Biomedical Engineering Challenges

Biomedical Engineering Challenges

A Chemical Engineering Insight

Edited by

Vincenzo Piemonte
Department of Engineering,
University Campus Biomedico of Rome,
Italy

Angelo Basile
Italian National Research Council,
Rende, Italy

Taichi Ito
Department of Bioengineering,
School of Medicine and Engineering,
University of Tokyo, Japan

Luigi Marrelli
Department of Engineering,
University Campus Biomedico of Rome,
Italy

WILEY

This edition first published 2018

Registered Office(s)
John Wiley & Sons, Inc., 111 River Street, Hoboken, NJ 07030, USA
John Wiley & Sons Ltd, The Atrium, Southern Gate, Chichester, West Sussex, PO19 8SQ, UK

Editorial Office
The Atrium, Southern Gate, Chichester, West Sussex, PO19 8SQ, UK

For details of our global editorial offices, customer services, and more information about Wiley products visit us at www.wiley.com.

Wiley also publishes its books in a variety of electronic formats and by print-on-demand. Some content that appears in standard print versions of this book may not be available in other formats.

Library of Congress Cataloging-in-Publication Data

Names: Piemonte, Vincenzo, editor. | Basile, Angelo, editor. | Ito, Taichi, editor. | Marrelli, Luigi, editor.
Title: Biomedical engineering challenges : a chemical engineering insight / edited by Vincenzo Piemonte, University Campus Biomedico of Rome, Italy, Angelo Basile, Italian National Research Council, Rende, Italy, Taichi Ito, University of Tokyo, Japan, Luigi Marrelli, University Campus Biomedico of Rome, Italy.
Description: First edition. | Hoboken, NJ : Wiley, 2018. | Includes bibliographical references and index. |
Identifiers: LCCN 2017054493 (print) | LCCN 2017056798 (ebook) | ISBN 9781119296027 (pdf) | ISBN 9781119296010 (epub) | ISBN 9781119296041 (hardback)
Subjects: LCSH: Biomedical engineering. | Biochemical engineering. | BISAC: SCIENCE / Biotechnology.
Classification: LCC R856 (ebook) | LCC R856 .B566 2018 (print) | DDC 610.28–dc23
LC record available at https://lccn.loc.gov/2017054493

Cover design by Wiley
Cover image: © Pobytov/iStockphoto

Set in 10/12pt Warnock by SPi Global, Pondicherry, India

10 9 8 7 6 5 4 3 2 1

Contents

List of Contributors *xi*
Preface *xiii*

1 Introduction *1*
Luigi Marrelli
References *6*

2 Artificial Kidney: The New Challenge *9*
Pasquale Berloco, Simone Novelli, and Renzo Pretagostini
2.1 Introduction *9*
2.2 Kidney Transplantation Statistics *11*
2.3 Transplantation Costs *12*
2.4 Post-Transplant Costs *12*
2.5 Renal Replacement Devices *13*
2.6 Implantable Artificial Kidney: Prototype Developments *16*
2.7 Kidney Tissue Engineering *17*
2.8 Next Steps *20*
2.9 Conclusion *21*
List of Acronyms *22*
References *23*

3 Current Status and New Challenges of the Artificial Liver *27*
Hiroshi Mizumoto, Nana Shirakigawa, and Hiroyuki Ijima
3.1 Introduction *27*
3.2 Non-Biological Artificial Liver *28*
3.2.1 Classification and Clinical Study *29*
3.2.2 PE and HDF *29*
3.2.2.1 High-Volume Therapeutic PE *29*
3.2.2.2 High-Flow Dialysate Continuous HDF *29*
3.2.2.3 PE with Online HDF *30*

3.2.3 Blood Purification with Albumin Dialysis *30*
3.2.3.1 Single-Pass Albumin Dialysis *30*
3.2.3.2 Molecular Adsorbent Recirculating System *31*
3.2.3.3 Fractionated Plasma Separation and Adsorption (Prometheus™) *32*
3.2.3.4 Hepa Wash *32*
3.2.4 Selective Plasma Filtration Therapy *32*
3.2.4.1 Biologic-Detoxifilter/Plasma Filter *32*
3.2.4.2 Selective Plasma-Exchange Therapy *32*
3.2.4.3 Plasma Filtration with Dialysis *33*
3.2.5 Clinical Observations of Various Combinations *33*
3.3 Bioartificial Liver *35*
3.3.1 Bioartificial Liver Support System *35*
3.3.2 Cell Source for BAL *37*
3.4 New Stream for Artificial Liver *40*
3.4.1 Tissue Engineering for Liver Construction *40*
3.4.2 Whole Organ Engineering for the Transplantable Artificial Liver *41*
3.5 Conclusion and Future Trends *43*
List of Acronyms *44*
References *45*

4 A Chemical Engineering Perspective on Blood Oxygenators *55*
Luisa Di Paola
4.1 Introduction *55*
4.2 A Historical Note *57*
4.3 Chemical Engineering Principles in Blood Oxygenators *60*
4.4 Chemical Engineering Process Analogues of ECMO Systems *65*
4.5 New Challenges *67*
4.6 Conclusion *69*
List of Symbols *69*
References *69*

5 Model Predictive Control for the Artificial Pancreas *75*
M. Capocelli, L. De Santis, A. Maurizi, P. Pozzilli, and Vincenzo Piemonte
5.1 Introduction *75*
5.2 Phenomenological Models *78*
5.2.1 Background and Two-Compartmental Models *78*
5.2.2 Three-Compartment Models *79*
5.3 Black-Block Approach *85*
5.4 Conclusions *90*
Nomenclature *91*
References *92*

6 Multiscale Synthetic Biology: From Molecules to Ecosystems *97*
Luisa Di Paola and Alessandro Giuliani
6.1 Introduction: An Historical-Epistemological Perspective *97*
6.2 Applications *99*
6.2.1 Protein Synthetic Biology *99*
6.2.2 Tissue Engineering and Artificial Organs *108*
6.2.3 Biotechnology and Ecology Applications *109*
6.3 Conclusions *111*
List of Symbols *112*
References *112*

7 Chemical Reaction Engineering Methodologies for Biomedical Imaging Analysis *119*
Masahiro Kawahara
7.1 Introduction *119*
7.2 Magnetic Resonance Imaging (MRI) *119*
7.2.1 ^{1}H-MRI *120*
7.2.2 ^{19}F-MRI *121*
7.2.3 MRI using Magnetization Transfer *122*
7.3 Positron Emission Tomography (PET) and Single Photon Emission Computed Tomography (SPECT) *123*
7.3.1 PET *123*
7.3.2 SPECT *125*
7.4 Fluorescence Imaging *126*
7.4.1 Fluorescent Proteins *126*
7.4.2 Small Organic Fluorophores *128*
7.5 Conclusion *131*
List of Abbreviations *131*
References *132*

8 Noninvasive and Label-Free Characterization of Cells for Tissue Engineering Purposes *145*
Shunsuke Tomita
8.1 Introduction *145*
8.2 Multivariate Analyses *146*
8.2.1 Principal Component Analysis (PCA) *147*
8.2.2 Linear Discriminant Analysis (LDA) *148*
8.2.3 Hierarchical Clustering Analysis (HCA) *148*
8.2.4 Other Multivariate Analyses *149*
8.3 Vibrational Spectroscopic Features *149*

8.3.1 Cell Characterization Based on Whole-Cell Analysis by Raman Spectroscopy *151*
8.3.2 Cell Characterization Based on Subcellular Analysis by Raman Spectroscopy *153*
8.3.3 Raman-Based Cell Characterization Toward Biomedical Applications *157*
8.4 Morphological Features *160*
8.4.1 Cell Characterization Based on Unstained Microscopic Images of Single Cells *160*
8.4.2 Cell Characterization Based on Unstained Microscopic Images of Cell Populations *162*
8.5 Secreted Molecule Features *165*
8.5.1 Cell Characterization Based on Response Signatures *165*
8.6 Conclusion and Outlook *167*
List of Acronyms *168*
References *168*

9 TMS-EEG: Methods and Challenges in the Analysis of Brain Connectivity *175*
Elisa Kallioniemi, Mervi Könönen, and Sara Määttä
9.1 Introduction *175*
9.1.1 Transcranial Magnetic Stimulation *175*
9.1.2 Electroencephalography *176*
9.1.3 Combined TMS and Electroencephalography *178*
9.1.4 Data Acquisition *178*
9.1.5 Artifacts and Their Prevention *180*
9.2 Signal Processing Methods *181*
9.2.1 Preprocessing *181*
9.2.2 Connectivity Analysis Methods in TMS-EEG *182*
9.2.3 Time Domain Methods *183*
9.2.4 Frequency Domain Methods *183*
9.3 TMS-EEG Applications in Studies of Connectivity *184*
9.3.1 General Aspects *184*
9.3.2 TMS-Evoked Potentials (TEPs) *185*
9.3.3 TMS-Induced Oscillations *186*
9.3.4 Clinical Perspectives *187*
9.3.4.1 Alzheimer's Disease *187*
9.3.4.2 Schizophrenia *188*
9.3.4.3 Disorders of Consciousness *189*
9.4 Conclusions and Future Trends *189*
List of Acronyms *190*
References *190*

10 Thermal Treatments of Tumors: Principles and Methods *199*
P. Saccomandi, E. Schena, M. Diana, J. Marescaux, and G. Costamagna
10.1 Introduction *199*
10.2 Effects of Temperature on Living Tissue *199*
10.2.1 Hyperthermal Tissue Destruction *200*
10.2.2 Cold Temperature for Tissue Destruction *202*
10.3 Physical Principles of Thermal Treatments *203*
10.3.1 Hyperthermal Treatments *203*
10.3.1.1 High-Intensity Focused Ultrasound Ablation *203*
10.3.1.2 Radiofrequency Ablation (RFA) *204*
10.3.1.3 Microwave Ablation (MWA) *205*
10.3.1.4 Laser Ablation (LA) *206*
10.3.2 Cryoablation *207*
10.4 Mathematical Modeling of Thermal Therapies *209*
10.5 Temperature Monitoring During Thermal Treatments *211*
10.5.1 Invasive (Contact) Thermometric Techniques *212*
10.5.2 Non-Invasive (Contactless) Thermometric Techniques *215*
10.6 Conclusions *218*
List of Acronyms *219*
List of Symbols *219*
References *220*

Index *229*

List of Contributors

Angelo Basile
Italian National Research Council, Rende, Italy

Pasquale Berloco
Università La Sapienza Roma, Dipartimento di Chirurgia Generale e Trapianti d'Organo, Rome, Italy

M. Capocelli
Unit of Process Engineering, Department of Engineering, Università Campus Bio-Medico di Roma, Italy

G. Costamagna
Institute of Image-Guided Surgery, IHU-Strasbourg, France

M. Diana
Institute of Image-Guided Surgery, IHU-Strasbourg, France

Alessandro Giuliani
Environment and Health Department, Istituto Superiore di Sanità, Rome, Italy

Hiroyuki Ijima
Department of Chemical Engineering, Faculty of Engineering, Kyushu University, Fukuoka, Japan

Taichi Ito
Department of Bioengineering, School of Medicine and Engineering, University of Tokyo, Japan

Elisa Kallioniemi
Diagnostic Imaging Center, Kuopio University Hospital, Finland

Masahiro Kawahara
Department of Chemistry and Biotechnology, Graduate School of Engineering, The University of Tokyo, Japan

Mervi Könönen
Diagnostic Imaging Center, Kuopio University Hospital, Finland

Sara Määttä
Diagnostic Imaging Center, Kuopio University Hospital, Finland

J. Marescaux
Institute of Image-Guided Surgery, IHU-Strasbourg, France

Luigi Marrelli
Department of Engineering, University Campus Biomedico of Rome, Italy

A. Maurizi
Unit of Endocrinology and Diabetes, Department of Medicine, Università Campus Bio-Medico di Roma, Italy

Hiroshi Mizumoto
Department of Chemical Engineering, Faculty of Engineering, Kyushu University, Fukuoka, Japan

Simone Novelli
Università La Sapienza Roma, Dipartimento di Ingegneria Meccanica e Aerospaziale, Rome, Italy

Luisa Di Paola
Unit of Chemical-Physics Fundamentals in Chemical Engineering, Department of Engineering, Università Campus Bio-Medico di Roma, Italy

Vincenzo Piemonte
Department of Engineering, University Campus Biomedico of Rome, Italy

P. Pozzilli
Unit of Endocrinology and Diabetes, Department of Medicine, Università Campus Bio-Medico di Roma, Italy

Renzo Pretagostini
Università La Sapienza Roma, Dipartimento di Chirurgia Generale e Trapianti d'Organo, Rome, Italy

P. Saccomandi
Unit of Measurements and Biomedical Instrumentation, Department of Engineering, Università Campus Bio-Medico di Roma, Italy

L. De Santis
Unit of Process Engineering, Department of Engineering, Università Campus Bio-Medico di Roma, Italy

E. Schena
Unit of Measurements and Biomedical Instrumentation, Department of Engineering, Università Campus Bio-Medico di Roma, Italy

Nana Shirakigawa
Department of Chemical Engineering, Faculty of Engineering, Kyushu University, Fukuoka, Japan

Shunsuke Tomita
Biomedical Research Institute, National Institute of Advanced Industrial Science and Technology, and DAILAB, Tsukuba, Ibaraki, Japan

Preface

Biomedical engineering has come a long way since Leonardo da Vinci (1452–1519) drew his revolutionary pictures of the skeleton and its musculature and studied the mechanics of bird flight. The modern era has seen the application of engineering in almost every branch of medicine, so much so that the practice of medicine is now completely dependent on the work and support of engineers.

Artificial organ technologies such as extracorporeal life support, dialysis and insulin control systems keep huge numbers of patients alive who await organ transplantation. A person with a new prosthetic arm can 'feel' things and thereby grip eggs without breaking them. Imaging techniques aim to see inside the intact human body. An important use of three-dimensional body imaging is incorporating biomechanical models of muscles and bones, enabling simulations of planned surgical procedures. Less invasive techniques and devices will lead to more comfortable recovery for patients, as well as faster rehabilitation and shorter stays in hospital, all of which also lead to a reduction in health care costs. These are just some examples of the amazing progress that has been made in biomedical engineering over the last few decades.

This book reports on some applications in the field of biomedical engineering ranging from artificial organs and tissue engineering to biomedical imaging.

In detail, after a short introductory chapter devoted to introducing the book and providing an overview of the topics covered, Chapter 2 (Berloco, Novelli and Pretagostini) deals with the artificial kidney. Nowadays, the only real dialysis substitute is kidney transplantation but the while scientific community is looking for a substitute, because of the lack of organs available and the complications that a transplant can cause. The understanding of the cellular and molecular bases of organ function will be translated into new therapeutic approaches, especially in the acute and chronic renal failure. The bases of these new therapies are cell therapy and tissue engineering with a final goal of introducing these cells into the patient either via extracorporeal circuits or as implantable constructs.

In Chapter 3 (Mizumoto, Shirakigawa and Ijima) a detailed survey of artificial liver is given. The development of a functional artificial liver has been desired to save patients' lives with serious liver failure. For the past 60 years, the removal rate of toxins in blood by artificial systems was improved. As a result, the effectiveness of recovery has been significantly improved by the non-biological artificial liver. On the other hand, the bioartificial liver (BAL), which is expected to compensate for important functions of the liver, has been developed as a practical treatment system by ingenious improvement of the devices. In the future, it is expected that high efficiency treatment will be developed by the combination of these systems.

Chapter 4 (Di Paola), dealing with blood oxygenators, traces out the parallel development of oxygenators and gas-liquid contactors, demonstrating in this emblematic case how chemical engineering principles contribute to biomedical engineering innovation.

Chapter 5 (Capocelli, De Santis, Maurizi, Pozzilli and Piemonte) provides an historical review of the predictive control models for artificial pancreas, individuating the drawbacks and the advantages of the most widespread class of algorithms. Moreover, the chapter describes the physical-mathematical fundamentals and the most important results of a three-compartment model recently proposed by the chapter's authors.

Chapter 6 (Di Paola and Giuliani) reports a perspective of actual discoveries and innovations in the field under the light of systems biology. Trying to translate knowledge from engineering to build up biological machines, there is a missing trick due to the strong, systemic nature of organisms, which challenges any interaction between machines and life. The systemic approach could drive a real revolution and provide devices that are compliant with life, but engineering itself could learn many lessons for instance from protein nanomachines about regulation, autoadaptive systems and modularity of smart systems.

Chapter 7 (Kawahara) focuses on the aspects of chemical reaction engineering in the molecular imaging techniques (MRI, PET/SPECT and fluorescence imaging) and summarizes their advances, capability and significance in the biomedical field.

Chapter 8 (Tomita) provides a brief overview of multivariate analyses as key statistical techniques that are commonly used in signature-based cell profiling. Subsequently, three approaches for noninvasive and label-free characterization of cultured cells are presented using selected examples.

Chapter 9 (Kallioniemi, Könönen and Määttä) introduces some of the central issues regarding theory, methods and applications of TMS-EEGin functional connectivity studies. Connectivity studies aim to evaluate and characterize the neural interactions between brain regions. The chapter begins with a presentation of the basic principles related to TMS-EEG methodology followed by an introduction to the most common data analysis methods in connectivity studies. Finally, the chapter discusses the applications related to TMS-EEG in studies of connectivity and presents some of the clinical perspectives in the field.

Finally, Chapter 10 (Saccomandi, Schena, Diana, Marescaux and Costamagna) discusses the application of thermal therapies to the treatment of tumors: the principles of working of the main thermal approaches are presented, together with the main techniques for the real-time monitoring of the outcome.

The Editors wish to take this opportunity to thank all the Authors for their hard work. Special thanks also to the editorial team at Wiley for their continued helpful support.

Prof. Vincenzo Piemonte
University Campus Biomedico of Rome
Italy

Prof. Angelo Basile
Italian National Research Council Rende
Italy

Prof. Taichi Ito
University of Tokyo
Japan

Prof. Luigi Marrelli
University Campus Biomedico of Rome
Italy

1
Introduction

Luigi Marrelli

Department of Engineering, University Campus Biomedico of Rome, Italy

Biomedical Engineering (BE) is a complex applied science that has applications in the fields of medicine concerning diagnosis, therapy and rehabilitation. As in all sectors of technology characterized by complexity, the capability of solving problems and developing new devices for therapy and rehabilitation or to devise innovative techniques for diagnosis and treatment of pathologies requires synergy of various forms of expertise. This need is especially felt in BE where the subject of research is the human body with its complex operations: molecular mechanisms, chemical and biochemical intracellular reactions, control systems, functions of human organs and so on.

Despite the complexity of the biological system that doctors and biomedical engineers must relate to, the elements of which the human body is composed and the functions they perform have a close affinity with the operations carried out in a chemical plant [1] where some raw materials undergo a series of transformations (reactions, separation operations, mass and heat exchanges, etc.) in order to obtain useful products and energy. In the last century, this analogy has led to striking graphic representations of the human body and of its functions as an industrial plant. It is worth mentioning the picture [2] named "Der Mensch als Industriepalast," a creation by Fritz Kahn, a German doctor, science writer and pioneer of information graphics.

Indeed, the human body is composed of a solid structure of support and a casing that encloses a series of organs with functions of mass exchange, synthesis and transformation, and are connected to each other by a network of ducts passed through by fluids. A pumping system equipped with valves has the task of ensuring blood circulation in the vascular circuit. The digestive system, through complex chemical reactions, transforms ingested raw materials into

Biomedical Engineering Challenges: A Chemical Engineering Insight, First Edition.
Edited by Vincenzo Piemonte, Angelo Basile, Taichi Ito, and Luigi Marrelli.

useful substances and energy needed for the operation of the whole system. The muscular system can be regarded as a set of actuators responsible for moving several parts of the body whereas the peripheral nervous system supplies sensory stimuli coming from the environment to the central nervous system that supervises the control and processing of various functions in a similar way to what happens in the system of sensors, monitoring and automatic control in an industrial plant.

Beyond this evocative picture, however, the analogy suggests the idea that many sophisticated techniques of chemical engineering (CE) could be usefully applied to face the technical challenges of BE. Since the beginnings of its history, CE has dealt with unit operations for the separation of mixtures, humidification of gases, chemical reactors, mass, heat and momentum transport, properties of materials and so on, on the basis of scientific fundamentals that are phase equilibria thermodynamics, chemical and biochemical kinetics, transport phenomena, automatic control and mathematical tools needed to better understand many complex phenomena and to represent the behavior of equipment through theoretical or semi-empirical models useful for the simulation and the optimization of the process. Therefore, CE can provide skills to the solution of problems that BE has to face and, vice versa by a cross-fertilization process, can receive from BE valuable input for the development of innovative methods and processes deduced from the behavior of biological systems.

The fields where CE can provide fundamental contributions are numerous and range from the macroscale of artificial and bio-artificial organs to the nanoscale of chemical-physical properties of materials of cell micro-reactors. Furthermore, also in the industrial biotechnology and pharmaceutical fields, CE points out its potentiality in the large-scale production of drugs and in sophisticated methods of targeted drug delivery.

Organs like the kidney, liver or heart-lung system have, among their functions, those of cleaning the blood from toxins or excesses of substances and of exchanging oxygen and carbon dioxide. When native organs are not able to correctly perform these functions for pathological reasons, an artificial kidney, artificial liver or lung oxygenation unit can substitute or at least support the damaged vital functions and allow the patient to stay alive indefinitely or, at least, long enough for the possibility of carrying out transplantation or revival of the native organ. Nowadays, ultrafiltration technology by selective membranes is an acquired asset in CE that can provide a key contribution to the development of increasingly effective and low-cost artificial organs. Some chapters of the present book are devoted to artificial organs and their behavior.

However, just separation operations and selective transport are not enough in many cases to mimic the functions of the native organ: complex organs such as the liver or pancreas carry out synthesis functions and biochemical reactions not currently reproducible by artificial systems. This need has led to the development of bio-artificial or hybrid organs in which the artificial component is coupled with a biological element; that is, a cell tissue able to perform functions not reproducible by a totally artificial system.

Therefore, hybrid organs are characterized by the presence of a kind of bioreactor where cells are kept in the optimal conditions for their survival and, in particular, to perform the functions of which they are responsible. The use of living cells as engineering materials is the basis of the so-called *tissue engineering* [3] where chemical engineering, material science and life sciences skills are involved. In the field of tissue engineering, the scaffold or support technology provides an important step in the growth and differentiation of the desired tissue. Even in this case, chemical reaction engineering and theory of reactors, the typical hallmarks of chemical engineers' activity, play a fundamental role in modeling, designing and properly running the bioreactor used for growing the new tissue.

A technology still under study is the development of an artificial pancreas. The purpose of this device is monitoring and properly releasing insulin, a hormone produced by β cells of the pancreatic islets of Langerhans, that regulates the absorption of glucose from blood and its conversion into glycogen or triglycerides. If β cells do not work, insulin can no longer be synthesized or secreted into the blood resulting in a high blood glucose concentration (type 1 diabetes). In order to solve this problem, chemical engineers are working on a computerized device able to monitor continuously blood glucose levels and to actuate micro-pumps for delivering insulin contained in a small reservoir. Chapter 5 of the present book is devoted to the artificial pancreas.

A well-known field of CE deals with scale-up techniques; that is, similar criteria needed to develop an industrial plant from information on the behavior of a pilot or bench scale plant. In the past, these techniques have already provided fundamental results for designing and running plants devoted to manufacturing products essential to human health. A typical early example is the process of producing penicillin on an industrial scale, suggested at the end of the World War II by Margaret Hutchinson Rousseau [4], a young chemical engineer. The industrial production process is based on an aerobic submerged fermentation. When penicillin was first made, the fungus *Penicillum notatum* was used and the yield of the process was about 1 mg/dm^3. Nowadays, using a different mold species (*Penicillum chrysogenum*) and by improving fermentation operating conditions and downstream processing, such as extraction techniques, a yield of 50 g/dm^3 is reached.

The opposite side of the coin is represented by scale-down techniques; that is, by the techniques to implement micro-devices [5, 6]. These devices are composed of a network of microchannels connecting micro-reactors, mixers, pumps and valves contained in vessels whose dimensions are in the order of micrometers with controlled volumes down to picolitres. At this scale, fluid transport in capillaries is laminar and the resulting very high surface area to volume ratio affects mass and heat transfer rates and catalytic reaction rates that depend on the interface area. Microfluidic devices (lab-on-chip assemblies) are increasingly used to carry out chemical and biochemical reactions for applications in the genomic field, immunoassays, sensors, drug discovery, new catalyst development and many other forthcoming uses.

Properties of materials is another field where CE together with material science can provide an important contribution to BE. Scaffolds used in tissue engineering have to be biocompatible and biodegradable [1] to allow their use in contact with biological material and their absorption by the surrounding tissues when scaffolds are used in implantable devices. In any case, the degradation rate of the support must be compatible with the rate of making new tissue and with the integration of this one with the surrounding tissues. A very important property of the scaffold is its porosity and the distribution of pore sizes to allow three-dimensional tissue growth. A fractal geometry approach has proven to be useful for the characterization of these properties.

The knowledge of rheological properties of biological fluids [7] is another essential requirement for a proper design of extracorporeal devices. Blood is a suspension made of an aqueous solution (plasma) of electrolytes, sugars and proteins and of a corpuscular part composed of erithrocytes, leucocytes and platelets. Plasma is a Newtonian fluid with a viscosity of $\mu_P = 1.16 \div 1.35 cp$ at 37 °C. Whole blood, on the other hand, shows non-Newtonian behavior ranging from Bingham to pseudoplastic fluid behavior depending on the value of the shear stress. Non-Newtonian characteristics are clearer in the thinnest ducts as capillaries.

These rheological properties make the modeling of blood flow particularly complex [8, 9] in the various blood ducts, especially if one wishes to account for the transient behavior of the flow due to heartbeat and for non-stiffness of blood vessels. Moreover, in the case of blood circulation in extracorporeal devices, a non-negligible feature to take into account is the possible damage that shear stresses in circulation ducts and even more in pumping systems can cause to corpuscular part of the blood and to proteins [10].

These features cannot properly be accounted for without the knowledge of momentum transport that is one of the typical expertise areas of CE.

Another significant CE contribution can be found in the increasingly sophisticated techniques that are being tuned for an effective and safe administration of drugs or for their targeted delivery. The development of controlled drug delivery strategies has the purpose of assuring a precise drug dosage and of avoiding the risk of insufficient dosing or of over-dosages that can appear in conventional methods of delivering (ingestion or injection). Some of these new methods have been developed at MIT in the 1970s by chemical engineer Robert Langer and his colleagues [11, 12] by exploiting the property of some highly porous structures (mixtures of hydrophobic polymers and proteins) impregnated with the drug to release it in slowly and controlled way under appropriate conditions. Such a device, in the form of an implantable wafer, has been successfully tested in delivering, with small side effects, a potent chemotherapy drug to brain cells attacked by glioblastoma.

Glucose-sensitive hydrogels can be used for the controlled release of insulin to diabetic patients. A system based on this kind of materials has been developed at the Purdue University by the team of chemical engineers coordinated by Nicholas Peppas [13–15].

Overdoses of many antitumor drugs are dangerous for healthy tissues and organs. This drawback has started up an intensive research activity in the community of chemical and biomedical engineers for the development of targeted delivery systems of chemotherapy drugs, that is for the drug delivery in the sick tissue. Attempts have been made by the encapsulation of the drug in carriers as liposomes, nanoscaled C_{60} fullerenes or entrapping it within a hydrogel matrix. These kinds of methods allow a reduction of drug degradation by the digestive system and avoid attack from the immune system if the carrier surface is properly masked. Using specific peptides or, in particular, monoclonal antibodies attached to the external carrier surface allows the drug to be led to target cells where it is released as a consequence of internal or external triggers such as pH, temperature, enzymes, ultrasound or magnetic signals. For example, hydrogels swell under different environmental conditions so that they can release a drug entrapped by changing the pH or the temperature. Furthermore, since hydrogels are stable in acidic environments, they do not undergo any degradation by passing through the stomach. Even in this BE field, CE fundamentals concerning mass transport phenomena and chemical and physical properties of polymers play a paramount role.

Finally, a few words about *synthetic biology* [16] to which a chapter of this book is devoted (Chapter 6). As the human body with its organs can be seen as an industrial plant, so cells can be considered as nanomachines on the molecular scale. The advances in biology indeed allow cell mechanisms to be described as an ensemble of biocatalytic reactions resulting in handling mass, energy and information. This consideration has induced researchers to use the molecular mechanisms of living matter for thinking up models, techniques and devices aimed to the development of new strategies of diagnosis and therapy more effective than the traditional ones. A typical example is the fight against infection, since increasing resistance to antibiotics has made their use partly ineffective.

This approach has led to the origin and evolution of the synthetic biology that is an interdisciplinary field of science based on the use of fundamentals of biology and engineering for the achievement of new and smart methods and devices useful in diagnosis and therapy.

Some examples in the field of therapy are:

- bacteriophage-based therapy that exploits the very specific action of the phages for killing infectious microorganisms;
- synthetic probiotic therapy based on the controlled and confined release under specific conditions of drug-molecules by engineered bacteria in the form of synthetic probiotics.

Similar strategies are evolving also in the field of diagnostics. In particular, a diagnostic technique is emerging based on the ability of specific bacteriophages to identify a particular bacterium on the basis of their ability to clear it.

These considerations clearly show that, as synergy between CE and BE has produced key contributions to the development of tools useful to human health by using typical methods of engineering, we are able to formulate promising achievable ideas. Ever-closer cooperation between these branches of engineering allows us to foresee that results obtained so far are only the tip of the iceberg and that, in the near future, currently seeming science fiction achievements can become reality.

References

1 AIChE. "Chemical engineers in action: Chemical engineers in the biomedical field: from discoveries to solutions." Online at: http://studylib.net/doc/10322128/chemical-engineers-in-the-biomedical-field--from-discoveries-to-solutions (accessed November 2017) (2009).
2 Cybernetic Zoo. Images of Fritz Kahn. Online at: http://cyberneticzoo.com/tag/fritz-kahn/ (accessed November 2017) (n.d.).
3 Hoque, M.E. *Bioartificial Liver Assist System: Tissue Engineering Challenges*, Lambert Academic Publishing (2010).
4 Maitland, G. Ten chemical engineers that shaped our world (Day 356). Online at: https//ichemeblog.org/ten-chemical-engineers-that-shaped-our-world (accessed October 2017). Institution of Chemical Engineers (2015).
5 Allen, R., "Editorial," *Ingenia On Line Magazine* (2001), 9, 40.
6 Stones, H.A. and S. Kim, "Microfluidics: basic issues, applications, and challenges," *AIChE Journal* (2001), 47(6), 1250.
7 Casentini, G., L. Di Paola, L. Marrelli, F. Palma "Rheological characterization of an artificial synovial fluid," *The International Journal of Artificial Organs* (2005), 28(7), 711.
8 Apostolidis, A.J. and A.N. Beris, "Modeling of the blood rheology in steady-state shear flows," *Journal of Rheology* (2016), 58, 607.
9 Apostolidis, A.J., M.J. Armstrong and A.N. Beris, "Modeling of human blood rheology in shear flows," *Journal of Rheology* (2016), 59, 275.
10 Marrelli, L., F. Vitale, and L. Di Paola, "Shear stress induced blood damage in artificial organs and haematica circuits" *Proceedings of Gr.I.C.U. Conference* (2008), 177.
11 Langer, R. and J. Folkman, "Polymers for the sustained release of proteins and other macromolecules," *Nature* (1976), 263, 797.
12 Bawa, R., R.A. Siegel, B. Marasca, M. Karel, and R. Langer, "An explanation for the controlled release of macromolecules from polymers," *Journal of Controlled Release* (1985), 1(4), 259.

13 Marek, S.R. and N.A. Peppas, "Insulin release dynamics from poly(diethylaminoethyl methacrylate) hydrogel systems," *AIChE Journal* (2013), 59(10), 3578.

14 Gaharwar, A.K., N.A. Peppas, and A. Khademhossein, "Nanocomposite hydrogels for biomedical applications," *Biotech. Bioeng.* (2014), 111(3), 441.

15 Peppas, N.A. and B. Narasimhan, "Mathematical models in drug delivery: how modeling has shaped the way we design new drug delivery systems," *J. Control. Release* (2014), 190, 75.

16 Braff, D., D. Shis, and J.J. Collins, "Synthetic biology platform technologies for antimicrobial applications," *Advanced Drug Delivery Reviews* (2016), 105, 35.

2

Artificial Kidney: The New Challenge

Pasquale Berloco[1], *Simone Novelli*[2], *and Renzo Pretagostini*[1]

[1] *Università La Sapienza Roma, Dipartimento di Chirurgia Generale e Trapianti d'Organo, Rome, Italy*
[2] *Università La Sapienza Roma, Dipartimento di Ingegneria Meccanica e Aerospaziale, Rome, Italy*

2.1 Introduction

In recent years, chronic kidney disease has seen an exponential increase worldwide. The epidemiologic data have caused considerable alarm and consequently politicians in many countries, including third world countries, have shown interest, as has the World Health Organisation (WHO).

All over the world, the number of people receiving dialysis treatment is increasing at a rate of 7% a year and projections see this figure doubled by 2020. Europe faces a similar situation where, in the 28 countries of The United Nations, 65 000 new patients are added to dialysis treatment every year. In other words, the incident rate is 137 with prevalence rate of 786 patients/million inhabitants, respectively.

Chronic kidney disease (CKD) is a major public health problem that is associated with an increased risk of kidney failure, cardiovascular events and all-cause mortality [1, 2]. However, the risk of mortality attributable to CKD in the general population is difficult to determine because the ultimate cause of death of CKD patients is often attributed to one of the complications induced by the condition, therefore obscuring the role played by the underlying disease [3]. Moreover, the impact of type of nephropathy and stage of the disease on the mortality of CKD patients has been little investigated at the population level [4]. The availability of clinically useful models for risk stratification of CKD patients according to their clinical and demographic characteristics and presence of major complications is important for informing treatment decisions [5] and planning dialysis and transplantation when

Biomedical Engineering Challenges: A Chemical Engineering Insight, First Edition.
Edited by Vincenzo Piemonte, Angelo Basile, Taichi Ito, and Luigi Marrelli.

needed. However, it requires large population-based databases to achieve the statistical power needed to set up predictive models.

In the USA, more than 400 000 Americans have end-stage renal disease, and over 300 000 of these patients require maintenance dialysis. Mortality rates remain above 20% per year with the use of dialysis, with more than half of the deaths related to cardiovascular disease [1].

The importance of kidney disease in Europe is similar to that in the USA (according to Thomas D. DuBose, President of the American Society of Nephrology, as stated in his Presidential Address in San Diego, California at the American Society of Nephrology Congress), and in Italy, today, there are 2 200 000 IRC (Chronic Renal Failure) patients not yet undergoing dialysis treatment. This is a high number. It has been estimated that there are 20 million people suffering from persistent proteinuria. The Faenza study [6] on the importance of kidney disease suggests that around 4% of the population could be suffering from kidney disease (based solely on the data for high plasmatic creatinine).

Dialysis, both haemodialysis and peritoneal, obliges the patient to undergo a 4-hour treatment three or four days a week. This affects their social life, their work and has negative psychological consequences. Depression is common in dialysis patients and has been shown to be associated with higher morbidity and mortality, but little is known about the course of symptoms over time. Depression and poor mental health are common in haemodialysis patients and are associated with higher risk of hospitalization and death [7].

If treatment lasts for years, as often happens, other organ damage occurs, especially to the heart, lungs and blood vessels. Dialysis also represents a considerable cost to national health systems both in America and in Europe.

In America, the annual direct medical costs for end-stage renal disease are nearly $23 billion. Although an estimated 8 million adults in the USA have CKD of at least stage 3 (as defined by an estimated glomerular filtration rate [GFR] of less than 60 ml per minute per 1.73 m^2 of body-surface area), less is known about the rates of death, cardiovascular disease and resource use among persons with a reduced estimated GFR who are not yet receiving maintenance dialysis [8].

In the EU in 2014, National Health costs expressed in Gross National Product percentages were the following: Europe 8.6%, EU (28 countries) 8.8%, Italy 8.5%, Germany 11%, France 9.5% and UK 7.8%. These figures show the incidence of dialysis costs on the National Health. It has been estimated that in Italy, dialysis costs for chronic patients (0.08% of the population) is 25 times that of National Health costs pro capita.

In Germany (0.08% of population undergoing dialysis), it is 20 times that of National Health costs pro capita, in France (0.05% of population undergoing dialysis), it is 20 times more and in the United Kingdom (0.03% of population undergoing dialysis) it is 30 times more. It is therefore apparent that patients undergoing dialysis have considerable incidence on health budgets in every country [9].

In Italy, there is no reliable data on National Health costs for these patients. However, if reimbursement rates for dialysis are applied and are then multiplied by the number of dialysis treatments and the number of patients, enormously high figures are obtained.

The average cost of one dialysis treatment is around €150. If we estimate 35 000 patients, the following figures are obtained only for dialysis treatment:

- Cost per patient: €22 800 per year
- Cost for patients in dialysis around €800 million.

To these we must add transport costs to and from home/hospital, drugs, hospitalization (on average one per year), invalid pensions and special needs accompanying pensions. This, of course excludes loss of work, which is often the result in permanent invalidity.

The other option for CKD is transplantation from a dead or live donor. This surgery procedure partly relieves the patient from social damaging consequences but obliges them to take anti-rejection drugs for life, which could cause infection or in rare cases, cancer. Kidney transplant is also of considerable cost to The National Health Service.

2.2 Kidney Transplantation Statistics

In Europe, the total numbers of patients on waiting lists reflect an impressive reality, even if they cover data from various transplant systems with different national policies and as well as evolving dynamics (see remarks that follow) and therefore only represent the tip of the iceberg.

A total of over 63 000 patients were officially placed on organ waiting lists on 31 December 2015 in the European Union; that is, for 508 million inhabitants (for comparison: 63 800 patients and 61 500 patients, respectively, were placed on waiting lists in the EU on 31 December 2013 and 2014). If patients from Iceland, Norway and Turkey are added: 86 000 patients were on the waiting lists (for a total population of 588 million inhabitants).

It is estimated that 4100 patients died while officially placed on these waiting lists in the course of 2013 in the 28 Member States of the European Union (respectively, 3780 and 5500 patients died while on waiting lists in 2012 and 2011). If patients from Iceland, Norway and Turkey are included, it can be estimated that 6000 patients died while on the waiting lists in 2015 [9].

In America, in 2015, 16 147 kidney transplants were carried out. this number includes all ages and transplants for non-US citizens. We predict increases in the numbers of some transplants from 2013 to 2014 because of projected population increases, even though the expected transplant rate per million people may decrease for various types of transplants [10].

2.3 Transplantation Costs

In Europe, kidney transplant costs vary from nation to nation, even if they are similar. Let us consider kidney transplant costs in Italy. Data analysis has allowed us to estimate National Health costs for patients who have undergone transplant compared to those that have not undergone transplant for the whole period of observation as both were included in our study. Methodology considered treatment length was equal to or above 3 years or equal to or below 4 years: all patients underwent transplant in 2008 and observation period continued to 31 December 2011. Every patient who did not undergo transplant was associated to a transplanted patient with analogical clinical and angraphical characteristics, and the observation period started at the moment the associated patient underwent transplant.

Total estimated cost for each transplant patient throughout observation period totalled €95 247. Of this, €52 543 were estimated costs for the transplant, equal to 55.2% of total relevant transplant costs, which correspond to surgery, pre-surgery and post-surgery hospitalization [11]. Post-transplant costs, for the whole period, which runs from first hospital discharge (in all cases in 2008) to 31 December 2011 are there for estimated at €42 704, equal to 44.8% of total estimated costs.

Detailed analysis of single elements in transplant costs highlights the fact that on average national levels, more significant costs (with similar amounts) are the costs for hospitalization in Kidney Transplant Centre (KTC) equal to €23 112 (44.1% of total transplant costs), and for surgery where estimated costs are on average €22 918 (equal to 43.6% of total).

We have taken into account that all surgical operations were not carried out in kidney centres but in other operative units (usually surgery units); therefore, we have also included costs referable to surgery staffing (personnel, drugs and units). Whereas, in the case of surgery units, these costs (all surgery costs except those of operating theatre) were included in the daily hospitalization costs [11].

2.4 Post-Transplant Costs

Analysis of single elements for the post-transplant period has highlighted that territorial immunosuppression, complications and other diseases on average have the highest costs and, once again, the figures at average national levels do not vary a great deal. Cost of territorial immunosuppressive therapy has been estimated on average at €12 419, while costs connected to complications and other diseases (above all post-transplant hospitalization) are estimated at an average of €1226 [10, 11].

Costs connected to death and organ failure (concerning only a small number of patients included in this study but which have in these few cases, incurred considerable costs) have on average (therefore divided among all patients included in the study) been estimated at €11 038. Follow-up costs, that is periodic patient check-ups after transplant hospital release, have an average estimated cost of €7020.

Drug therapy, already included in relevant estimated transplant and post-transplant costs, was also subject to deeper specific analysis, which allowed us to observe its itemization. Relative kidney transplant costs in America have been listed in Table 2.1.

As we have shown, both dialysis and kidney transplant provide two good solutions for chronic kidney failure but both have considerable costs and, in different ways, both have negative social and psychological side effects,

In order to create treatment that is free of or partially free of social and psychological problems, and in order for patients suffering from kidney failure to go back to a normal working and social life, there have been considerable investments (in the order of millions of Euros) in the worldwide scientific community aimed at obtaining portable or implantable devices or to create more artificial organs through tissue engineering [12].

CKD is more widespread than acute kidney disease and is constantly on the increase all over the world, as we have seen. The main cures that have been trialled over the last 10 years are kidney dialysis and kidney transplant. However, the potential for using stem or progenitor cells for tissue engineering and renal replacement as novel therapies are slowly being realized. Beginning with the principles of developmental biology and knowledge of factors that specify renal stem cells, investigators are applying new principles to the kidney in the hopes of enhancing the regenerative capacity and improving long-term function.

2.5 Renal Replacement Devices

In chronic kidney failure, haemodialysis increases both morbidity and mortality rates in patients when compared to the general population. In fact, several statistics show how adjusted rates of all-cause mortality are up to eight times greater for dialysis patients compared with age-matched individuals in the general population [13, 14].

Although long-term chronic renal replacement therapy with either haemodialysis or peritoneal dialysis has dramatically changed the prognosis of renal failure, it is not complete replacement therapy, because it provides only the filtration function (usually on an intermittent basis) and does not replace the homeostatic,

Table 2.1 Relative kidney transplant costs in the US in 2015.

Total estimated number of transplants	30-day pre-transplant	Procurement	Hospital transplant admission	Physician during transplant	180 days post-transplant discharge	Op immuno-suppressants and other rx	Total
16 147	$23 200.00	$84 400.00	$119 600.00	$20 500.00	$66 800.00	$19 800.00	$334 300.00

Source: UNOS website (https://optn.transplant.hrsa.gov/data/organ-datasource/kidney/).

Figure 2.1 First experimental dialysis (1913). *Source:* Journal of Pharmacology and Experimental Therapeutics January 1914, 5 [3] 275–316, http://jpet.aspetjournals.org/content/5/3/275. (*See insert for color representation of the figure.*)

regulatory, metabolic and endocrine functions of the kidney (Figure 2.1). Tissue engineering of an implantable artificial kidney composed of both biological and synthetic components could result in substantial benefits for patients [15].

In the field of acute and chronic renal failure the rapid understanding of the cellular and molecular bases of organ function and disease processes will be translated in the next decade into new therapeutic approaches. Central to these new therapies are the developing technologies of cell therapy and tissue engineering, which are based on the ability to expand stem or progenitor cells in tissue culture to perform differentiated tasks and to introduce these cells into the patient either via extracorporeal circuits or as implantable constructs. Cell therapy devices are currently being developed to replace the filtrative, metabolic and endocrinological functions of the kidney lost in both acute and chronic renal failure.

These devices have the promise to be combined to produce a wearable or implantable bioartificial kidney for full renal replacement therapy that may significantly diminish morbidity and mortality in patients with acute or chronic renal disease [12].

There is thus a critical unmet need for new dialysis technologies that offer patients an enhanced spectrum of choices and may address some of the key limitations of the current end-stage renal disease (ESRD) treatment paradigm [16]. An alternative with the potential to improve on current haemodialysis technologies is a continuously operating and wearable artificial kidney (WAK) designed to provide continuous solute clearance and ultrafiltration capacity [16]. This device, which utilizes dialysate-regenerating sorbent technology paired with a miniaturized, dual-channel, battery-operated, pulsatile pump for driving both blood and dialysate simultaneously. This pulsatile pump generates a unique flow pattern that enhances convective transfer of solutes across the dialyser membrane. Previous pilot studies of the WAK have demonstrated its short-term safety and efficacy in solute clearance and fluid removal but have been limited to treatment durations of less than 8h [17, 18].

The WAK is designed to be worn and used by patients for up to 24h per day and, thus, represents a wearable form of extended-hours haemodialysis in which patients receive substantially longer treatments than available with conventional thrice-weekly haemodialysis. Extended-hours therapies have been shown to lead to improved control of electrolyte homeostasis, including reductions in serum phosphorus [19, 20]. Extended-hours therapies also allow for slower and more physiological rates of volume removal via ultrafiltration, which may reduce the frequency of intradialytic hypotensive events and intradialytic myocardial stunning [21, 22]. Additionally, longer weekly treatment times have been shown to result in increased dialytic clearance of larger-molecular weight and protein-bound uremic retention solutes, which may at least partly underlie the improved outcomes observed among patients undergoing extended-hours regimens [8, 23].

In conclusion, the WAK is a dialysis technology with the potential to radically change the way dialysis therapy is delivered. As demonstrated in this FDA-approved exploratory clinical trial, 24-hour treatment of individuals with ESRD with a wearable, miniaturized, sorbent-based haemodialysis system is well-tolerated and results in effective electrolyte homeostasis, solute clearance and volume removal.

The possible complications that we encountered were: Cumulative occurrence of device-related technical problems, which included excessive carbon dioxide bubbles in the dialysate circuit that exceeded degassing capacity, tubing kinks and variable pump function, resulting in fluctuating blood and dialysate flow rate.

Despite a variety of device-related technical complications encountered, this study serves as proof of concept of the WAK as an innovative alternative dialysis technology that has the potential to change the way ESRD is treated through enhancing patient choice and optimizing health-related quality of life for patients with ESRD [1].

2.6 Implantable Artificial Kidney: Prototype Developments

Researchers in San Francisco University (California) have stated that they have set up the first implantable artificial kidney in collaboration with many other research groups. According to researchers, a usable prototype in medicine could be available in five to seven years, which is the necessary time to increase miniaturization.

The whole set up has been projected to obtain most of the benefits obtained with kidney transplants, stated Shuyo Roy, coordinator of research groups involved in the project [23]. As there are only a limited number of donors a year, most patients have to resort to dialysis, where current protocols foresee three 3–5-hour treatments a week. This treatment replaces only 13% kidney function and has negative long-term consequences on patient health [24].

The system has two phases. Particular microscopic filters remove metabolic toxins from the blood in the first phase, while in the second, tubuli kidney cells, which will imitate kidney action in balancing electrolytes, metabolites and water and will allow the equipment to carry out other biological healthy kidney functions, are produced inside a bioreactor. Progress in tissue engineering has made this possible [14, 16, 19].

Therefore, Shuvo et al. [24] posed a fundamentally different approach using microelectromechanical systems (MEMS) fabrication techniques to create thin-flat sheets of silicon-based membranes (SNM) for implantable or portable haemodialysis applications. The SNM have biomimetic slit-pore geometry and uniform pores size distribution that allow for exceptional permeability and selectivity. A quantitative diffusion model identified structural limits to diffusive solute transport and motivated a new microfabrication technique to create SNM with enhanced diffusive transport. Consequently, conventional polymer haemodialysis membranes are likely limited in their capacity to provide an implantable or portable device that requires long-term continuous blood exposure [24].

In contrast, MEMS technology allows for fabrication of membranes with reproducible and uniform pore size distribution [25–27] with tunable pore geometry that can be optimized for molecular transport and permeability [25]. Microfabricated silicon membranes also enable surface chemistry modification that can limit immunologic reaction and protein fouling, while enhancing selectivity based on electrostatic charge [28, 29]. Furthermore, MEMS enables device miniaturization not feasible with current polymer haemodialysers given their inherently large package size and high internal flow resistance.

Shuvo et al. [30] used MEMS technology to pioneer a novel and mechanically robust silicon nanopore membrane (SNM) for implantable renal replacement therapy.

2.7 Kidney Tissue Engineering

The critical elements of renal function must be replaced, including the excretory, regulatory transport and endocrinology functions. The functioning excretory unit of the kidney is composed of the filtering unit, the glomerulus and the regulatory or transport unit, the tubule. Therefore, a bioartificial kidney requires two main units, the glomerulus and the tubule, to replace excretory and metabolic functions of the kidney. As to the development of a tissue-engineered kidney, a step function approach towards a fully implantable bioartificial kidney (BAK) has been adopted. An initial focus was to develop an extracorporeal bioartificial kidney comprised of a conventional synthetic haemofilter with a renal tubule cell assist device (RAD) in an acute extracorporeal blood circuit (see Figure 2.2). With success of this formulation, proof of concept for a wearable bioartificial kidney (WEBAK) combining Peritoneal dialysis (PD) with a bioartificial renal epithelial cell system (BRECS) has been recently achieved [31, 32], but a key critical hurdle for any bioartificial organ programme, including kidney replacement, is identifying a robust cell source for the replacement device. In this regard, a specific cell-based therapy requires several important methodologic choices and the solution of a number of technological problems. For cell-based therapies, large quantities of cells need to be expanded pathogen free, while maintaining uniformity in activity [12]. Current approaches to ensure robust cell expansion and uniformity requirements are dependent on either stem/ progenitor cells or

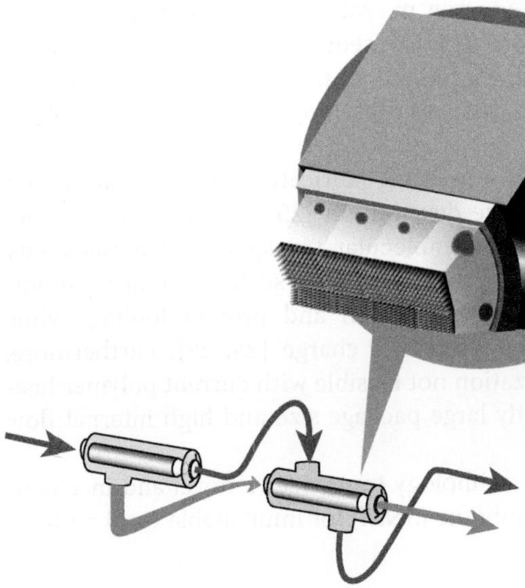

Figure 2.2 Schematic of an extracorporeal circuit of a bioartificial kidney used to treat patients with acute renal failure. The first cartridge is a hemofiltration cartridge in series with a renal tubule cell device (RAD). The ultrafiltrate is delivered to the luminal compartment of the RAD which contains the cells and the post-filtered blood is pumped into the extracapillary space of the RAD. The processed luminal ultrafiltrate from the RAD is discarded to waste and the processed blood is returned to the patient. *Source:* Principles of Tissue Engineering, Robert Lanza, Robert Langer, Joseph Vacanti (2013) 1142. (*See insert for color representation of the figure.*)

transformed cells. The use of human embryonic stem (ES) cells versus adult stem cells is under rigorous societal debate, with the current political environment strongly favouring adult stem cell processes [33, 34]. Although cell-based therapy has substantial technological, regulatory and ethical barriers, the potential to develop innovative treatments for a large number of clinical disorders, including acute and chronic renal diseases, is expanding rapidly. Progress in this field is highly dependent on an interdisciplinary approach at the interface of a number of scientific disciplines. This approach is at times empiric rather than reductive in nature, without full understanding of the manner in which cells may alter the complex pathophysiology of a systemic disorder. This limitation should not preclude continued efforts in this field. In fact, this empiric approach may result in unanticipated insights into basic biology, similar to those obtained in immunology as the field of human solid organ transplantation evolved. Cell therapy has the potential to creatively leverage nature's ability to provide new and much needed treatments to patients with acute and chronic diseases. For renal tubule epithelial cells, cell sourcing issues from human sources has been overcome [31]. This recent development has demonstrated a dramatic increase in the yield of renal epithelial cells, exceeding 1016 cells/g of kidney cortex from a healthy human source. This enhanced propagation methodology can easily produce over 100 000 devices per kidney and allow for autologous use of renal cells from a biopsy specimen from an individual [12].

The final developmental step based upon prior success of the RAD in acute disorders and the WEBAK in preclinical models, is the current approach to design, fabricate and test in preclinical models a functional fully implantable bioartificial kidney (IBAK) [35] (Figure 2.3).

The ultimate goal of tissue engineering, however, is the growth of fully functional and fully integrated replacement organs [15]. Thus, another approach towards the achievement of improved renal function involves the augmentation renal tissue with kidney cell expansion in vitro and subsequent autologous transplantation. The feasibility of achieving renal cell growth, expansion and in vivo reconstitution with the use of tissue-engineering techniques has been explored. Most recently, an attempt was made to harness the reconstitution of renal epithelial cells for the generation of functional nephron units. Renal cells were harvested and expanded in culture. The cells were seeded onto a tubular device constructed from a polycarbonate membrane, connected at one end with a silastic catheter that terminated into a reservoir. The device was implanted in athymic mice. Histologic examination of the implanted devices over time revealed extensive vascularization with formation of glomeruli and highly organized tubule-like structures. Immunocytochemical staining confirmed the renal phenotype. Yellow fluid was collected from inside the implant, and the fluid retrieved was consistent with the makeup of dilute urine in its creatinine and uric acid concentrations [36, 37]. Further

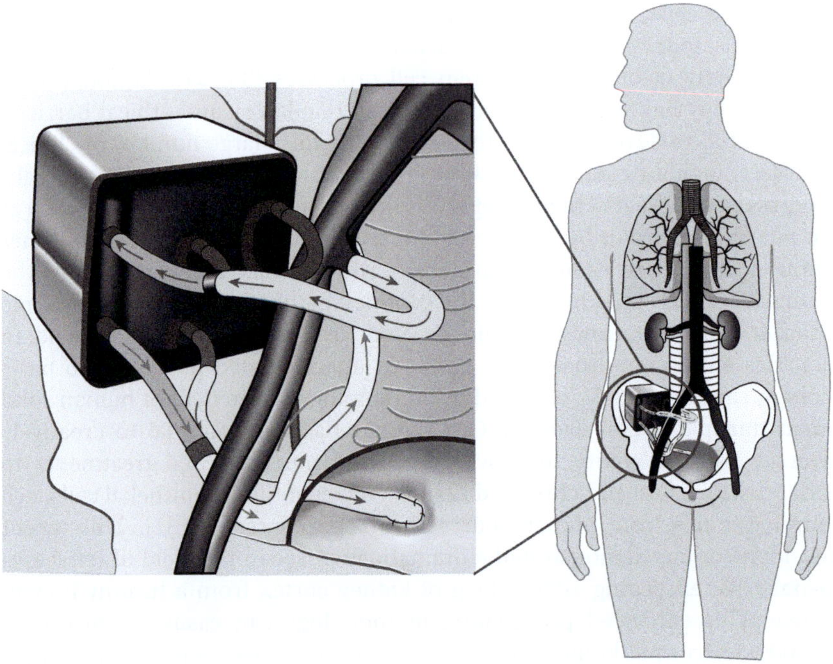

Figure 2.3 Concept illustration of implantable bioartificial kidney. *Source:* Principles of Tissue Engineering, by Robert Lanza, Robert Langer, Joseph Vacanti (2013), p. 1150. (*See insert for color representation of the figure.*)

studies have been performed showing the formation of renal structures in cows using nuclear transfer techniques [38]. Challenges await this technology, including the expansion of this system [15].

2.8 Next Steps

Significant preliminary data have been published documenting each of the three platform technologies necessary for an implantable bio artificial device that can provide enough small solute clearance to allow the patient with failing kidneys to avoid dialysis. Significant engineering challenges remain, but these challenges are not new to bioengineering. First and foremost, just as haemodialysis in the treatment of renal failure is more than the hollow fibre dialyser alone, the design of the implanted device will have to be guided by the planned therapeutic strategy [39]. Specific means for patients to self-monitor and reprogramme device function will be critical for truly independent self-care using an implanted device. Lifecycle management of the device, including recognition of impending device failure and a minimally invasive approach to renewing or replacing failed components, modules or cartridges seems essential. Selection of implant site will be guided by the paramount need to preserve vascular sites for future allografts. These challenges can be met and answered, just as they have been for other implanted devices, from corneas to ventricular assist devices [40, 41]. The remaining work to be completed is more than justified by the overwhelming burden of illness faced by patients with renal failure, the financial burden placed on payers for dialysis, and the revolutionary innovations already achieved in SNM technology, polymer chemistry and cell biology [42].

This technique involves sowing live cells in a biocompatible and biodegradable scaffold and the following crop in a bioreactor in order to encourage cell proliferation and tissue expansion [43]. In spite of many tissue-engineering successes, there are still some limitations including the choice of ideal biodegradable material for specific cell types and if high cell density is unobtainable, the constructs tend to degrade over time. One promising approach, which allows us to overcome these limitations, seems to be bio-printing. This is an engineering approach that uses the same auto-assembly techniques as cell units in order to encourage tissue fusion thus allowing cells to organise themselves without aid. The final model will therefore, have had no external interference. The first step in producing an organ or three-dimensional tissue through bio-printing is imaging, used in medicine to obtain necessary cell, tissue and organ information for the engineering model. Non-invasive techniques are used such as computerized tomography (CT) and magnetic resonance imaging (MRI) [44]. These images are then elaborated on a computer and mathematical models help create the required tissue or organ models. Bio-ink is then deposited on bio-paper using a bio-printer [43, 45–47].

2.9 Conclusion

Despite all the advances in renal replacement therapies, a portable, continuous, dialysate-free artificial kidney remains the holy grail of renal tissue engineering [48, 49]. Future research in renal tissue engineering will need to focus on reproducing mechanisms of whole-body homeostasis. A high priority must be given to sensing and regulating extracellular fluid volume, even if only at the crude level of having the patient weigh him- or herself daily and adjust ultrafiltration and reabsorption by the bioartificial kidney. Chemical-field effect transistors (ChemFETs) offer the possibility of measuring electrolyte levels in a protein-free ultrafiltrate and reading out the potassium level to the patient, who could then alter diet or treat him- or herself with potassium absorbing resins [15, 50]. The critical building blocks of an autonomous bioartificial kidney are advancing rapidly with revolutionary clinical trials currently underway at multiple medical centres. The technology with which to adapt these advances to a more autonomous, dialysate-free system is under development. In addition, progress has been made in the field of cryopreservation and thus the ability to manufacture, store and distribute bioartificial organs is advancing. The next decade, like the previous, will likely see quantum advances in renal tissue engineering [51, 52].

The final and most ambitious objective is, however, to be able to produce complete, functional tissue and organs from autologous patient cells thus allowing reject risks to become minimal. In applying the bottom up procedure, we try to replica and take advantage of natural dynamics in order to produce tissue and organs, which are as similar as possible to the original. In a context where, according to estimates of The International Monetary Fund, health costs will increase by 50% without innovation in therapeutic approaches and where, according to The World Economic Forum, only in Europe 12 people on the organ waiting list will die, the fundamental importance of this new technique becomes clear. It is fast, safe and reproducible while at the same time being the most precise available.

Although there are still many hurdles to overcome in order to reach the final objective, such as problems connected to vascularization and those concerning reproduction of scaffolds with complex geometries the idea of being able to produce whole, 3D organs using a 3D printer no longer represents a Utopistic dream.

List of Acronyms

BAK	BioArtificial Kidney
BRECS	Bioartificial Renal Epithelial Cell System
ChemFets	Chemical-Field Effect Transistors
CKD	Chronic Kidney Disease
CT	Computerized Tomography
ES	Embryonic Stem
ESRD	End-Stage Renal Disease
FDA	Food and Drug Administration
GFR	Glomerular Filtration Rate
IRC	Chronic Renal Failure
KTC	Kidney Transplant Centre
MEMS	MicroElectromMechanical Systems
MRI	Magnetic Resonance Imaging
PD	Peritoneal Dialysis
RAD	Renal tubule cell Assist Device
SNM	Silicon Nanopore Membranes
WAK	Wearable Artificial Kidney
WEBAK	WEarable BioArtificial Kidney

References

1 Go AS, Chertow GM, Fan D, McCulloch CE, Hsu CY. Chronic kidney disease and the risks of death, cardiovascular events, and hospitalization. *N Engl J Med* 2004;351:1296–1305.

2 Neovius M, Jacobson SH, Eriksson JK, Elinder CG, Hylander B. Mortality in chronic kidney disease and renal replacement therapy: a population-based cohort study. *BMJ Open* 2014;4:e004251.

3 Elie C, De Rycke Y, Jais J, Landais P. Appraising relative and excess mortality in population-based studies of chronic diseases such as end-stage renal disease. *Clin Epidemiol* 2011;3:157–169.

4 Thomas R, Kanso A, Sedor JR. Chronic kidney disease and its complications. *Prim Care* 2008;35:329–344, vii.

5 Tangri N, Kitsios GD, Inker LA, Griffith J, Naimark DM, Walker S, Rigatto C, et al. Risk prediction models for patients with chronic kidney disease: a systematic review. *Ann Intern Med* 2013;158:596–603.

6 Faenza A, Fuga G, Nardo B, Donati G, Cianciolo G, Scolari MP, Stefoni S. Metabolic syndrome after kidney transplantation. *Transplant Proc* 2007;39:1843–1846.

7 White Y, Grenyer BF. The biopsychosocial impact of end-stage renal disease: the experience of dialysis patients and their partners. *J Adv Nurs* 1999;30:1312–1320.

8 Greene T, Daugirdas JT, Depner TA, Gotch F, Kuhlman M, Frequent hemodialysis network study G, Solute clearances and fluid removal in the frequent hemodialysis network trials. *Am J Kidney Dis* 2009;53:835–844.

9 Public Health Europe. Organ donation and transplantation. Recent Facts & Figures. Available online at: http://ec.europa.eu/health/sites/health/files/blood_tissues_organs/docs/ev_20141126_factsfigures_en.pdf; 26 November 2014 (last accessed November 2017).

10 UNOS. Website. Data. UUnfO. Available at: http://www.unos.org/data/ (accessed November 2017).

11 Ministero della Salute, Trapianti CN. Website, available at: http://www.trapianti.salute.gov.it/cnt/cnt.htm (accessed November 2017).

12 Lanzer R, Langer R, Vacanti J. Principles of Tissue Engineering. Elsevier, 2013.

13 Grams ME, Juraschek SP, Selvin E, Foster MC, Inker LA, Eckfeldt JH, et al. Trends in the prevalence of reduced GFR in the United States: a comparison of creatinine- and cystatin C-based estimates. *Am J Kidney Dis* 2013;62:253–260.

14 Chronic Kidney Disease Prognosis Consortium, Matsushita K, van der Velde M, Astor BC, Woodward M, Levey AS, de Jong PE, et al. Association of estimated glomerular filtration rate and albuminuria with all-cause and cardiovascular mortality in general population cohorts: a collaborative meta-analysis. *Lancet* 2010;375:2073–2081.

15 Atala A. Tissue engineering and regenerative medicine: concepts for clinical application. *Rejuvenation Res* 2004;7:15–31.

16 Gura V, Rivara MB, Bieber S, Munshi R, Smith NC, Linke L, et al. A wearable artificial kidney for patients with end-stage renal disease. *JCI Insight* 2016;1.

17 Gura V, Davenport A, Beizai M, Ezon C, Ronco C. Beta2-microglobulin and phosphate clearances using a wearable artificial kidney: a pilot study. *Am J Kidney Dis* 2009;54:104–111.

18 Davenport A, Gura V, Ronco C, Beizai M, Ezon C, Rambod E. A wearable haemodialysis device for patients with end-stage renal failure: a pilot study. *Lancet* 2007;370:2005–2010.

19 Tentori F, Zhang J, Li Y, Karaboyas A, Kerr P, Saran R, Bommer J, et al. Longer dialysis session length is associated with better intermediate outcomes and survival among patients on in-center three times per week hemodialysis: results from the Dialysis Outcomes and Practice Patterns Study (DOPPS). *Nephrol Dial Transplant* 2012;27:4180–4188.

20 Walsh M, Culleton B, Tonelli M, Manns B. A systematic review of the effect of nocturnal hemodialysis on blood pressure, left ventricular hypertrophy, anemia, mineral metabolism, and health-related quality of life. *Kidney Int* 2005;67:1500–1508.

21 Jefferies HJ, Virk B, Schiller B, Moran J, McIntyre CW. Frequent hemodialysis schedules are associated with reduced levels of dialysis-induced cardiac injury (myocardial stunning). *Clin J Am Soc Nephrol* 2011;6:1326–1332.

22 Saran R, Bragg-Gresham JL, Levin NW, Twardowski ZJ, Wizemann V, Saito A, et al. Longer treatment time and slower ultrafiltration in hemodialysis: associations with reduced mortality in the DOPPS. *Kidney Int* 2006;69:1222–1228.

23 Cornelis T, van der Sande FM, Eloot S, Cardinaels E, Bekers O, Damoiseaux J, et al. Acute hemodynamic response and uremic toxin removal in conventional and extended hemodialysis and hemodiafiltration: a randomized crossover study. *Am J Kidney Dis* 2014;64:247–256.

24 Kim S, Feinberg B, Kant R, Chui B, Goldman K, Park J, et al. Diffusive silicon nanopore membranes for hemodialysis applications. *PLoS One* 2016;11:e0159526.

25 Fissell WH, Dubnisheva A, Eldridge AN, Fleischman AJ, Zydney AL, Roy S. High-performance silicon nanopore hemofiltration membranes. *J Memb Sci* 2009;326:58–63.

26 Kanani DM, Fissell WH, Roy S, Dubnisheva A, Fleischman A, Zydney AL. Permeability - selectivity analysis for ultrafiltration: Effect of pore geometry. *J Memb Sci* 2010;349:405.

27 Siddiqui MU, Arif AF, Bashmal S. Permeability-selectivity analysis of microfiltration and ultrafiltration membranes: effect of pore size and shape distribution and membrane stretching. *Membranes (Basel)* 2016;6.

28 Li L, Marchant RE, Dubnisheva A, Roy S, Fissell WH. Anti-biofouling sulfobetaine polymer thin films on silicon and silicon nanopore membranes. *J Biomater Sci Polym Ed* 2011;22:91–106.
29 Popat KC, Desai TA. Poly(ethylene glycol) interfaces: an approach for enhanced performance of microfluidic systems. *Biosens Bioelectron* 2004;19:1037–1044.
30 Kim S, Roy S. Microelectromechanical systems and nephrology: the next frontier in renal replacement technology. *Adv Chronic Kidney Dis* 2013;20:516–535.
31 Buffington DA, Pino CJ, Chen L, Westover AJ, Hageman G, Humes HD. Bioartificial renal epithelial cell system (BRECS): A compact, cryopreservable extracorporeal renal replacement device. *Cell Med* 2012;4:33–43.
32 Johnston KA, Westover AJ, Rojas-Pena A, Buffington DA, Pino CJ, et al. Development of a wearable bioartificial kidney using the Bioartificial Renal Epithelial Cell System (BRECS). *J Tissue Eng Regen Med* 2016; doi:doi.org/10.1002/term.2206.
33 Brower, V. Human ES cells: Can you build a business around them? *Nature Biotech* 1999;17.
34 Wertz DC. Embryo and stem cell research in the USA: a political history. *Trends Mol Med* 2002;8:143–146.
35 Roy S, Goldman K, Marchant R, Zydney A, Brown D, Fleischman A, et al. Implanted renal replacement for end-stage renal disease. *Panminerva Med* 2011;53:155–166.
36 Atala A. Recent developments in tissue engineering and regenerative medicine. *Curr Opin Pediatr* 2006;18:167–171.
37 Upputuri PK, Sivasubramanian K, Mark CS, Pramanik M. Recent developments in vascular imaging techniques in tissue engineering and regenerative medicine. *Biomed Res Int* 2015;2015:783983.
38 Lanza RP, Chung HY, Yoo JJ, Wettstein PJ, Blackwell C, Borson N, et al. Generation of histocompatible tissues using nuclear transplantation. *Nat Biotechnol* 2002;20:689–696.
39 Koh CJ, Atala A. Tissue engineering, stem cells, and cloning: opportunities for regenerative medicine. *J Am Soc Nephrol* 2004;15:1113–1125.
40 Orlando G, Baptista P, Birchall M, De Coppi P, Farney A, Guimaraes-Souza NK, et al. Regenerative medicine as applied to solid organ transplantation: current status and future challenges. *Transpl Int* 2011;24:223–232.
41 Sadri-Ardekani H, Atala A. Regenerative medicine for the treatment of reproductive system disorders: current and potential options. *Adv Drug Deliv Rev* 2015;82–83:145–152.
42 Demirtas TT, Irmak G, Gumusderelioglu M. Bioprintable form of chitosan hydrogel for bone tissue engineering. *Biofabrication* 2017;13;9(3):03500.
43 Pati F, Cho DW. Bioprinting of 3D tissue models using decellularized extracellular matrix bioink. *Methods Mol Biol* 2017;1612:381–390.

44 Ducouret E, Loriaut P, Boyer P, Perozziello A, Pesquer L, Mounayer C, Dallaudiere B. Tunnel positioning assessment after anterior cruciate ligament reconstruction at 12 months: comparison between 3d CT and 3d MRI. A Pilot Study. *Orthop Traumatol Surg Res* 2017;103(6):937–942.

45 Fricain JC, De Olivera H, Devillard R, Kalisky J, Remy M, Keriquel V, et al. [3D bioprinting in regenerative medicine and tissue engineering]. *Med Sci (Paris)* 2017;33:52–59.

46 Zhu W, Qu X, Zhu J, Ma X, Patel S, Liu J, et al. Direct 3D bioprinting of prevascularized tissue constructs with complex microarchitecture. *Biomaterials* 2017;124:106–115.

47 Colosi C, Costantini M, Barbetta A, Dentini M. Microfluidic bioprinting of heterogeneous 3D tissue constructs. *Methods Mol Biol* 2017;1612:369–380.

48 Sadri-Ardekani H, Atala A. Regenerative medicine. *Methods* 2016;99:1–2.

49 Orlando G, Wood KJ, De Coppi P, Baptista PM, Binder KW, Bitar KN, et al. Regenerative medicine as applied to general surgery. *Ann Surg* 2012;255:867–880.

50 Baptista PM, Atala A. Regenerative medicine: the hurdles and hopes. *Transl Res* 2014;163:255–258.

51 Trantidou T, Friddin M, Elani Y, Brooks NJ, Law RV, Seddon JM, Ces O. Engineering compartmentalized biomimetic micro- and nanocontainers. *ACS Nano* 2017; 25;11(7):6549–6565.

52 Moradi M, Hood B, Moradi M, Atala A. The potential role of regenerative medicine in the management of traumatic patients. *J Inj Violence Res* 2015;7:27–35.

3

Current Status and New Challenges of the Artificial Liver

Hiroshi Mizumoto, Nana Shirakigawa, and Hiroyuki Ijima

Department of Chemical Engineering, Faculty of Engineering, Kyushu University, Fukuoka, Japan

3.1 Introduction

The liver is the main metabolic organ *in vivo*. Therefore, severe liver dysfunction results in serious diseases with high mortality rates. Since Starzl et al. reported the first liver transplantation in a human in 1963, orthotropic liver transplantation has evolved by improving quality control, immune inhibition, and infection prevention of the donor's liver. As the result, liver transplantation became the most effective treatment for severe liver failure in patients, causing many lives to be saved. According to data from 2016, the number of patients waiting for liver transplantation in the USA was 14540 and 7841 patients received liver transplantation. However, 1240 patients died waiting for liver transplantation (United Network for Organ Sharing (UNOS). Available from: https://optn.transplant.hrsa.gov/data/view-data-reports/). In other words, donor shortage is a severe problem.

Therefore, an artificial liver (also called an artificial liver support system) can be expected to be a temporary substitute while a patient awaits transplantation. Furthermore, it has the potential to eliminate the need for liver transplantation by promoting liver regeneration and functional recovery. The necessary alternative function for treating liver failure is removal of toxins in blood. Based on this view point, the development of the artificial liver was considered to begin from Abel's report in 1914. He performed dialysis with colloid membranes (Abel et al., 1914). However, practical development and many reports have been produced since the 1950s (Kiley

Biomedical Engineering Challenges: A Chemical Engineering Insight, First Edition.
Edited by Vincenzo Piemonte, Angelo Basile, Taichi Ito, and Luigi Marrelli.

et al., 1958), about a half century after the first report. Hemodialysis with various types of membrane and hemoperfusion by using charcoal or synthetic resin has been carried out. These are classified as non-biological (non-bio) artificial livers. On the other hand, bioartificial livers (BAL) aim to compensate for the essential liver function by using biological components including whole livers or liver cells. The early clinical studies of BAL systems included cross-hetero-hemodialysis using xenogeneic animals or livers (Kimoto et al., 1959, Ozawa et al., 1982), extracorporeal liver perfusion (Eiseman et al., 1965, Sen et al., 1966), and an extracorporeal bioreactor with suspension hepatocytes (Matsumura et al., 1987). However, the outcome of these classical treatments was not satisfactory enough to save the patients' lives.

Based on these backgrounds, the artificial liver has been developed and has become an effective treatment in clinical use. In this chapter, the current status and the future vision of non-bio and bioartificial livers are reviewed. Furthermore, tissue- and organ-engineered livers are introduced as a new stream of liver failure treatments. Finally, the future vision of liver failure treatment is summarized.

3.2 Non-Biological Artificial Liver

The non-bio artificial liver aims to purify the blood by removing toxins. Historically, dialysis has been used to remove small molecular weight (MW) substances (Kiley et al., 1956), while an activated charcoal column has been used because of its excellent ability to adsorb middle MW substances (Yatzidis, 1964). Plasma exchange (PE) consisting of a hollow-fiber module has been used to remove albumin-bound toxins and other medium to large MW substances (Pineda and Taswell, 1981). However, dialysis cannot be used to help awaken patients from coma. In addition, activated charcoal cannot prevent capture/denaturation of blood cell components and the development of cerebral edema or coma. Additionally, the toxin removal ratio of PE is low, which decreases its effectiveness in awakening patients from coma.

The basic mechanism of each non-bio artificial liver has not changed recently, but good clinical data have been reported by combining existing systems or creating new systems. For example, hemodiafiltration (HDF), which removes moisture by ultrafiltration with hemodialysis and supplying an electrolyte solution, has a good ability to remove small to large MW substances (Yoshiba et al., 1996). HDF and high-flow HDF, which combines high-flow dialysis with PE, have reportedly significant effectiveness to aid with coma recovery and the potential to save the lives of patients with acute liver failure (ALF) (Yoshiba et al., 1996).

3.2.1 Classification and Clinical Study

The non-bio artificial liver in clinical use combines some systems. Each artificial liver system is described in Table 3.1.

3.2.2 PE and HDF

3.2.2.1 High-Volume Therapeutic PE

High-volume therapeutic PE (HVP) (Figure 3.1a) involves the replacement of the patient's plasma with fresh frozen plasma (FFP). Larsen defined HVP as the PE of 8 – 12% or 15% of one's ideal body weight with FFP (Larsen et al., 2016). They then treated ALF patients with at least grade 2 hepatic encephalopathy. Patients >18 years of age (N = 182) were randomly divided into groups: 90 patients who received standard medical therapy (SMT) and 92 patients who received SMT plus HVP (SMT + HVP) for 3 days. The HVP was performed by plasma replacement amounting to 15% of the ideal body weight at the rate of 1 – 2 L/h. After 84 days, viability was 58.7% in the SMT + HVP-treated patients and 47.8% in the SMT–treated patients.

3.2.2.2 High-Flow Dialysate Continuous HDF

The dialysate used for high-flow dialysate continuous HDF (CHDF) is delivered at a high flow rate (Figure 3.1b). Yokoi reported that HDF was performed at a dialysate flow rate of 300 – 500 mL/min and blood flow of 200 – 250 mL/min (Yokoi et al., 2009). Slow PE was also performed depending on each patient's prothrombin time. In the circuit, both PE and HDF are shown but the treatments were performed individually. The data from 90 patients with fulminant

Table 3.1 Overview of non-biological artificial liver systems.

System	Mechanism
Plasma exchange	Plasma separation from patient's blood and replacement with fresh frozen plasma
Albumin dialysis	Purification of patient's blood by dialysis with human albumin solution to remove albumin-bound toxins
Hemofiltration (HF)	Ultrafiltration of patient's blood by pressure without use of dialysate to remove target molecules
Hemodialysis (HD)	Dialysis using diffusion by connecting the dialysate with the blood via semipermeable membrane
Hemodiafiltration	Combination of HD (main function) and HF
Selective plasma separation	Plasma separation using special membranes that separate specific fractions

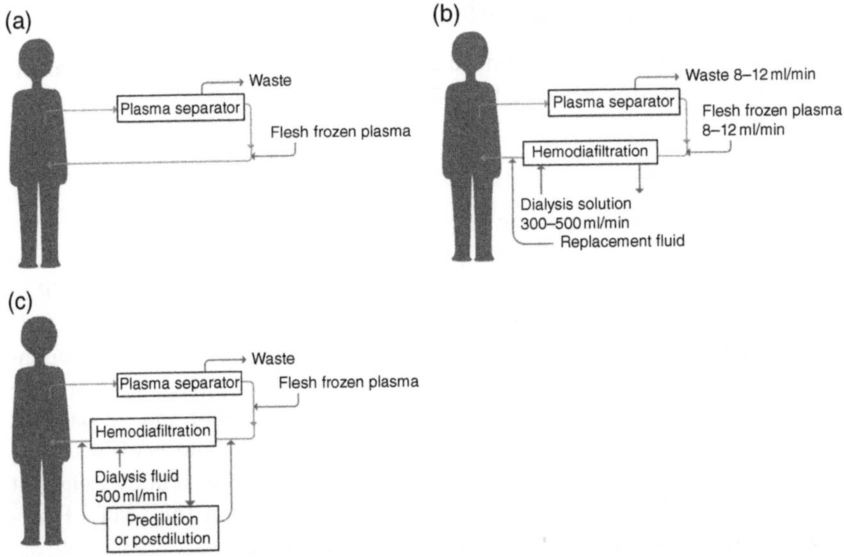

Figure 3.1 System images of non-biological artificial livers using plasma exchange and hemodiafiltration. (a) Plasma exchange, (b) plasma exchange and high-flow hemodiafiltration, and (c) plasma exchange with online hemodiafiltration. (*See insert for color representation of the figure.*)

hepatic failure, occurring after 1990, were analyzed. The rate of recovery from coma of the high-flow CHDF–treated group was 70.2% (33 of 47 patients) (Shinozaki et al., 2010; Yokoi et al., 2009). This value was significantly higher than that of the slow PE plus normal CHDF (47.1%, 16/34).

3.2.2.3 PE with Online HDF

This technique combines PE with FFP and HDF with pre- or post-dilution (Figure 3.1c). In this circuit, the substitution fluid is produced online from the dialysate through three ultrafilters (Arata et al., 2010). Arata reported treating 17 patients with ALF daily with online HDF and PE. Of them, 16 (94.1%) reportedly completely recovered from hepatic encephalopathy after approximately five online HDF sessions.

3.2.3 Blood Purification with Albumin Dialysis

3.2.3.1 Single-Pass Albumin Dialysis

In the single-pass albumin dialysis (SPAD) method (Figure 3.2a), 4 – 5% human albumin is used as a dialysate, which is not reused. Kreymann reported the use of SPAD in a patient with fulminant Wilson disease effectively removed the copper and bilirubin (Kreymann et al., 1999). In this case, the albumin solution was not recycled because the copper could not be removed by the

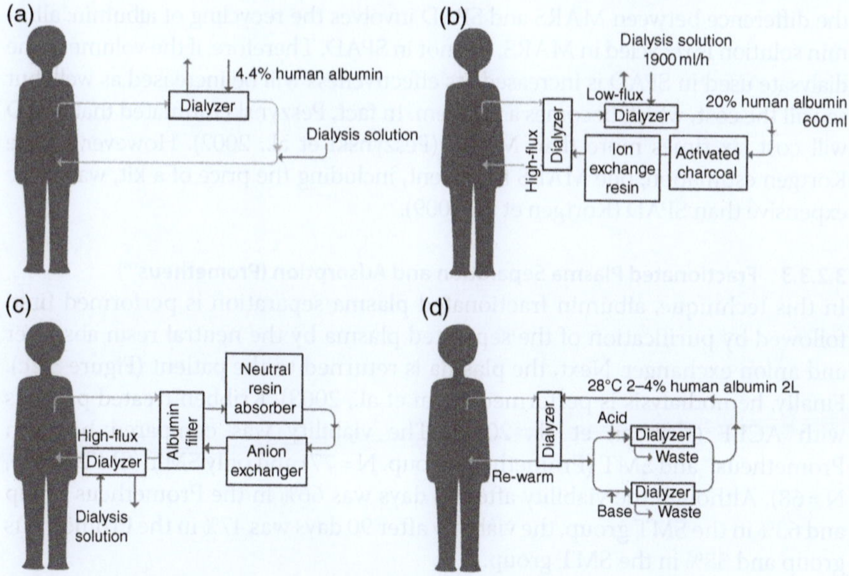

Figure 3.2 System images of non-biological artificial livers using albumin dialysis. (a) Single-pass albumin dialysis, (b) molecular adsorbent recirculating system, (c) fractionated plasma separation and adsorption, and (d) Hepa Wash. (*See insert for color representation of the figure.*)

charcoal cartridge and the bilirubin absorber. Additionally, Kıhtır reported that SPAD effectively treated a patient with phenobarbital poisoning (Kıhtır et al., 2016). Caution should be taken with regard to metabolic derangements and electrolyte disturbances, particularly when SPAD uses regional citrate anti-coagulation (Sponholz et al., 2015).

3.2.3.2 Molecular Adsorbent Recirculating System

The molecular adsorbent recirculating system (MARS) consists of a flux dialyzer with that recycles 600 mL of a 20% human albumin solution (Figure 3.2b). The albumin solution is regenerated by the dialysis and flowing through the activated charcoal and the ion exchange resin (Sauer et al., 2004; Tan, 2004; Annesini et al., 2005, 2008; Novelli et al., 2009a, 2009b; Annesini et al., 2010).

Sponholz compared MARS with SPAD. In SPAD, 4% albumin was used at a flow rate of 700 mL/h (Sponholz et al., 2015). A total of 32 patients with ALF, acute-on-chronic liver failure (ACLF), or liver graft failure were treated. The plasma bilirubin levels were reduced using both systems. However, only MARS affected other paraclinical parameters (i.e., serum bile acids, albumin-binding capacity, and creatinine and urea levels). On the contrary, Piechota mentioned that flow rate is key because the bile acid and ammonia levels were reduced when SPAD was performed at a flow rate of 1000 mL/h for 10 h (Piechota and Piechota, 2016). In other words, the difference between MARS and SPAD involves the recycling of albumin; albumin solution is recycled in MARS, but not in SPAD. Therefore, if the volume of the dialysate used in SPAD is increased, its effectiveness will be increased as well, but so will the cost, which becomes a problem. In fact, Peszynski estimated that SPAD will cost six times more than MARS (Peszynski et al., 2002). However, in the Kortgen estimation, one MARS treatment, including the price of a kit, was more expensive than SPAD (Kortgen et al., 2009).

3.2.3.3 Fractionated Plasma Separation and Adsorption (Prometheus™)

In this technique, albumin fractionated plasma separation is performed first, followed by purification of the separated plasma by the neutral resin absorber and anion exchanger. Next, the plasma is returned to the patient (Figure 3.2c). Finally, hemodialysis is performed (Rifai et al., 2003). Kribben treated patients with ACLF (Kribben et al., 2012). The viability was compared between Prometheus™ and SMT (Prometheus group, N = 77) and only SMT (SMT group, N = 68). Although the viability after 28 days was 66% in the Prometheus group and 63% in the SMT group, the viability after 90 days was 47% in the Prometheus group and 38% in the SMT group.

3.2.3.4 Hepa Wash

Hepa Wash involves blood dialysis using a 2 – 4% human albumin solution that is recycled (Figure 3.2d). The dialysate of human albumin solution is regenerated by dialysis and filtration with adding acid and base and circulated (Huber et al., 2017). Huber reported treating cases of ACLF (n = 9) and secondary liver failure (n = 5) with Hepa Wash several times, after which point serum bilirubin, creatinine, and blood urea nitrogen levels were significantly decreased.

3.2.4 Selective Plasma Filtration Therapy

3.2.4.1 Biologic-Detoxifilter/Plasma Filter

In the biologic-detoxifilter/plasma filter (Biologic-DT plus PF) technique, the patient's plasma is separated, purified by flowing through powdered charcoal and cation exchangers, and returned to the patient (Biologic-DT) (Figure 3.3a). The plasma is then separated again, purified by powdered charcoal with silica, and returned (plus PF) (Ash et al., 2001). Ash reported the treatment of eight adult ICU patients with both systemic inflammatory response syndrome and multiple organ failure. The four patients were treated with Biologic-DT and the four patients were treated with Biologic-DT plus PF. The sepsis was resolved in five of eight patients.

3.2.4.2 Selective Plasma-Exchange Therapy

In selective plasma-exchange therapy (SEPET), the plasma is separated by a SEPET cartridge and replaced with FFP (Figure 3.3b). The separated plasma

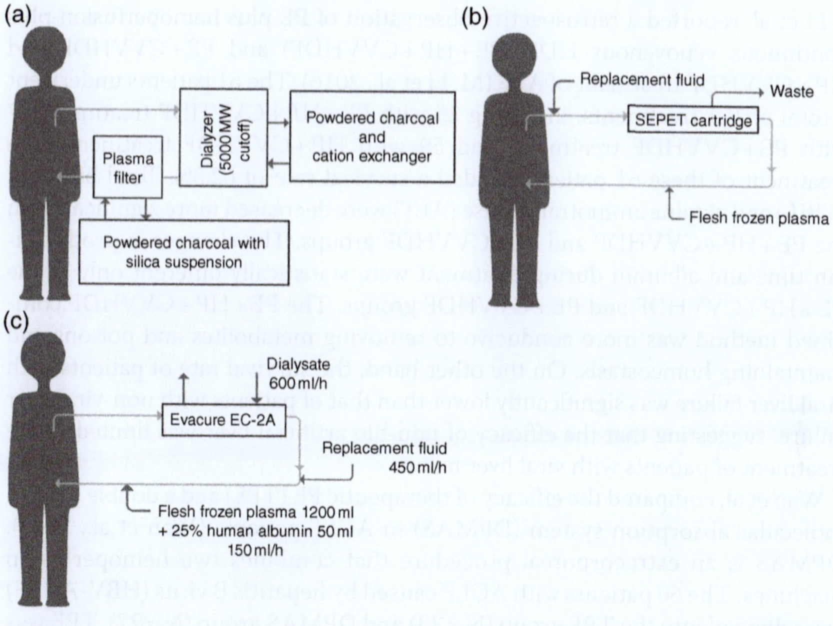

Figure 3.3 System images of non-biological artificial livers using selective plasma filtration. (a) Biologic-detoxifilter/plasma filter, (b) selective plasma-exchange therapy, and (c) plasma filtration with dialysis. (*See insert for color representation of the figure.*)

was replaced by 2.5 L of FFP over 6 h in a study by Rozga (Rozga et al., 2006). Rozga described treating pigs with experimentally induced fulminant hepatic failure and reported that SEPET was safe and effective for extending survival time and arresting brain swelling.

3.2.4.3 Plasma Filtration with Dialysis

In plasma filtration with dialysis (PDF), the dialysate is directed through a plasma separator and separated by control of its flow volume (Figure 3.3c). Dialysis is simultaneously performed. After that, FFP and 25% human albumin solution is returned to the patient (Yamamoto et al., 2015). Yamamoto treated 17 patients with PDF and PE. The plasma adiponectin level was increased after PDF but decreased after PE.

3.2.5 Clinical Observations of Various Combinations

Although the different non-bio artificial liver models – including PE, blood or plasma perfusion, and continuous blood dialysis filtration – have different principles and functions, their combined applications can be used to achieve effective treatment.

Li et al. reported a retrospective observation of PE plus hemoperfusion plus continuous venovenous HDF (PE + HP + CVVHDF) and PE + CVVHDF and HP + CVVHDF treatment of ALF (M. Li et al., 2016). The 61 patients underwent a total of 171 treatments, including 25 with PE + HP + CVVHDF treatment, 87 with PE + CVVHDF treatment, and 59 with HP + CVVHDF treatment. The treatment of these 61 patients yielded a survival rate of 62.3%. Total bilirubin (TBIL) and alanine aminotransferase (ALT) were decreased more significantly in the PE + HP + CVVHDF and PE + CVVHDF groups. The changes in prothrombin time and albumin during treatment were statistically different only in the PE + HP + CVVHDF and PE + CVVHDF groups. The PE + HP + CVVHDF combined method was more conducive to removing metabolites and poisons and maintaining homeostasis. On the other hand, the survival rate of patients with viral liver failure was significantly lower than that of patients with non-viral liver failure, suggesting that the efficacy of non-bio artificial liver was limited in the treatment of patients with viral liver failure.

Wan et al. compared the efficacy of therapeutic PE (TPE) and a double plasma molecular absorption system (DPMAS) in ACLF patients (Wan et al., 2017). DPMAS is an extracorporeal procedure that combines two hemoperfusion machines. The 60 patients with ACLF caused by hepatitis B virus (HBV-ACLF) were divided into the TPE group (N = 33) and DPMAS group (N = 27). TPE was more efficient at eliminating protein-bound toxins (i.e., TBIL, direct bilirubin [DBIL], total bile acid), and decreasing serum highly sensitive C-reactive protein (hsCRP) levels. On the other hand, albumin levels were remarkedly reduced after TPE and DPMAS, with a higher reduction rate in the former than the latter. However, no difference in 12-week survival was detected between the TPE and DPMAS treatments. Therefore, DPMAS and TPE are both effective and safe therapeutic approaches for HBV-ACLF patients. Despite TPE resulting in higher removal rates of TBIL, DBIL, and hsCRP, it required large volumes of FFP, carried the risk of plasma allergies and blood-borne disease transmission, and had higher loss rate of albumin than DPMAS.

Gong et al. evaluated the efficacy of a new artificial liver support system based on fractionated plasma separation and adsorption integrated with continuous venovenous hemofiltration (FPSA-CVVH, similar to the Prometheus system) compared with a traditional plasma separation and adsorption (PSA) system (Gong et al., 2013). Sixteen patients with ALF each received the first treatment session using the PSA. They then received a second session using the FPSA-CVVH, in which albumin-rich plasma separated using a fraction plasma separator was ultrafiltered using a hemofilter and perfused through an adsorber before being returned to the blood. The FPSA-CVVH had a higher clearance of bilirubin and slower decline of clearance over time. In particular, the decline in clearance was much slower than that seen with PSA, MARS, and the Prometheus system. The FPSA-CVVH also showed a reduced prolongation of coagulation parameters and decline in albumin level compared to PSA. Therefore, the authors concluded that FPSA-CVVH is superior to PSA with regard to toxin elimination and side effects on serum proteins.

Zhou et al. assessed overall survival and the possible prognostic predictors in a cohort of ACLF patients who underwent PE or plasma bilirubin adsorption (PBA) combined with PE (Zhou et al., 2015). The mean survival times of patients treated with PBA + PE and patients treated with PE alone were significantly different (531 d vs 343 d, respectively; $P = 0.012$). PE and PBA have similar ability to lower bilirubin and inflammatory cytokine levels, but PBA has the advantages of not transmitting blood-borne diseases or featuring rare side effects. The combination therapy of PBA + PE can reduce the amount of plasma required for treatment compared to PE alone.

As a novel approach for artificial liver support, Defterevos et al. evaluated the curative effect of peritoneal albumin dialysis using a porcine acute hepatic failure model (Defterevos et al., 2016). They claimed that the concept of peritoneal albumin dialysis was cheaper, was more readily available, did not require special equipment or trained personnel, and avoided the use of extracorporeal circulation.

The artificial liver support system can rescue patients with liver failure by optimizing treatment combinations in properly selected patients. However, larger prospective randomized controlled trials are needed before these treatments can be recommended as standard clinical practice.

3.3 Bioartificial Liver

3.3.1 Bioartificial Liver Support System

BAL is an extracorporeal circulation system that comprises a bioreactor containing immobilized functional liver cells. Since they use living functional hepatocytes, BAL systems are expected to compensate for important functions of the liver. BAL devices generally consist of a bioreactor including functional liver cells in various culture configurations. The BAL system type has been extensively reviewed elsewhere (Park et al., 2005) and several of them have been evaluated in clinical trials.

The extracorporeal liver assist device (ELAD) is based on a conventional hollow-fiber bioreactor loaded with C3A cells derived from human hepatoblastoma HepG2. In the first phase I trial, 11 patients were treated with the ELAD and no short-term safety problems were noted (Sussman et al., 1994). Metabolic support was documented in 10 of the 11 patients, and six reached a successful end-point. In the pilot controlled study, 24 patients were divided into two groups – 17 with a potentially recoverable lesion (group I) and seven listed for transplantation (group II) – and then randomly allocated to ELAD hemoperfusion or control (Ellis et al., 1996). The median period of ELAD hemoperfusion was 72 hours. In group I, seven of nine (78%) ELAD-treated patients survived compared to six of eight (75%) controls. In group II, one of three (33%) ELAD-treated patients survived versus one of four (25%) controls. Both survivors in this group underwent transplantation. A modified version of the ELAD was also evaluated (Millis et al., 2002).

Most recently, a phase III trial of 96 patients with alcohol-induced liver decompensation concluded that ELAD treatment might be favorable in a subgroup of younger patients with sufficient renal function (serum creatinine < 1.5 mg/dL) and less severe coagulopathy (international normalized ratio (INR) < 2.5); both components are included in the Model for End-stage Liver Disease (MELD) score (Wiegand et al., 2016).

Demetriou et al. developed a BAL system combining a perfusion circuit containing cellulose-coated activated charcoal columns and a hollow-fiber module with porcine hepatocytes on microcarriers in the extracapillary space (HepatAssist®). In the first clinical trial, 10 patients were treated with the BAL system and eight underwent orthotopic liver transplantation (OLT) following BAL treatment (Demetriou et al., 1995). Later, the system was evaluated in a large randomized trial including 171 patients with fulminant (FHF)/subfulminant (SHF) hepatic failure and primary nonfunction (PNF) following liver transplantation (Demetriou et al., 2004). The patients were randomly assigned to the BAL (N = 85) or control (N = 86) group. The 30-day patient survival rates were 71% for BAL versus 62% for control (P = 0.26). When confounding factors were accounted for in the analysis, there was no statistically significant difference in survival between the two groups in the entire patient population. However, when PNF patients were excluded, FHF/SHF patients treated with BAL had a statistically significant survival advantage compared with controls (44% reduction in mortality).

The hybrid liver support system (LSS) and modular extracorporeal liver support system (MELS) were developed by Gerlach and coworkers. Their device consists of several bundles of interwoven hollow-fiber capillaries with liver cells in the intercapillary space (Sauer et al., 2001). This system utilizes porcine or human hepatocytes. For porcine hepatocytes, seven patients in coma stage II–IV were treated with a BAL system for 8–46 hours (Mundt et al., 2002). All patients were bridged successfully to liver transplantation. In a clinical trial using human hepatocytes, the bioreactor (CellModule) – immobilized primary human hepatocytes were combined with two detoxification modules for albumin dialysis and CVVH if required (MELS) (Sauer et al., 2002). Eight patients were treated for 7–144 hours. In all cases, neurological status improved during treatment and no adverse events were observed.

The AMC-BAL was a BAL system devised at the Academic Medical Center of Amsterdam. The bioreactor is equipped with a non-woven polyester matrix for hepatocyte adhesion and oxygen-permeable capillaries to enhance the oxygen supply at the site of the hepatocytes. In the first phase I clinical trial, seven patients were treated with the AMC-BAL system. The total treatment duration was 8–35 hours (van de Kerkhove et al., 2002). Five additional patients were included in the phase I trial (for a total of 12 patients) (van de Kerkhove et al., 2004). Eleven patients were successfully bridged to OLT, while the other patient survived without OLT. Four patients died after OLT due to disease or OLT-related problems. The eight other patients (66%) are now in good health with post-BAL survival times of 6–30 months.

As described previously, many studies reported that the clinical condition of patients improved remarkably with BAL treatment and allowed successful bridging to liver transplantation or their own liver regeneration. However, an analysis of the effect of BAL treatment on survival in controlled and randomized clinical trials including large and uniform patient groups is required to yield statistically significant outcomes. Simultaneously, the issue of cell source should be addressed.

3.3.2 Cell Source for BAL

Cell source is one of the most important components of the BAL system. Various cell types are being investigated as candidate sources of hepatocytes for BAL, including primary porcine hepatocytes, isolation and preservation of primary human hepatocytes, liver cell lines, and stem cells. Most clinical BAL studies have used primary porcine hepatocytes. Porcine hepatocytes are readily obtainable in large quantities and express liver-specific functions comparable to human hepatocytes. Despite these advantages, several drawbacks hamper the widespread use of porcine hepatocytes for BAL application. One of the major concerns is related to zoonosis. Porcine endogenous retrovirus (PERV) may display pig-to-human xenozoonosis. Numerous studies about PERV infection of human cells in vitro and rodents in vivo have been reported (Wilson, 2008). However, no clinically important zoonosis or PERV transmission has been reported in the treatment of BAL systems using porcine hepatocytes. Di Nicuolo et al. reported that no cases of PERV infection were found in long-term immunosuppressed patients who received porcine hepatocyte-based BAL after 8.7 years of follow-up (Di Nicuolo et al., 2010). However, the potential for PERV infection remains during BAL treatment. Yang et al. reported that microencapsulated primary porcine hepatocytes have a risk of PERV transmission (Yang et al., 2010). Thus, the utilization of primary porcine hepatocytes for BAL is prohibited in several European countries (Chamuleau et al., 2006).

Primary human hepatocytes are the most desirable cell source for BAL. However, liver donors for the isolation of primary human hepatocytes, as well as for liver transplantation, are scarce. Currently, human livers for hepatocyte isolation are commonly obtained from rejected livers for transplantation or non-transplantable liver organs, such as from non-heart-beating donors (Baccarani et al., 2003; Hughes et al., 2006). Since primary human hepatocytes have limited proliferative activity, an approach for preserving hepatocyte viability and phenotype for use in BAL is needed.

It has been widely suggested that approximately 20–30% of the total liver volume is required for survival, indicating that a large number of hepatocytes (about tens of billions) would be required for clinical-scale BAL (Chamuleau et al., 2006; Sakiyama et al., 2017). Because of the limited proliferative activity of primary hepatocytes, hepatic cell lines, which have a vigorous proliferative activity, are a potential cell source for BAL application. The human hepatoma cell line HepG2 is one of the representative human hepatoma cell lines, while C3A, a subclone of HepG2, has been used in ELAD systems in clinical trials. On the other hand, some liver cell lines have been developed by gene transfer of the simian virus 40 large tumor antigen gene (SV40 LT) or human telomerase reverse transcriptase such as immortalized human hepatocyte cell lines (Kobayashi et al., 2000b; Li et al., 2005; Tsuruga et al., 2008). Because these types of cell lines carry a risk of tumorigenesis, researchers have adopted a reversible immortalization system that uses the Cre-loxP site-specific recombination reaction to remove the targeting gene in genetically engineered hepatocytes (Kobayashi et al., 2000a).

In recent years, there has been a focus on deriving human hepatocytes from stem cells. Stem cells have the potential to differentiate into functional hepatocytes, including embryonic stem cells (ESCs), induced pluripotent stem cells (iPSCs), mesenchymal stem cells (MSCs), and liver progenitor cells (LPCs). ESCs are pluripotent stem cells isolated from the inner cell mass of blastocysts (Evans and Kaufman, 1981; Martin, 1981). ESCs have unlimited proliferative capability and the potential to differentiate into a variety of cell lineages. However, they are created from fertilized eggs, which creates bioethical and potentially legal issues. In 2006, Yamanaka and coworkers published the first report of iPSCs in a mouse, followed by the successful generation of human iPSCs the next year (Takahashi and Yamanaka, 2006, Takahashi et al., 2007). Such iPSCs were established from terminally differentiated somatic cells by the transfection of several genes, and they have the same qualities as ESCs but do not pose the same issues. Thus, iPSCs are expected to be a candidate cell source for BAL instead of ESCs. Many groups have reported the differentiation protocol of ESCs and iPSCs into hepatic lineages; that is, maturation into functional hepatocytes. ESC- and iPSC-derived hepatocytes expressed some liver-specific genes and functions (Cai et al., 2007; Kondo et al., 2014; Si-Tayeb et al., 2010; Song et al., 2009; Takata et al., 2011).

MSCs were originally reported as adult stem cells with the capacity to differentiate into mesodermal lineages such as osteocytes, adipocytes, and chondrocytes (Barry and Murphy, 2004). Later, researchers reported that MSCs have the potential to differentiate into ectodermal and endodermal lineages (Lee et al., 2004). Human MSCs were isolated from bone marrow, adipose tissue, and cord blood (Mo et al., 2016). MSCs have the advantage of relatively easy availability. On the other hand, their proliferation capacity is limited. Thus, the establishment of a large-scale preparation process is needed for their use in BAL.

In the context of embryonic development, hepatic progenitor cells found in the early fetal liver bud, called hepatoblasts, are characterized by their bipotentiality for generating both hepatocytes and cholangiocytes (Schmelzer et al., 2007). Several liver progenitor cells have been isolated and characterized from fetal and adult liver. These cells have the capacity to differentiate into hepatocyte-like cells in vitro (Turner et al., 2011). The human hepatoma cell line HepaRG is reportedly a bipotent liver progenitor cell line that differentiates into hepatocytes (Gripon et al., 2002). Thus, HepaRG may be a promising candidate for BAL application (Hoekstra et al., 2011).

In past clinical trials, each cell source had different issues. Several genetically engineered liver cell lines and immortalized human hepatocytes have been evaluated in large-animal experiments (Kobayashi, 2009; Zhao et al., 2012). The establishment of a high-performing non-tumorigenic liver cell line would help provide a stable source of hepatocytes for BAL treatment. On the other hand, stem cells with the capability to differentiate into functional hepatocytes are also expected to be used as cell sources of BAL.

Several researchers reported the expansion and differentiation (or maturation) of stem cells into the bioreactors. Some liver stem cells, fetal hepatocytes, and pluripotent stem cells proliferated and differentiated in the various bioreactors (Amimoto et al., 2011; Fonsato et al., 2010; Iwamuro et al., 2012; Miki et al., 2011; Ring et al., 2010). Moreover, some researchers evaluated small-scale BAL devices with stem cells in animal experiments (Mizumoto et al., 2012; Nibourg et al., 2012). Most recently, Shi et al. reported a novel BAL system with generated human hepatocytes by lineage conversion (hiHep-BAL) at a preclinical scale (Shi et al., 2016). They converted human fibroblasts into hepatocytes (hiHep) by using a direct reprogramming approach similar to that reported previously (Huang et al., 2011; Sekiya et al., 2011). The hiHep cells were expanded to a large scale and a BAL system using three billion hiHep cells was evaluated in a porcine ALF model. The hiHep-BAL treatment restored liver function, corrected blood levels of ammonia and bilirubin, and prolonged survival.

In the use of stem cells, it is essential to establish the efficient expansion and differentiation process at low cost. Furthermore, since stem cell-derived hepatocytes are thought to have the same properties as mature primary hepatocytes, it is important to devise culture methods to maintain their long-term function.

3.4 New Stream for Artificial Liver

3.4.1 Tissue Engineering for Liver Construction

Back in 1993, Langer and Vacanti first propounded tissue engineering for the reconstruction of injured tissue or organs *in vivo*. Tissue engineering is a concept in which a tissue is constructed using cells, a scaffolding, and cytokines to promote cell growth or functional expression (Figure 3.4). Thereafter, many researchers have attempted to achieve tissue reconstruction, which has resulted in the creation of skin, cornea, and cartilage using a patient's own cells (Japan Tissue Engineering Co., Ltd., and Cornea Biosciences™) (Chua et al., 2016; Ghezzi et al., 2015; Liu et al., 2017).

On the other hand, practical studies of the construction of an organ with a rich metabolism or thick structure are scarce because it is difficult to maintain a thick structure in a created tissue due to the need for a mass transfer of oxygen or inadequate scaffold strength. Although there are many *in vitro* studies, *in vivo* studies of liver tissue creation using animals are rare, some of which we describe here.

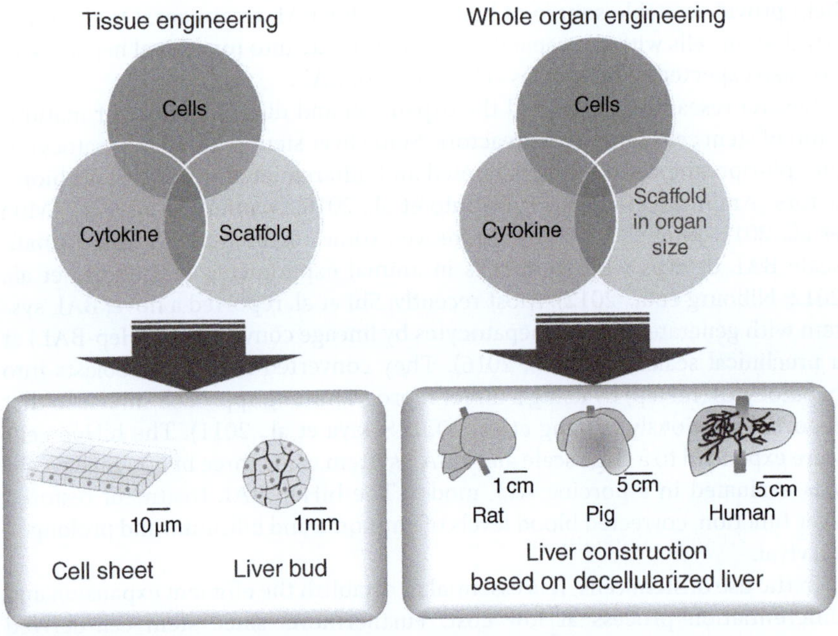

Figure 3.4 Concepts of tissue engineering and whole-organ engineering. (*See insert for color representation of the figure.*)

One study used a cell sheet to construct liver tissue (Ohashi et al., 2007). The cell sheet was obtained by the culturing of cells in a dish coated with a temperature-sensitive polymer. Once the cells reach confluence, the temperature is changed and the cell sheet detaches from the dish. Using this method, hepatocyte sheets were obtained and used to construct the liver tissue. Ohashi reported that the transplanted hepatocytes grew within a neovascularized location in mice treated with partial hepatectomy (Ohashi et al., 2007). Additionally, Nagamoto (2016) reported that hepatocyte sheets could be obtained using hepatocyte-like cells created by human iPS cells (iPS-HLCs) (Nagamoto et al., 2016). The iPS-HLC sheets can then be transplanted to the surface of a mouse liver from which the capsule had been removed. Human albumin production was then detected in the blood of the transplanted mice. Furthermore, the viability of ALF mice induced by carbon tetrachloride was improved by iPS-HLC sheet transplantation.

On the other hand, Hou reported that a heparin-conjugated matrix used to immobilize growth factors improved transplanted hepatocyte viability (Hou et al., 2012). Nakamura also reported that the liver-specific matrix obtained from native liver improved liver function (Nakamura and Ijima, 2013). These reports suggest that the cell–cell interaction is important to transplanted hepatocyte survival. They also suggested that a rich vascular location or cell comfortable environment is important for hepatocyte transplantation. However, in all reports, the thickness of the constructed liver tissue was <1 mm, likely due to a limited oxygen supply, as the required human liver tissue thickness for clinical treatment is in the order of centimeters of thickness. Scaling up of liver tissue thickness is needed for clinical use.

In contrast, another report stated that a liver bud was obtained by self-organization after the combination of hepatic endoderm cells (obtained by inducing human iPS cells), human umbilical vein endothelial cells, and human MSCs (Takebe et al., 2013). The transplanted liver bud induced vascularization and connected its internal vascular network to the original vascular network in the transplant location. Although the mixing of undifferentiated cells carries the potential risk of teratoma formation, the authors reported that the liver bud could construct a few millimeters' worth of liver tissue by the transplantation. In the future, the authors of this study also hope to scale up the results to enable liver functional assist or replacement.

3.4.2 Whole Organ Engineering for the Transplantable Artificial Liver

Whole organ engineering was recently propounded (Song and Ott, 2011). Although the explanation of this concept varies among researchers, "cell," "scaffold at organ scale," and "cytokine that improves cell growth or function" are listed as three elements (Figure 3.4). The decellularized organ obtained by removing cells from a native organ has been studied as an organ-scale scaffold by many researchers. The construction of various types of organs such as the heart, lung, liver, kidney, and pancreas using decellularized organs has been studied (Song and Ott, 2011). Some studies have focused on liver construction using decellularized liver obtained using detergents. Such treatment is not dependent on a donor liver.

Uygun reported that decellularized and recellularized rat liver was heterotopic-transplanted by rerouting of the body's blood supply from the kidney to the recellularized liver for 8 h (Uygun et al., 2010). Shirakigawa reported a different decellularized method of creating a rat liver that retains a fine vascular structure (Shirakigawa et al., 2012, 2013) (Figure 3.5). They aimed to construct a liver by reconstructing the vascular network by seeding liver cells in a suitable environment between the vascular networks. The decellularization of porcine or human liver has also been reported (Mazza et al., 2015; Yagi et al., 2013).

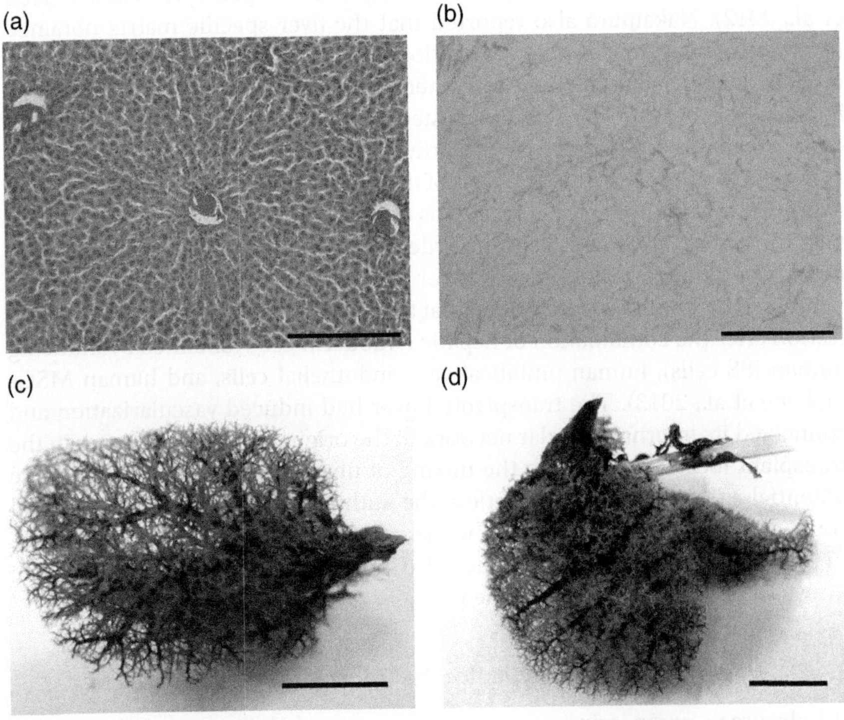

Figure 3.5 Histological analysis using hematoxylin and eosin staining (a, b) and molded resin of vascular structure (c, d). native liver (a, c) and decellularized liver (b, d). Scale bars indicate 100 μm (a, b) and 5 mm (c, d), respectively (reproduced from Shirakigawa et al., 2013 with permission from Elsevier) (Shirakigawa et al., 2013). (*See insert for color representation of the figure.*)

However, there are no reports about transplantation via the connection of blood vessels over long periods. Ko reported on the decellularization of porcine liver and recellularization with endothelial cells (Ko et al., 2015). One-day transplantation was performed and blood was observed flowing in the transplanted organ. Liver cells were not seeded, so no reference was made to liver function. However, this report suggested that the vascularization of decellularized tissue is important for long-term transplantation.

As described earlier, the creation of fine vascular networks is needed for the construction of liver tissues and organs to provide sufficient oxygen to each liver cell. We suppose that both blood vessels that can endure the blood flow and pressure and the fine vascular network that attaches to liver cells are needed in the liver construction process. These studies will achieve liver construction and enable transplantation of the constructed liver instead of a donor liver. This will resolve the severe liver donor shortage problem and pave the way for other uses of the constructed liver such as new drug development.

3.5 Conclusion and Future Trends

In this chapter, the current status and new challenges of artificial liver were introduced. The classical non-bio artificial liver is reportedly effective for improving patient conditions such as awakening from coma but did not improve the survival ratio. However, the use of non-bio artificial liver recently achieved a significant survival ratio by the combination of some systems or improvement of the removal ratio of toxic substrates. On the other hand, although some excellent bioartificial livers have been reported, its clinical use is severely limited because of the cell source problem. In other words, the most important problem of the bioartificial liver for clinical use is establishing a method of obtaining a safe and functional cell source. In the future, the establishment of a cell source and combined use with a non-bio artificial liver may improve patient conditions and survival ratios. Also, as a new challenge, tissue- and organ-engineered livers have been studied. These techniques aim to construct functional living liver tissue and are hoped to become fundamental treatments for full recovery. However, their construction requires much time, effort, and cost. As described previously, we propose the ideal treatment of artificial liver assist as follows. First, treatment using non-bio artificial liver assistance will save patient lives. Second, conventional liver transplantation from the donor or the induction of regeneration of the patient's liver by bioartificial liver assistance will aid in full recovery. And third, the organ-engineered liver will be transplanted to circumvent the donor shortage. In conclusion, the improvements of all types of artificial liver are needed and their combination use will provide ideal therapeutic treatment for severe liver failure patients.

List of Acronyms

ACLF	acute-on-chronic liver failure
ALF	acute liver failure
ALT	alanine aminotransferase
AMC-BAL	BAL system devised at the Academic Medical Center of Amsterdam
BAL	bioartificial liver
Biologic-DT plus PF	biologic-detoxifilter/plasma filter
CHDF	continuous hemodiafiltration
CVVH	continuous venovenous hemofiltration
CVVHDF	continuous venovenous hemodiafiltration
DBIL	direct bilirubin
DPMAS	double plasma molecular absorption system
ELAD	extracorporeal liver assistance device
ESCs	embryonic stem cells
FFP	fresh frozen plasma
FHF	fulminant hepatic failure
FPSA-CVVH	fractionated plasma separation and adsorption integrated with continuous venovenous hemofiltration
HBV-ACLF	patients with ACLF caused by hepatitis B virus
HD	hemodialysis
HDF	hemodiafiltration
HF	hemofiltration
HP	hemoperfusion
hsCRP	highly sensitive C-reactive protein
HVP	high-volume therapeutic PE
INR	international normalized ratio
iPSCs	induced pluripotent stem cells
LPCs	liver progenitor cells
LSS	liver support system
MARS	molecular adsorbent recirculating system
MELD	Model for End-stage Liver Disease
MELS	modular extracorporeal liver support system
MSCs	mesenchymal stem cells
MW	molecular weight
non-bio	non-biological
OLT	orthotopic liver transplantation
PBA	plasma bilirubin adsorption
PDF	plasma filtration with dialysis
PE	plasma exchange
PERV	porcine endogenous retrovirus
PNF	primary nonfunction

PSA	plasma separation and adsorption
SEPET	selective plasma-exchange therapy
SHF	subfulminant hepatic failure
SMT	standard medical therapy
SPAD	single-pass albumin dialysis
TBIL	total bilirubin
TPE	therapeutic plasma exchange

References

Abel, J. J., Rowntree, L. G., and Turner, B. B. (1914). On the removal of diffusible substances from the circulating blood of living animals by dialysis. *Journal of Pharmacology and Experimental Therapeutics*, *5*(3), 275 LP-316. Retrieved from http://jpet.aspetjournals.org/content/5/3/275.abstract.

Amimoto, N., Mizumoto, H., Nakazawa, K., Ijima, H., Funatsu, K., and Kajiwara, T. (2011). Hepatic differentiation of mouse embryonic stem cells and induced pluripotent stem cells during organoid formation in hollow fibers. *Tissue Engineering. Part A*, *17*(15–16), 2071–2078. https://doi.org/10.1089/ten.TEA.2010.0689.

Annesini, M. C., Di Carlo, C., Piemonte, V., and Turchetti, L. (2008). Bilirubin and Tryptophan Adsorption in Albumin-Containing Solutions: I. Equilibrium Isotherms on Activated Carbon. *Biochemical Engineering Journal*, *40*, 205–210. http://dx.doi.org/10.1016/j.bej.2007.12.010.

Annesini, M. C., Di Paola, L., Marrelli, L., Piemonte, V., and Turchetti, L. (2005). Bilirubin removal from albumin containing solution by adsorption on polymer resin. *International Journal of Artificial Organs*, *28*(7), 686–693.

Annesini, M. C., Piemonte, V., and Turchetti, L. (2010). Albumin-bound Toxins Removal from Albumin-containing Solutions: Tryptophan Fixed-bed Adsorption on Activated Carbon. *Chemical Engineering Research and Design*, *88*, 1018–1023. http://dx.doi.org/10.1016/j.cherd.2010.01.022.

Arata, S., Tanaka, K., Takayama, K., Moriwaki, Y., Suzuki, N., Sugiyama, M., and Aoyagi, K. (2010). Treatment of hepatic encephalopathy by on-line hemodiafiltration: a case series study. *BMC Emergency Medicine*, *10*, 10. https://doi.org/10.1186/1471–227X-10–10.

Ash, S. R., Steczko, J., Levy, H., Blake, D. E., and Carr, D. J. (2001). Treatment of systemic inflammatory response syndrome by push-pull powdered sorbent pheresis: A Phase 1 clinical trial. *Ther. Apher.*, *5*(6), 497–505.

Baccarani, U., Sanna, A., Cariani, A., Sainz-Barriga, M., Adani, G. L., Zambito, A. M., et al. (2003). Isolation of human hepatocytes from livers rejected for liver transplantation on a national basis: Results of a 2-year experience. *Liver Transplantation*, *9*(5), 506–512. https://doi.org/10.1053/jlts.2003.50087.

Barry, F. P., and Murphy, J. M. (2004). Mesenchymal stem cells: clinical applications and biological characterization. *The International Journal of Biochemistry & Cell Biology, 36*(4), 568–584. https://doi.org/10.1016/j.biocel.2003.11.001.

Cai, J., Zhao, Y., Liu, Y., Ye, F., Song, Z., Qin, H., et al. (2007). Directed differentiation of human embryonic stem cells into functional hepatic cells. *Hepatology, 45*(5), 1229–1239. https://doi.org/10.1002/hep.21582.

Chamuleau, R. A. F. M., Poyck, P. P. C., and van de Kerkhove, M.-P. (2006). Bioartificial liver: its pros and cons. *Therapeutic Apheresis and Dialysis: Official Peer-Reviewed Journal of the International Society for Apheresis, the Japanese Society for Apheresis, the Japanese Society for Dialysis Therapy, 10*(2), 168–174. https://doi.org/10.1111/j.1744–9987.2006.00359.x.

Chua, A. W. C., Khoo, Y. C., Tan, B. K., Tan, K. C., Foo, C. L., and Chong, S. J. (2016). Skin tissue engineering advances in severe burns: review and therapeutic applications. *Burns & Trauma, 4*(1), 3. https://doi.org/10.1186/s41038–016–0027-y.

Defterevos, G., Nastos, C., Papalois, A., Kalimeris, K., Margelos, V., Fragulidis, G., Pafiti, A., Mikrovas, A., Nomikos, T., Smyrniotis, V., and Arkadopoulos, N. (2016). Peritoneal Albumin Dialysis as a Novel Approach for Liver Support: Study in a Porcine Model of Acute Hepatic Failure. *Artificial Organs, 40*(8), 755–764. https://doi.org/10.1111/aor.12687.

Demetriou, A. A., Brown, R. S., Busuttil, R. W., Fair, J., McGuire, B. M., Rosenthal, P., et al. (2004). Prospective, randomized, multicenter, controlled trial of a bioartificial liver in treating acute liver failure. *Ann Surg, 239*(5), 660–670. https://doi.org/10.1097/01.sla.0000124298.74199.e5.

Demetriou, A. A., Rozga, J., Podesta, L., Lepage, E., Morsiani, E., Moscioni, A. D., et al. (1995). Early clinical experience with a hybrid bioartificial liver. *Scand J Gastroenterol Suppl, 208*(S208), 111–117. https://doi.org/10.3109/00365529509107771.

Di Nicuolo, G., D'Alessandro, A., Andria, B., Scuderi, V., Scognamiglio, M., Tammaro, A., et al. (2010). Long-term absence of porcine endogenous retrovirus infection in chronically immunosuppressed patients after treatment with the porcine cell-based Academic Medical Center bioartificial liver. *Xenotransplantation, 17*(6), 431–439. https://doi.org/10.1111/j.1399–3089.2010.00617.x.

Eiseman, B., Liem, D. S., and Raffucci, F. (1965). Heterologous liver perfusion in treatment of hepatic failure. *Annals of Surgery, 162*(3), 329–345. Retrieved from www.ncbi.nlm.nih.gov/pmc/articles/PMC1476914/.

Ellis, A. J., Hughes, R. D., Wendon, J. A., Dunne, J., Langley, P. G., Kelly, J. H., et al. (1996). Pilot-controlled trial of the extracorporeal liver assist device in acute liver failure. *Hepatology, 24*(6), 1446–1451. https://doi.org/10.1053/jhep.1996.v24.pm0008938179.

Evans, M. J., and Kaufman, M. H. (1981). Establishment in culture of pluripotential cells from mouse embryos. *Nature*. https://doi.org/10.1038/292154a0.

Fonsato, V., Herrera, M. B., Buttiglieri, S., Gatti, S., Camussi, G., and Tetta, C. (2010). Use of a rotary bioartificial liver in the differentiation of human liver stem cells. *Tissue Engineering. Part C, Methods, 16*(1), 123–132. https://doi.org/10.1089/ten.TEC.2008.0634.

Ghezzi, C. E., Řnjak-Kovacina, J., and Kaplan, D. L. (2015). Corneal tissue engineering: recent advances and future perspectives. *Tissue Engineering. Part B, Reviews, 21*(3), 278–287. https://doi.org/10.1089/ten.TEB.2014.0397.

Gong, D., Ji, D., Zhu, D., Xu, B., and Liu, Z. (2013). Efficient removal of serum ilirubin by a novel artificial liver support system using albumin convection: A pilot study. *Blood Purification, 34*(3–4), 201–208. https://doi.org/10.1159/000342111.

Gripon, P., Rumin, S., Urban, S., Le Seyec, J., Glaise, D., Cannie, I., et al. (2002). Infection of a human hepatoma cell line by hepatitis B virus. *Proceedings of the National Academy of Sciences of the United States of America, 99*(24), 15655–15660. https://doi.org/10.1073/pnas.232137699.

Hoekstra, R., Nibourg, G. A. A., Van Der Hoeven, T. V., Ackermans, M. T., Hakvoort, T. B. M., et al. (2011). The HepaRG cell line is suitable for bioartificial liver application. *International Journal of Biochemistry and Cell Biology, 43*(10), 1483–1489. https://doi.org/10.1016/j.biocel.2011.06.011.

Hou, Y. T., Ijima, H., Shirakigawa, N., Takei, T., and Kawakami, K. (2012). Development of growth factor-immobilizable material for hepatocyte transplantation. *Biochemical Engineering Journal, 69*, 172–181. https://doi.org/10.1016/j.bej.2012.09.007.

Huang, P., He, Z., Ji, S., Sun, H., Xiang, D., Liu, C., et al. (2011). Induction of functional hepatocyte-like cells from mouse fibroblasts by defined factors. *Nature, 475*(7356), 386–389. https://doi.org/10.1038/nature10116.

Huber, W., Henschel, B., Schmid, R., and Al-Chalabi, A. (2017). First clinical experience in 14 patients treated with ADVOS: a study on feasibility, safety and efficacy of a new type of albumin dialysis. *BMC Gastroenterology, 17*(1), 32. https://doi.org/10.1186/s12876–017–0569-x.

Hughes, R. D., Mitry, R. R., Dhawan, A., Lehec, S. C., Girlanda, R., Rela, M., et al. (2006). Isolation of hepatocytes from livers from non-heart-beating donors for cell transplantation. *Liver Transplantation: Official Publication of the American Association for the Study of Liver Diseases and the International Liver Transplantation Society, 12*(5), 713–717. https://doi.org/10.1002/lt.20732.

Iwamuro, M., Shiraha, H., Nakaji, S., Furutani, M., Kobayashi, N., Takaki, A., and Yamamoto, K. (2012). A preliminary study for constructing a bioartificial liver device with induced pluripotent stem cell-derived hepatocytes. *BioMedical Engineering Online, 11*(1), 1–12. https://doi.org/10.1186/1475–925X-11–93.

Kıhtır, H. S., Yildirim, H. M., Yesilbas, O., Duramaz, B. B., and Sevketoglu, E. (2016). Single-pass albumin dialysis in a child aged six months with phenobarbital poisoning. *Turk Pediatri Arsivi, 51*(4), 228–230. https://doi.org/10.5152/TurkPediatriArs.2016.2335.

Kiley, J. E., Welch, H. F., Pender, J. C., and Welch, C. S. (1956). Removal of blood ammonia by hemodialysis. *Proceedings of the Society for Experimental Biology and Medicine. Society for Experimental Biology and Medicine (New York, N.Y.), 91*(3), 489–490.

Kiley, J. E., Pender, J. C., Welch, H. F., and Welch, C. S. (1958). Ammonia intoxication treated by hemodialysis. *The New England Journal of Medicine, 259*(24), 1156–1161.

Kimoto, S. (1959). The artificial liver experiments and clinical application. *Asaio J, 5*(1), 102–112.

Ko, I. K., Peng, L., Peloso, A., Smith, C. J., Dhal, A., Deegan, D. B., et al. (2015). Bioengineered transplantable porcine livers with re-endothelialized vasculature. *Biomaterials, 40*, 72–79. https://doi.org/10.1016/j.biomaterials.2014.11.027.

Kobayashi, N. (2009). Life support of artificial liver: Development of a bioartificial liver to treat liver failure. *Journal of Hepato-Biliary-Pancreatic Surgery, 16*(2), 113–117. https://doi.org/10.1007/s00534–008–0022–1.

Kobayashi, N., Fujiwara, T., Westerman, K. A., Inoue, Y., Sakaguchi, M., Noguchi, H., et al. (2000a). Prevention of acute liver failure in rats with reversibly immortalized human hepatocytes. *Science (New York, N.Y.), 287*(5456), 1258–1262.

Kobayashi, N., Miyazaki, M., Fukaya, K., Inoue, Y., Sakaguchi, M., Uemura, T., et al. (2000b). Transplantation of highly differentiated immortalized human hepatocytes to treat acute liver failure. *Transplantation, 69*(2), 202–207.

Kondo, Y., Yoshihashi, S., Mimori, K., Ogihara, R., Kanehama, Y., Maki, Y., et al. (2014). Selective culture method for hepatocyte-like cells differentiated from human induced pluripotent stem cells. *Drug Metabolism and Pharmacokinetics, 29*(5), 407–413. https://doi.org/10.2133/dmpk.DMPK-14-RG-022.

Kortgen, A., Rauchfuss, F., Götz, M., Settmacher, U., Bauer, M., and Sponholz, C. (2009). Albumin dialysis in liver failure: Comparison of molecular adsorbent recirculating system and single pass albumin dialysis – A retrospective analysis. *Therapeutic Apheresis and Dialysis, 13*(5), 419–425. https://doi.org/10.1111/j.1744-9987.2009.00760.x.

Kreymann, B., Seige, M., Schweigart, U., Kopp, K. F., and Classen, M. (1999). Albumin dialysis: Effective removal of copper in a patient with fulminant Wilson disease and successful bridging to liver transplantation: A new possibility for the elimination of protein-bound toxins. *Journal of Hepatology, 31*(6), 1080–1085. https://doi.org/10.1016/S0168–8278(99)80322–5.

Kribben, A., Gerken, G., Haag, S., Hergetrosenthal, S., Treichel, U., Betz, C., et al. (2012). Effects of fractionated plasma separation and adsorption on survival in patients with acute-on-chronic liver failure. *Gastroenterology, 142*(4), 782–789. e3. https://doi.org/10.1053/j.gastro.2011.12.056.

Langer, R., and Vacanti, J. P. (1993). Tissue engineering. *Science, 260*(May), 920–926. https://doi.org/10.1126/science.8493529.

Larsen, F. S., Schmidt, L. E., Bernsmeier, C., Rasmussen, A., Isoniemi, H., Patel, V. C., et al. (2016). High-volume plasma exchange in patients with acute liver failure: An open randomised controlled trial. *Journal of Hepatology, 64*(1), 69–78. https://doi.org/10.1016/j.jhep.2015.08.018.

Lee, K.-D., Kuo, T. K.-C., Whang-Peng, J., Chung, Y.-F., Lin, C.-T., Chou, S.-H., et al. (2004). In vitro hepatic differentiation of human mesenchymal stem cells. *Hepatology, 40*(6), 1275–1284. https://doi.org/10.1002/hep.20469.

Li, J., Li, L.-J., Cao, H.-C., Sheng, G.-P., Yu, H.-Y., Xu, W., and Sheng, J.-F. (2005). Establishment of highly differentiated immortalized human hepatocyte line with simian virus 40 large tumor antigen for liver based cell therapy. *ASAIO Journal (American Society for Artificial Internal Organs: 1992), 51*(3), 262–268.

Li, M., Sun, J., Li, J., Shi, Z., Xu, J., Lu, B., et al. (2016). Clinical observation on the treatment of acute liver failure by combined non-biological artificial liver. *Experimental and Therapeutic Medicine, 12*(6), 3873–3876. https://doi.org/10.3892/etm.2016.3887.

Liu, Y., Zhou, G., and Cao, Y. (2017). Recent progress in cartilage tissue engineering – our experience and future directions. *Engineering, 3*(1), 28–35. https://doi.org/10.1016/J.ENG.2017.01.010.

Martin, G. R. (1981). Isolation of a pluripotent cell line from early mouse embryos cultured in medium conditioned by teratocarcinoma stem cells. *Proceedings of the National Academy of Sciences, 78*(12), 7634–7638. https://doi.org/10.1073/pnas.78.12.7634.

Matsumura, K. N., Guevara, G. R., Huston, H., Hamilton, W. L., Rikimaru, M., Yamasaki, G., and Matsumura, M. S. (1987). Hybrid bioartificial liver in hepatic failure: Preliminary clinical report. *Surgery, 101*(1).

Mazza, G., Rombouts, K., Rennie Hall, A., Urbani, L., Vinh Luong, T., Al-Akkad, W., et al. (2015). Decellularized human liver as a natural 3D-scaffold for liver bioengineering and transplantation. *Scientific Reports, 5*, 13079. https://doi.org/10.1038/srep13079.

Miki, T., Ring, A., and Gerlach, J. (2011). Hepatic differentiation of human embryonic stem cells is promoted by three-dimensional dynamic perfusion culture conditions. *Tissue Engineering Part C: Methods, 17*(5), 557–568. https://doi.org/10.1089/ten.tec.2010.0437.

Millis, J. M., Cronin, D. C., Johnson, R., Conjeevaram, H., Conlin, C., Trevino, S., and Maguire, P. (2002). Initial experience with the modified extracorporeal liver-assist device for patients with fulminant hepatic failure: system modifications and clinical impact. *Transplantation, 74*(12), 1735–1746. https://doi.org/10.1097/01.TP.0000038483.93833.21.

Mizumoto, H., Hayashi, S., Matsumoto, K., Ikeda, K., Kusumi, T., Inamori, M., et al. (2012). Evaluation of a hybrid artificial liver module based on a spheroid culture system of embryonic stem cell-derived hepatic cells. *Cell Transplantation, 21*(2–3), 421–428. https://doi.org/10.3727/096368911X605321.

Mo, M., Wang, S., Zhou, Y., Li, H., and Wu, Y. (2016). Mesenchymal stem cell subpopulations: phenotype, property and therapeutic potential. *Cellular and Molecular Life Sciences, 73*(17), 3311–3321. https://doi.org/10.1007/s00018-016-2229-7.

Mundt, A., Puhl, G., Muller, A., Sauer, I., Muller, C., Richard, R., et al. (2002). A method to assess biochemical activity of liver cells during clinical application of extracorporeal hybrid liver support. *Int J Artif Organs, 25*(6), 542–548.

Nagamoto, Y., Takayama, K., Ohashi, K., Okamoto, R., Sakurai, F., Tachibana, M., et al. (2016). Transplantation of a human iPSC-derived hepatocyte sheet increases survival in mice with acute liver failure. *Journal of Hepatology, 64*(5), 1068–1075. https://doi.org/10.1016/j.jhep.2016.01.004.

Nakamura, S., and Ijima, H. (2013). Solubilized matrix derived from decellularized liver as a growth factor-immobilizable scaffold for hepatocyte culture. *Journal of Bioscience and Bioengineering, 116*(6), 746–753. https://doi.org/10.1016/j.jbiosc.2013.05.031.

Nibourg, G. A. A., Chamuleau, R. A. F. M., van der Hoeven, T. V., Maas, M. A. W., Ruiter, A. F. C., Lamers, W. H., et al. (2012). Liver progenitor cell line HepaRG differentiated in a bioartificial liver effectively supplies liver support to rats with acute liver failure. *PloS One, 7*(6), e38778. https://doi.org/10.1371/journal.pone.0038778.

Novelli, G., Annesini, M. C., Morabito, V., Cinti, P., Pugliese, F., Novelli, S., Piemonte, V., Turchetti, L., Rossi, M., and Berloco, P. B. (2009). Cytokines Levels Modifications. Molecular Adsorbent Recirculating System Versus Standard Medical Therapy. *Transplantation Proceedings, 41*(4), 1243–1248. http://dx.doi.org/10.1016/j.transproceed.2009.03.035.

Novelli, G., Annesini, M. C., Morabito, V., Ferretti, G., Lai, Q., Pugliese, F., Novelli, S., Piemonte, V., Turchetti, L., Rossi, M., and Berloco, P. B. (2009). Predictive criteria for the outcome of patients with Acute Liver Failure treated with albumin dialysis Molecular Adsorbent Recirculating System. *Terapeutic Apheresis and Dialysis, 13*(5), 404–412. (DOI: 10.1111/j.1744-9987.2009.00759.x).

Ohashi, K., Yokoyama, T., Yamato, M., Kuge, H., Kanehiro, H., Tsutsumi, M., et al. (2007). Engineering functional two- and three-dimensional liver systems in vivo using hepatic tissue sheets. *Nature Medicine, 13*(7), 880–885. https://doi.org/10.1038/nm1576.

Ozawa, K., Kamiyama, Y., Kimura, K., Ukikusa, M., Kono, Y., Yamato, T., et al. (1982). clinical experience of postoperative hepatic failure treatment with pig or baboon liver cross-hemodialysis with an interposed membrane.

Artificial Organs, 6(4), 433–446. https://doi.org/10.1111/j.1525–1594.1982.tb04141.x

Park, J., and Lee, D. (2005). Bioartificial liver systems: current status and future perspective. *Journal of Bioscience and Bioengineering, 99*(4), 311–319. https://doi.org/10.1263/jbb.99.311.

Peszynski, P., Klammt, S., Peters, E., Mitzner, S., Stange, J., and Schmidt, R. (2002). Albumin dialysis: single pass vs. recirculation (MARS). *Liver, 22* Suppl 2, 40–2. https://doi.org/10.1034/j.1600–0676.2002.00007.x.

Piechota, M., and Piechota, A. (2016). An evaluation of the usefulness of single pass albumin dialysis: key role of dialysate flow rate. *Critical Care*, 5–7. https://doi.org/10.1186/s13054-016-1287-4.

Pineda, A. A., and Taswell, H. F. (1981). Selective plasma component removal: alternatives to plasma exchange. *Artificial Organs, 5*(3), 234–240.

Rifai, K., Ernst, T., Kretschmer, U., Bahr, M. J., Schneider, A., Hafer, C., et al. (2003). Prometheus?? – A new extracorporeal system for the treatment of liver failure. *Journal of Hepatology, 39*(6), 984–990. https://doi.org/10.1016/S0168-8278(03)00468-9.

Ring, A., Gerlach, J. C., Peters, G., Pazin, B. J., Minervini, C. F., Turner, M. E., et al. (2010). Hepatic maturation of human fetal hepatocytes in four-compartment three-dimensional perfusion culture. *Tissue Engineering. Part C, Methods, 16*(5), 835–845. https://doi.org/10.1089/ten.tec.2009.0342.

Rozga, J., Umehara, Y., Trofimenko, A., Sadahiro, T., and Demetriou, A. A. (2006). A novel plasma filtration therapy for hepatic failure: Preclinical studies. *Therapeutic Apheresis and Dialysis, 10*(2), 138–144. https://doi.org/10.1111/j.1744–9987.2006.00355.x

Sakiyama, R., Blau, B. J., and Miki, T. (2017). Clinical translation of bioartificial liver support systems with human pluripotent stem cell-derived hepatic cells. *World Journal of Gastroenterology, 23*(11), 1974–1979. https://doi.org/10.3748/wjg.v23.i11.1974.

Sauer, I. M., Goetz, M., Steffen, I., Walter, G., Kehr, D. C., Schwartlander, R., et al. (2004). In vitro comparison of the molecular adsorbent recirculation system (MARS) and single-pass albumin dialysis (SPAD). *Hepatology, 39*(5), 1408–1414. https://doi.org/10.1002/hep.20195.

Sauer, I. M., Obermeyer, N., Kardassis, D., Theruvath, T., and Gerlach, J. C. (2001). Development of a hybrid liver support system. *Annals of the New York Academy of Sciences, 944*, 308–319. https://doi.org/10.1111/j.1749–6632.2001.tb03843.x.

Sauer, I. M., Zeilinger, K., Obermayer, N., Pless, G., Grünwald, A., Pascher, A., et al. (2002). Primary human liver cells as source for modular extracorporeal liver support – a preliminary report. *International Journal of Artificial Organs, 25*(10), 1001–1005.

Schmelzer, E., Zhang, L., Bruce, A., Wauthier, E., Ludlow, J., Yao, H., et al. (2007). Human hepatic stem cells from fetal and postnatal donors. *The Journal of Experimental Medicine, 204*(8), 1973–1987. https://doi.org/10.1084/jem.20061603.

Sekiya, S., and Suzuki, A. (2011). Direct conversion of mouse fibroblasts to hepatocyte-like cells by defined factors. *Nature, 475*(7356), 1–6. https://doi.org/10.1038/nature10263

Sen, P. K., Bhalerao, R. A., Parulkar, G. P., Samsi, A. B., Shah, B. K., and Kinare, S. G. (1966). Use of isolated perfused cadaveric liver in the management of hepatic failure. *Surgery, 59*(5), 774–781. Retrieved from www.ncbi.nlm.nih.gov/pubmed/5931492.

Shi, X.-L., Gao, Y., Yan, Y., Ma, H., Sun, L., Huang, P., et al. (2016). Improved survival of porcine acute liver failure by a bioartificial liver device implanted with induced human functional hepatocytes. *Cell Research, 26*(26), 1–11. https://doi.org/10.1038/cr.2016.6.

Shinozaki, K., Oda, S., Abe, R., Tateishi, Y., Yokoi, T., and Hirasawa, H. (2010). Blood purification in fulminant hepatic failure. *Contributions to Nephrology, 166*, 64–72. https://doi.org/10.1159/000314854.

Shirakigawa, N., Ijima, H., and Takei, T. (2012). Decellularized liver as a practical scaffold with a vascular network template for liver tissue engineering. *Journal of Bioscience and Bioengineering, 114*(5), 546–551. https://doi.org/10.1016/j.jbiosc.2012.05.022.

Shirakigawa, N., Takei, T., and Ijima, H. (2013). Base structure consisting of an endothelialized vascular-tree network and hepatocytes for whole liver engineering. *Journal of Bioscience and Bioengineering, 116*(6), 740–745. https://doi.org/10.1016/j.jbiosc.2013.05.020.

Si-Tayeb, K., Noto, F. K., Nagaoka, M., Li, J., Battle, M. A., Duris, C., et al. (2010). Highly efficient generation of human hepatocyte-like cells from induced pluripotent stem cells. *Hepatology, 51*(1). https://doi.org/10.1002/hep.23354.

Song, J. J., and Ott, H. C. (2011). Organ engineering based on decellularized matrix scaffolds. *Trends in Molecular Medicine, 17*(8), 424–432. https://doi.org/10.1016/j.molmed.2011.03.005.

Song, Z., Cai, J., Liu, Y., Zhao, D., Yong, J., Duo, S., et al. (2009). Efficient generation of hepatocyte-like cells from human induced pluripotent stem cells. *Cell Research, 19*(11), 1233–42. https://doi.org/10.1038/cr.2009.107.

Sponholz, C., Matthes, K., Rupp, D., Backaus, W., Klammt, S., Karailieva, D., et al. (2015). Molecular adsorbent recirculating system and single-pass albumin dialysis in liver failure – a prospective, randomised crossover study. *Critical Care, 20*(1), 2. https://doi.org/10.1186/s13054–015–1159–3.

Starzl, T. E., Marchioro, T. L., Kaulla, K. N. Von, Hermann, G., Brittain, R. S., Waddell, W. R., and Denver, F. A. C. S. (1963). Homotranspantation of the liver in humans. *Surg Gynecol Obstet., 117*(May 1963), 659–676. https://doi.org/10.1097/00007890–196707001–00003.

Sussman, N. L., Gislason, G. T., Conlin, C. A., and Kelly, J. H. (1994). The Hepatix extracorporeal liver assist device: Initial clinical experience. *Artificial Organs, 18*(5), 390–396. https://doi.org/10.1111/j.1525–1594.1994.tb02221.x.

Takahashi, K., Tanabe, K., Ohnuki, M., Narita, M., Ichisaka, T., Tomoda, K., and Yamanaka, S. (2007). Induction of pluripotent stem cells from adult human fibroblasts by defined factors. *Cell, 131*(5), 861–872. https://doi.org/10.1016/j.cell.2007.11.019.

Takahashi, K., and Yamanaka, S. (2006). Induction of pluripotent stem cells from mouse embryonic and adult fibroblast cultures by defined factors. *Cell, 126*(4), 663–676. https://doi.org/10.1016/j.cell.2006.07.024.

Takata, A., Otsuka, M., Kogiso, T., Kojima, K., Yoshikawa, T., Tateishi, R., et al. (2011). Direct differentiation of hepatic cells from human induced pluripotent stem cells using a limited number of cytokines. *Hepatology International, 5*(4), 890–898. https://doi.org/10.1007/s12072-011-9251-5.

Takebe, T., Sekine, K., Enomura, M., Koike, H., Kimura, M., Ogaeri, T., et al. (2013). Vascularized and functional human liver from an iPSC-derived organ bud transplant. *Nature, 499*(7459), 481–484. https://doi.org/10.1038/nature12271.

Tan, H. K. (2004). Molecular adsorbent recirculating system (MARS). *Annals of the Academy of Medicine Singapore, 33*(3).

Tsuruga, Y., Kiyono, T., Matsushita, M., Takahashi, T., Kasai, H., Matsumoto, S., and Todo, S. (2008). Establishment of immortalized human hepatocytes by introduction of HPV16 E6/E7 and hTERT as cell sources for liver cell-based therapy. *Cell Transplantation, 17*(9), 1083–1094.

Turner, R., Lozoya, O., Wang, Y., Cardinale, V., Gaudio, E., Alpini, G., et al. (2011). Human hepatic stem cell and maturational liver lineage biology. *Hepatology, 53*(3), 1035–1045. https://doi.org/10.1002/hep.24157.

Uygun, B. E., Soto-Gutierrez, A., Yagi, H., Izamis, M.-L., Guzzardi, M. A., Shulman, C., et al. (2010). Organ reengineering through development of a transplantable recellularized liver graft using decellularized liver matrix. *Nature Medicine, 16*(7), 814–820. https://doi.org/10.1038/nm.2170.

van de Kerkhove, M. P., Di Florio, E., Scuderi, V., Mancini, A., Belli, A., Bracco, A., et al. (2002). Phase I clinical trial with the AMC-bioartificial liver. *International Journal of Artificial Organs, 25*(10), 950–959. https://doi.org/0391-3988/950-10

van de Kerkhove, M. P., Hoekstra, R., Chamuleau, R. A. F. M., and van Gulik, T. M. (2004). Clinical application of bioartificial liver support systems. *Annals of Surgery, 240*(2), 216–230. https://doi.org/10.1097/01.sla.0000132986.75257.19.

Wan, Y.-M., Li, Y.-H., Xu, Z.-Y., Yang, J., Yang, L.-H., Xu, Y., and Yang, J.-H. (2017). Therapeutic plasma exchange versus double plasma molecular absorption system in hepatitis B virus-infected acute-on-chronic liver failure treated by entercavir: A prospective study. *Journal of Clinical Apheresis*, 1–9. https://doi.org/10.1002/jca.21535.

Wiegand, J., Van Bömmel, F., Duarte-Rojo, A., Altamirano, J., Abraldes, J. G., Villanueva, A., and Berg, T. (2016). Clinical Trial Watch: Reports from the Liver Meeting®, AASLD, San Francisco, November 2015. *Journal of Hepatology, 64*(6), 1428–1445. https://doi.org/10.1016/j.jhep.2016.02.020.

Wilson, C. A. (2008). Endogenous retroviruses: Porcine endogenous retroviruses and xenotransplantation. *Cellular and Molecular Life Sciences, 65*(21), 3399–3412. https://doi.org/10.1007/s00018-008-8498-z

Yagi, H., Fukumitsu, K., Fukuda, K., Kitago, M., Shinoda, M., Obara, H., et al. (2013). Human-scale whole-organ bioengineering for liver transplantation: a regenerative medicine approach. *Cell Transplantation, 22*(2), 231–242. https://doi.org/10.3727/096368912X654939.

Yamamoto, H., Nakae, H., Uji, Y., Maeda, K., Tani, T., and Eguchi, Y. (2015). Plasma adiponectin levels in acute liver failure patients treated with plasma filtration with dialysis and plasma exchange. *Therapeutic Apheresis and Dialysis, 19*(4), 349–354. https://doi.org/10.1111/1744-9987.12344.

Yang, Q., Liu, F., Pan, X. P., Lv, G., Zhang, A., Yu, C. B., and Li, L. (2010). Fluidized-bed bioartificial liver assist devices (BLADs) based on microencapsulated primary porcine hepatocytes have risk of porcine endogenous retroviruses transmission. *Hepatology International, 4*(4), 757–61. https://doi.org/10.1007/s12072-010-9210-6.

Yatzidis, H. (1964). Research on extrarenal purification with the aid of activated charcoal. *Nephron, 1*, 310–312.

Yokoi, T., Oda, S., Shiga, H., Matsuda, K., Sadahiro, T., Nakamura, M., and Hirasawa, H. (2009). Efficacy of high-flow dialysate continuous hemodiafiltration in the treatment of fulminant hepatic failure. *Transfusion and Apheresis Science, 40*(1), 61–70. https://doi.org/10.1016/j.transci.2008.11.006.

Yoshiba, M., Inoue, K., Sekiyama, K., and Koh, I. (1996). Favorable effect of new artificial liver support on survival of patients with fulminant hepatic failure. *Artif. Organs, 20*(11), 1169–1172. Retrieved from www.ncbi.nlm.nih.gov/pubmed/8908326.

Zhao, L., Li, J., Lv, G., Zhang, A., Zhou, P., Yang, Y., et al. (2012). Evaluation of a reversibly immortalized human hepatocyte line in bioartificial liver in pigs. *African Journal of Biotechnology, 11*(17), 4116–4126. https://doi.org/10.5897/AJB11.3411.

Zhou, P., Zheng, S., Yu, M., He, S., and Weng, Z. (2015). Prognosis of acute-on-chronic liver failure patients treated with artificial liver support system, *World J Gastroenterol., 21*(32), 9614–9622. https://doi.org/10.3748/wjg.v21.i32.9614.

4

A Chemical Engineering Perspective on Blood Oxygenators

Luisa Di Paola

Unit of Chemical-Physics Fundamentals in Chemical Engineering, Department of Engineering, Università Campus Bio-Medico di Roma, Italy

4.1 Introduction

Oxygenators for extra-corporeal blood oxygenation have become one of the most applied and crucial biomedical devices in surgery and clinical practice for respiratory and cardiac pathologies. Today, the widest application of blood oxygenation regards Extra-Corporeal Life Support in cardiopulmonary bypass surgery.

Application has steeply increased in recent years (see Figure 4.1, from ref. [1]) and attracted the focus of research as well: searching literature for "ECMO" (Extra-Corporeal Membrane Oxygenators) as the key words and limiting the search to documents in English in the last 10 years generates the report data shown in Figure 4.2.

This huge interest is grounded in the growing application rate of ECMO methodologies for respiratory assistance too, not only cardiac surgery, and to a more encouraging percentage rate of survival, with respect to previous years (now around 60%) [2, 3]. However, while ECMO technology is consolidated, it still requires many improvements: perspective developments point to wearable devices and to microfluidic design to improve usability and comfort to patients [4, 5], thus expanding areas of application (for instance, to severe chronic respiratory pathologies or as bridge for lung transplantation).

The ECMO technology development has experienced (and is actually still experiencing) a strong interdisciplinary effort, massively interacting with developing parallel technologies in other fields (chemical process, air upgrading). From a chemical engineering point of view, blood oxygenators are gas-liquid

Biomedical Engineering Challenges: A Chemical Engineering Insight, First Edition.
Edited by Vincenzo Piemonte, Angelo Basile, Taichi Ito, and Luigi Marrelli.

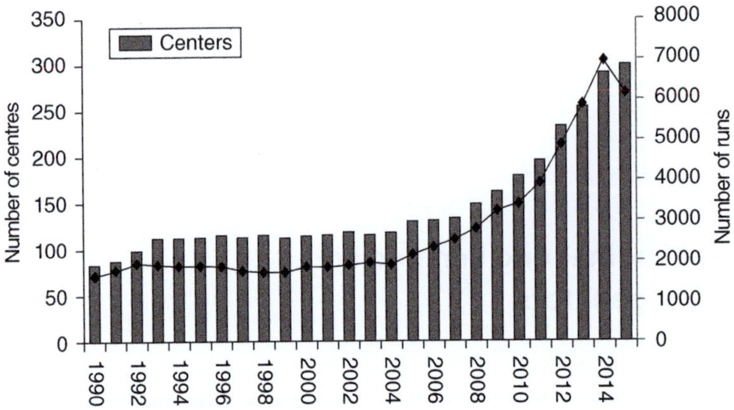

Figure 4.1 Number of ECMO active centers and runs worldwide [1]. *Source:* Butt, https://www.ncbi.nlm.nih.gov/pmc/articles/PMC4847564/. Used under CC-By 4.0 https://creativecommons.org/licenses/by/4.0/. (*See insert for color representation of the figure.*)

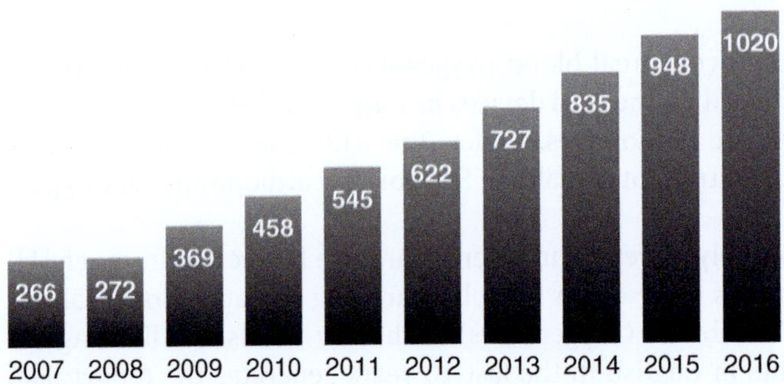

Figure 4.2 Number of papers regarding ECMO during years (*Source:* PubMed, language English). (*See insert for color representation of the figure.*)

contactors, which can be designed and optimized following the general approach of unit operations [6]. A major target in blood oxygenator design is to reach a good level of carbon dioxide removal, which is the most challenging task for such systems [6, 7]. On the other hand, carbon dioxide removal is a major task for exhaust flue gas treatment, a hot topic in recent years [8].

In this chapter, I draw a general sketch of blood oxygenators, starting from their history and then highlighting the crucial role chemical engineering fundamentals have played in developing actual and future devices. It is interesting to note (and somehow not easy to define the borders) how ECMO technology development was flanked by the parallel development of gas-liquid contactors for key applications in the field of chemical engineering process and environment protection. For this reason, I devote a section to parallel technologies in industrial applications, with special attention paid to historical timelines, in order to better appreciate how interdisciplinary exchange has provided multiple advantages and outstanding results in this field. Finally, a perspective section will demonstrate how chemical engineering principles may also help in developing new technologies (microfluidics, improved gas exchange) in order to reach the target of wearable and compatible artificial lungs.

4.2 A Historical Note

To better understand how modern blood membrane oxygenators work, it is useful to follow their history and perceive the effort to overcome all hurdles to extra-corporeal life support for blood oxygenation.

In the path of the development of modern oxygenators, three types of oxygenators have been designed [9]:

1) Film-type oxygenators: the contact between air and blood is direct, the gas exchange occurs at the surface of a thin blood film (the disperse phase in the direct contact is the liquid into air);
2) Bubble-type oxygenators: gas is introduced into the blood as bubbles (in this case, the disperse phase is the gas into the blood); the gas exchange occurs at the wide bubble surface. However, mechanical stress at the gas-liquid interface results in the poorest blood compatibility among oxygenators. An additional issue is the complete removal of bubbles from blood to avoid embolism;
3) Membrane-type oxygenators: the contact between air and blood is indirect through a membrane. Blood is not affected by trauma due to direct contact with air, so the membrane oxygenator is considered the less traumatic blood oxygenator. The main issue with membrane oxygenators is compensation for the additional resistance to respiratory gas exchange due to the membrane; other factors influencing the membrane oxygenator are the membrane materials affecting both blood compatibility and performance [10], the uneven transfer of oxygen and carbon dioxide, and the residual shear stress trauma on blood cells.

The main aim in designing blood oxygenators is to mimic natural lungs, in which gas and blood exchange respiratory gases (oxygen and carbon dioxide) through the alveolar capillary wall. The efficiency of lungs is due to the high surface extension per unit volume of the alveolar surface, wet by a very thin film of blood.

So far, it has been not possible to replicate the key features of lung structure (very thin blood film and high specific surface area).

Early studies on blood perfusion and oxygenation were lead in the nineteenth century, resulting in a successful perfusion of organs through direct contact between air and blood in film or bubble oxygenators [11]. The experimental data collected in the latter years of the nineteenth and the first decade of the twentieth century, however, did not result in clinical application, due to the evident issue of blood coagulation. In 1916 the discovery of heparin by Jay Maclean paved the way for clinical extra-corporeal blood treatments [12]. However, it was only in 1953 that Gibbon successfully applied an extra-corporeal blood oxygenation in a cardiac surgery for the first time: the first model was a film oxygenator developed over 20 years of experiments [13] (Figure 4.3).

In this circuit there were two key elements: the rotating film oxygenator, composed by metal sheets on which a continuous film of blood formed (J in Figure 4.3) and the atraumatic blood pumps (D, E, and P in Figure 4.3), specifically designed by DeBakey in 1934 to reduce the impact of high shear rate on blood cells [14].

In 1955 R.A. De Wall and C.W. Lillehei introduced their bubble oxygenator, representing the benchmark for two decades in open-heart surgery. Later in 1956, Kay and Cross presented a rotating disk bubble oxygenator, an improved version of the Wall–Lillehei device. However, both film and bubble direct

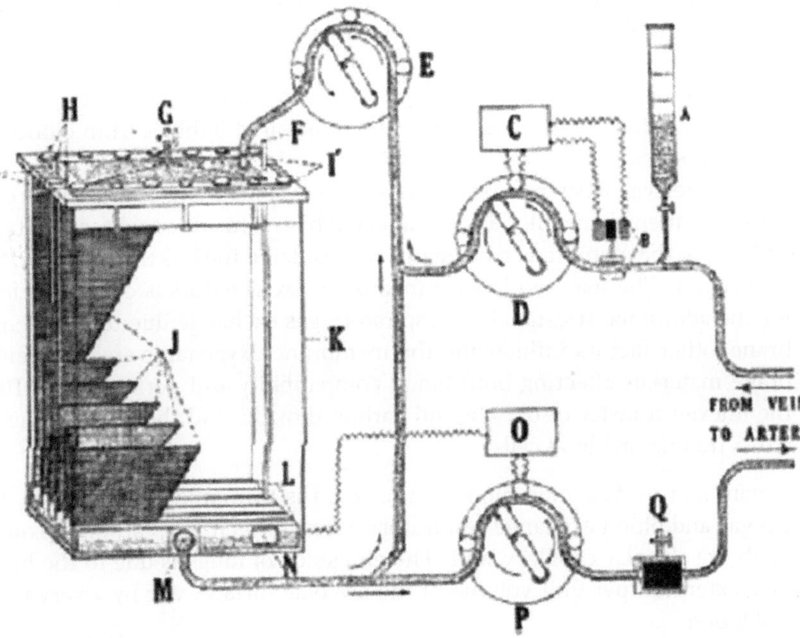

Figure 4.3 A schematic picture of the extra-corporeal circuit developed by Gibbon: the key elements are the film oxygenator (on flat metal sheets J) and the peristaltic pumps D, E, and P (reprinted with permission from Iwahashi, H., Yuri, K., and Nosé, Y. (2004) Development of the oxygenator: Past, present, and future. *J. Artif. Organs*, **7** (3), 111–120).

contact oxygenators had to cope with a major issue: in a short time (within few hours) the direct contact between blood and air causes an irreversible damage to blood components (proteins, blood cells, and platelets) at the air-blood interface, causing disorders such as blood coagulation, destruction of blood cells and platelets, and protein denaturation.

To solve this main issue, in 1956 membrane blood oxygenators were introduced and now they represent the benchmark for ECLS [15]. With respect to bubble and film oxygenators, membrane oxygenators present two main advantages: (1) they avoid direct blood-air contact; and (2) it is not required to remove gas from the blood. The disadvantages with respect to direct contact oxygenators are the larger priming volume of blood required for efficient oxygenation and the additional resistance to mass transfer due to the membrane (which, in turn, forces a large priming volume for efficient oxygenation).

The first configuration of a membrane oxygenator comprised a supported sheet cellulose membrane, proposed by GHA Clowes [15] in 1956. In the same year, Kolff's team introduced a coil-type disposable oxygenator with polyethylene membranes [16]: in this configuration the blood has a spiral motion, while the air flows parallel to the cylinder axis. In 1971, Kolobov and coworkers introduced a significant innovation in the coil-type membrane oxygenator: they introduced a silicone rubber membrane, which dramatically improved gas exchange efficiency [17]. This device was applied for a long time in cardiac surgery practice and for respiratory assistance. The hollow-fiber membrane oxygenator was developed in the early 1970s but the first commercial hollow-fiber oxygenator, Capiox, was placed on the market on 1981 [18]. The core of this device was the microporous polyethylene membrane developed by the Terumo Corporation and K. Suma. The high efficiency and blood compatibility of Capiox allowed them to reach a market share of 60% in USA. To further improve blood compatibility, polyethylene membranes were covered by a silicone rubber thin film, which also reduced plasma leakage, a typical problem for microporous membranes. These features also allowed the application of Capiox in long-term extra-corporeal respiratory assistance.

Another key point is the design of proper blood pumps to reduce blood distress in ECMO applications due to shear stress. This effect, known since the early days of ECLS practice [19] is generally due to different factors:

1) Interactions with solid surface;
2) Centrifugal forces;
3) Hemolysis at air-liquid interface (when present);
4) Cell-cell interactions;
5) Pressure-drop effects.

Pumps must accomplish some general tasks: provide a flow appropriate for the patient close to the physiological blood flow (75–150 mL/kg/min in pediatric applications) with pressure within a given range to avoid pressure hemolysis. Low inlet pressures (suction) threaten blood components by cavitation. High outlet pressures, on the other hand, may compromise the circuitry integrity [20].

In the first oxygenators, blood was moved by roller pumps: these pumps have a low inlet pressure, thus cavitation was more than a mere risk provoking extensive hemolysis. Another risk factor is the mechanical action of the pulsatile tube occlusion leading to cell damage [21]. Current centrifugal pumps work with a magnetic actuator, which strongly reduces the blood trauma [22]. The configuration of modern oxygenators does not differ much from the early ECMOs (in Figure 4.4 a typical modern ECMO circuit is shown [20]) and the technology is widely consolidated for cardiopulmonary bypass surgery, but still some issues are present:

1) the membrane material must achieve a trade-off between blood compatibility, plasma leakage, and gas exchange capability;
2) blood pumps are still too stressful on blood components, so blood disorders remain a major collateral effect of ECLS;
3) the quest for a wearable artificial lung is still open (and far from being closed).

All these issues form the basis of a hot topic in research, notwithstanding the practice of ECMO in cardiopulmonary surgery for more than 30 years.

4.3 Chemical Engineering Principles in Blood Oxygenators

Respiratory gases (O_2 and CO_2) interact with hemoglobin, a protein abundant in blood red cells [23]. The equilibrium between these gases and hemoglobin is at the very basis of blood interaction with air, in natural (lungs) and artificial (oxygenators) systems, and is represented by means of dissociation curves. This mechanism is quite complex, since the binding of the two ligands (O_2 and CO_2) with hemoglobin (a protein complex made up of four chains) induces a conformational shift needed for regulatory purposes (the Haldane and Bohr effect [24]) that can be interpreted in terms of allosteric models: in 1965, the Monod–Wyman–Changeux model was proposed to describe the oxygen-blood dissociation curve and now it represents the benchmark to describe similar systems [25, 26]. In Figure 4.5 the dissociation curves of oxygen at different partial pressures of CO_2 are reported, showing the interlaced role of oxygen and carbon dioxide in hemoglobin binding.

From a thermodynamic point of view, oxygen and carbon dioxide transfer in blood follows the rule of gas-liquid equilibrium in the presence of reactions. In detail, oxygen dissociation curves can be quantitatively interpreted in terms of thermodynamic equilibrium between the bound and unbound species and the different conformational states, and, according to this thermodynamic scheme,

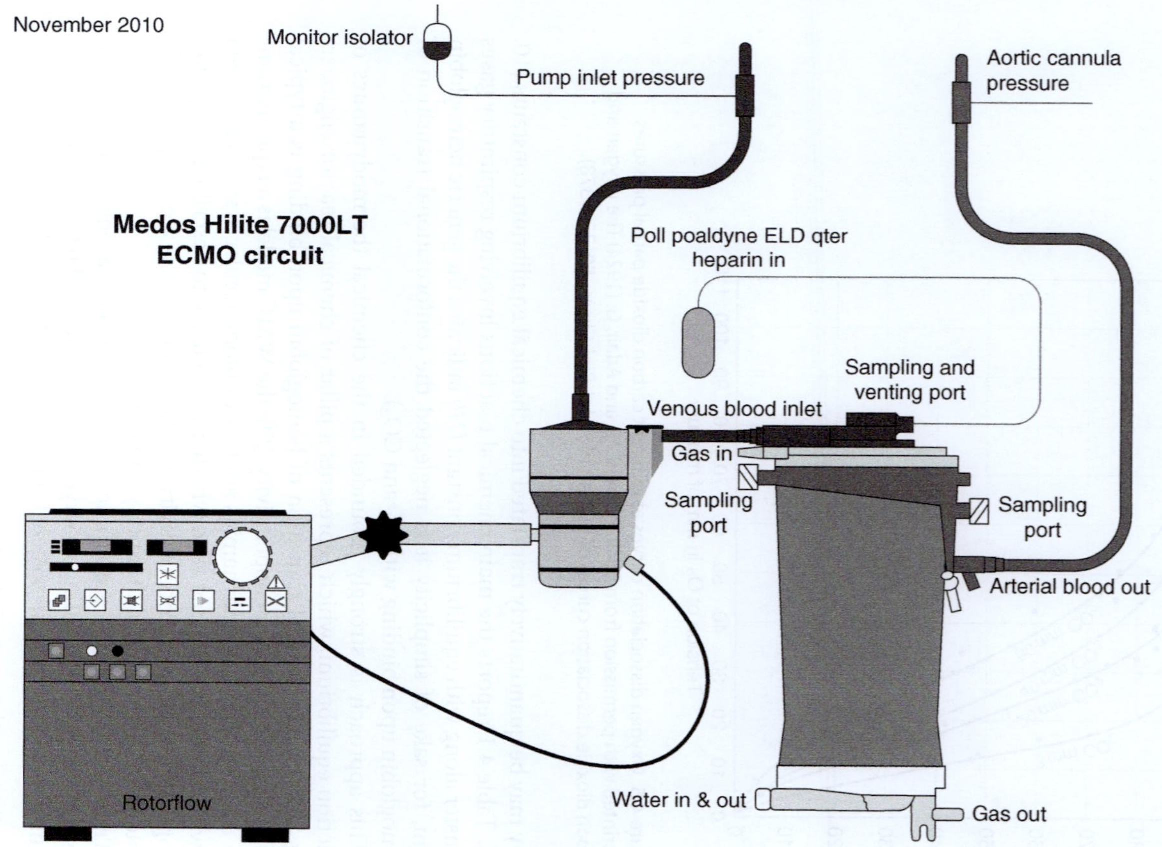

Figure 4.4 A modern ECMO circuit (reprinted with from Lequier, L., Horton, S., McMullan, D., and Bartlett, R.H. (2013) Extracorporeal membrane oxygenation circuitry. *Pediatr. Crit. care ...*, **14**, 1–10). (*See insert for color representation of the figure.*)

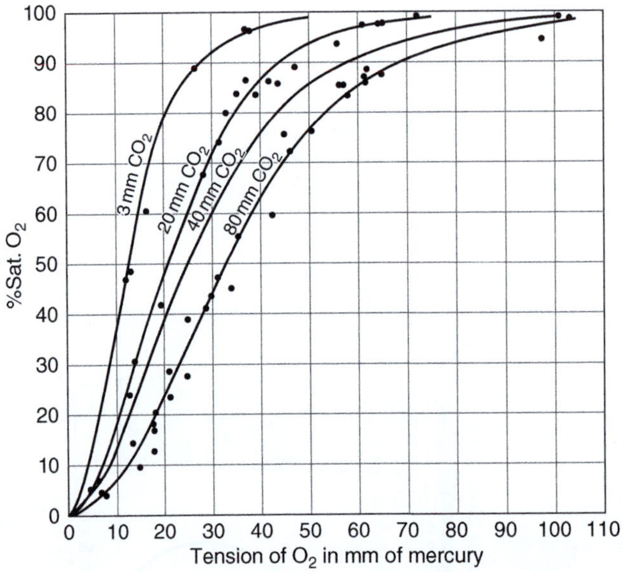

Figure 4.5 Oxygen dissociation curves at different carbon dioxide partial pressures (reprinted with permission from Bock, A., Field, H., and Adair, G. (1924) The oxygen and carbon dioxide dissociation curves of human blood. *J. Biol. Chem.*, **59**, 353–378).

they may be quantitatively translated into chemical equilibrium constants [10, 28]. Table 4.1 reports the main chemical reactions involving respiratory gases transfer along with equilibrium constant (*Hb* indicates the generic hemoglobin form, for sake of simplicity it is neglected the conformational transition of hemoglobin upon binding with O_2 and CO_2).

This approach is strongly grounded in the chemical thermodynamics of reaction equilibrium, which represents a pillar of chemical reactor engineering. The conformational transition of hemoglobin upon binding is a typical mechanism of molecular cooperativity [29]: the WMC model is a typical example of molecular thermodynamics applied to biomolecular systems and even now the debate is still open as to whether it is reliable to describe hemoglobin cooperativity [30, 31]. Cooperativity, on the other hand, is a concept strongly grounded in biochemical thermodynamics [32].

Transport phenomena are another pillar of the chemical engineering fundamentals largely contributing to oxygenator design [6]. Mass and momentum transfer play a key role in oxygenator design, aimed at [33, 34]:

- optimizing the gas transfer per unit volume;
- reducing the exchange surface to improve economical and health performances of the device;
- improving the blood flow in terms of impact on blood components.

Table 4.1 Main chemical reactions occurring in blood during gas exchange; chemical equibrium constants are evaluated at room temperature.

Reactions	Equilibrium Constants
$Hb + O_2 \rightleftharpoons Hb-(O_2)_m$	$K_O = \frac{[Hb-(O_2)_m]}{[Hb]\cdot[O_2]} = 4.7\cdot10^{-6}\ M^{-1}, m = 2.7$
$CO_2 + H_2O \rightleftharpoons HCO_3^{-}H^{+}$	$K_{a1} = \frac{[HCO_3^-]\cdot[H^+]}{[CO_2]\cdot[H_2O]} = 4.3\cdot10^{-7}\ M$
$HCO_3^- \rightleftharpoons CO_3^{2-} + H^+$	$K_{a2} = \frac{[CO_3^{2-}]\cdot[H^+]}{[HCO_3^-]} = 5.61\cdot10^{-11}\ M$
$CO_2 + nHb \rightleftharpoons Hb-(CO_2)_n$	$K_C \ll K_O\ (negligible)$

(*Source:* Di Paola, L., Terrinoni, A.R., and Vitale, F. (2012) Extracorporeal membrane blood oxygenators: effect of membrane wetting on gas transfer and device performance. Asia-Pacific J. Chem. Eng., 7 (S3), S348–S355)

As for the first point, hollow-fiber oxygenators achieve the highest surface per unit volume of the device, providing by design the highest flux of gas exchange per unit volume. The exchange surface is the result of a trade-off between high exchange rates and plasma leakage: indeed, in microporous membranes plasma permeation from blood to gas is a major issue, but presents strong higher permeances with respect to dense membranes. Moreover, the addition of silicon rubber layers reduces plasma leakage and improves blood compatibility, but also reduces gas fluxes [35]. Blood pumps harm blood components through mechanical stress, thus a careful analysis of the momentum transfer may help innovative pump design with reduced shear stress on blood components and more compatibility [36]. The Computational Fluid Dynamics (CFD) approach provides insightful indications for blood pump design [37]. The key points in blood oxygenator design are:

- oxygen and carbon dioxide transfer through membranes is uneven, since less carbon dioxide is transferred through membranes; for this reason, mass transfer effort focuses on carbon dioxide, set as the limiting component for transfer [38];
- blood flow in oxygenators induces blood trauma [39]; to reduce this detrimental effect, particularly evident in long-term ECLS applications, the pumping system and device fluid dynamics must be carefully designed;
- membrane materials and structure strongly affect blood compatibility [35], noxious side effects – membrane leakage – and gas transfer performance [40].

Most mass transfer studies in ECMO only include physical (transport across membranes) mechanisms [33, 41]: however, other studies explicitly include the effect on transport of chemical reactions occurring in blood [10]. CFD studies now solve mass and momentum balances in unsteady conditions providing the most complete description of ECMO systems; however, this approach is has a heavy computational burden and requires devoted CFD software [42].

Mass transfer analysis is the basis for blood oxygenator design and optimization [6]. In their key works [43, 44], Wickramasinghe and coworkers reported empirical correlations between adimensional numbers *Re, Sh,* and *Sc* to evaluate the mass and momentum transfer in ECMO, with blood flowing in the fibers. To generalize the mass transfer correlations to all flow regimes it is useful to introduce the Graetz number, defined for mass transfer as:

$$Gr = \frac{d_H}{L} \cdot Re \cdot Sc \tag{4.1}$$

d_H and L are, respectively, the fiber diameter and length. Table 4.2 reports correlations for adimensional numbers for mass transfer in hollow-fiber oxygenators.

As previously stated, CO_2 removal is a main target of ECMO system optimization [38]: carbon dioxide undergoes chemical reactions in blood, so its flux must account also for chemical reaction resistance. To improve CO_2 removal, membranes with entrapped carbonic anhydrase have been tested and demonstrated to be effective in long-term respiratory assistance [45].

Finally, a main issue for membrane contactor performance is porous membrane wetting (also promoting plasma leakage), which dramatically impairs gas transfer [10, 46, 47]: indeed, in dry membranes usually the controlling resistance is the blood film, while in wetted membranes the controlling resistance shifts to that of the wetted membrane (liquid entrapped into the microporous membrane) with a sharp decrease of the overall transfer

Table 4.2 Mass transfer empirical correlations in hollow-fiber oxygenators in the range.

Flow geometry	Flow range	Correlation
Flow inside fibers	*any*	$Sh = 1.62 \cdot Gr^{0.33}$
Flow outside and parallel to fibers	$Gr < 60$	$Sh = 0.019 \cdot Gr^{1.0}$
Flow outside and across fibers	$Re > 2.5$	$Sh = 0.154 \cdot Re^{0.8} \cdot Sc^{0.33}$
	$Re < 2.5$	$Sh = 0.122 \cdot Re^{1.0} \cdot Sc^{0.33}$

(Adapted from Wickramasinghe, S.R., Semmens, M.J., and Cussler, E.L. (1992) Mass transfer in various hollow fiber geometries. J. Membr. Sci., 69 (3), 235–250).

coefficients for both gases. In general, wetting for microporous membranes depends on different properties of the system [10, 48]:

- membrane materials (whether hydrophobic or polar) and structural features (porosity, mean pore radius, and tortuosity);
- liquid composition, affecting surface tension with respect to gas;
- fluid dynamics;
- pressure drop between the two compartments separated by the membrane.

The structural features of pores determine the critical pressure for leakage, called breakthrough pressure and defined according to the Young–Laplace equation [49]:

$$\Delta p_{BT} = p_{gas} - p_{liquid} = \frac{2 \cdot \sigma \cdot \cos(\theta)}{r} \tag{4.2}$$

being σ the gas/liquid surface tension, θ the material contact angle with regards to the liquid and the gas and r the pore radius. Hydrophobic and superhydrophobic membranes ($\theta > 150°$) accomplish the task of favorable $\Delta p_{BT} < 0$ and also present anti-clotting properties, thus improving blood compatibility [50]. For all these reasons, silicone rubber [35] and polypropylene [20] are widely used materials for ECMO membranes.

This quick perspective on ECMO demonstrates the high impact that chemical engineering science has had on the development of modern ECMO, starting from materials to fluid dynamics. This large amount of technological know-how stemmed from classical chemical engineering devices, intended to favor gas-liquid transfer. In the next section, I will show this similarity through a very quick perspective on gas-liquid contactors and their applications in chemical industrial processes.

4.4 Chemical Engineering Process Analogues of ECMO Systems

Mass transfer in gas-liquid systems is a very common task in chemical engineering processes [51]. The equipment devoted to this application spans from very classical, huge devices (plate columns) to more specialized and compact systems, the proper equipment for reduced capacity (hollow-fiber modules).

Gas-liquid equipment depend on different factors, the main ones being gas and liquid flows, the dispersed phase, and mass transfer per unit volume. For instance, classical systems handling very high gas and flow rates are plate columns, meant for gas absorption or liquid stripping. When the gas and liquid flow rates are rather smaller, more specialized systems come in play.

The evolution of such systems proceeded in the direction of increasing interface area per unit volume of the device.

Tracing a parallel with earlier oxygenators, wetted wall columns realize the gas-liquid contact by means of a liquid film flowing down the inner wall of the column [52] (Figure 4.6a). This configuration, while simple, is quite ineffective, since the interfacial gas-liquid area per unit volume is very low; thus this device complies only for absorption of a very soluble gas or, conversely, for stripping a poorly soluble gas.

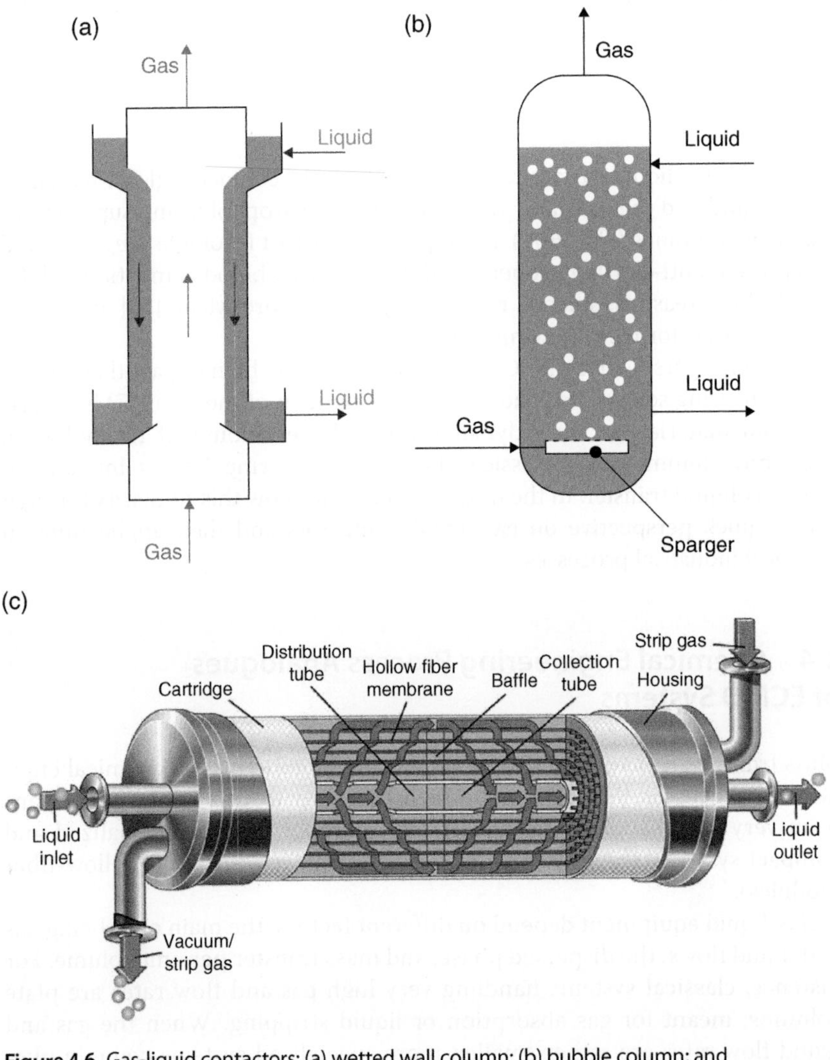

Figure 4.6 Gas-liquid contactors: (a) wetted wall column; (b) bubble column; and (c) gas-liquid hollow-fiber gas-liquid contactor. (*See insert for color representation of the figure.*)

Bubble columns (often reactors) are meant for gas-liquid contact in situations where pneumatic agitation may be a proper solution [53] (Figure 4.6b). As matter of fact, bubbling induces a good level of agitation, with no need for mechanical agitation. Gas-liquid interface per unit volume reaches high values for high flow rates per unit volume of liquid. A major issue with bubble contactors is foaming where the liquid contains tensioactive components, which strongly impairs gas-liquid separation stages.

Hollow-fiber membrane gas-liquid contactors allow the avoidance of direct contact between gas and liquid and achievement of a huge gas-liquid interface per unit volume [54] (Figure 4.6c). Another interesting feature of these devices is their modularity, so they can also be applied to a wide range of fluid flow rates.

Applications of gas-liquid contactors analogous to blood gas exchange are oxygenations in bioreactors and gas upgrading (removal of SO_X and CO_2). In aerobic microbial industrial processes, oxygen transfer is a key step in designing bioreactors. Oxygen, as in the human body, is required by aerobic microbes for growth, maintenance, and metabolite production, thus oxygen supply strongly affects whole process performance [55]. A key parameter for design is the Oxygen Transfer Rate (OTR), which is determined experimentally or evaluated *in silico* [56]. The mass transfer of oxygen in aerated bioreactors is evaluated in terms of liquid side mass transfer coefficient *kl*, which is generally measured along with the interfacial specific area *a* [57].

Bubble column bioreactors are the typical aerobic fermenters, also exploiting air bubbling for pneumatic agitation [58]: in these systems, gas is dispersed in liquid and the mass transfer coefficient $k_l a$ depends on liquid properties (dynamic and cinematic viscosity) and on the gas flow rate (flow regimes of bubbles in the liquid). Stirred bioreactors couple mechanical agitation by impellers with air sparging [59].

Finally, an application of hollow-fiber gas-liquid contactors is the removal of pollutant gases from air [60]: generally the application is motivated by specific liquid absorbents that react with the gases to be removed, enhancing low soluble gas mass transfer rates. A wide field of application is the upgrading of flue gases with acid gases (CO_2, SO_x) removal [60]. In this field, carbon dioxide removal is very similar to carbon dioxide exchange in blood oxygenators: liquid is generally an aqueous solution of ammines, which react with carbon dioxide producing soluble carbamate ions [61].

4.5 New Challenges

The original purpose of ECMO systems is to accomplish extra-corporeal blood flow to allow open-heart surgery for few hours. In this application, blood flow is kept high to allow gas exchange and the corresponding hemolysis was recognized in this condition only as a low risk side effect. In recent years, ECLS has been adopted more and more as option for respiratory assistance to mechanical ventilation, since it damages lungs on long-term application. In intensive care practice, patients affected by acute respiratory distress syndrome (ARDS) are treated by ventilation combined with ECLS to reduce side effects of ventilation [62, 63]. At this purpose, ECMO systems must be adapted essentially to accomplish CO_2 removal at a low blood flow (1–1.5 L/min, Extra-Corporeal CO_2 Removal = $ECCO_2R$). Pumpless systems (adopting a passive arterial-venous shunt) have been developed and have come in clinical practice in the last decade [2, 7, 64].

The increasingly frequent application in ARDS intensive care has required some technological improvement:

1) membrane materials: modern oxygenators have membranes in polymethyl pentene, which show low resistance and improved compatibility; gas exchange capability is higher than in silicon rubber oxygenators; finally, they reduce the impact on coagulation factors with resulting lower hemolysis [65];
2) centrifugal pumps: modern oxygenators include centrifugal pumps strongly reducing hemolysis due to classical roller pumps and this factor is greater in long-term applications;
3) miniaturization: the application in intensive care rather than in surgery has required simpler and less bulky equipment.

In this framework, ECMO systems play the role of long-term artificial organs, whose imagined development is the wearable version. The wearable artificial lung, an immediate result of the miniaturization of the ECMO systems for ARDS care, is meant as a bridge to organ transplantation to ensure more comfort to patients once confined to bed [66].

In developing a wearable artificial lung, some major issues emerge [67]: adapting the gas exchange capability to meet the variable need; reducing stress on blood to avoid blood clotting (a combined action on materials and fluid dynamics design); reducing the stress on heart, being a pumpless device. Microfluidics provides the framework to develop a real wearable artificial lung [3, 4]: the growing tools in this field for the first time promote a brand new perspective in artificial lungs, supplying technology for, finally, mimicking the unique structure of alveoli in natural lungs. The gas exchange capability for unit volume is tremendously higher than in modern classical oxygenators [5]. At the moment, these systems are at lab level, but promise to become the benchmark for all clinical applications, due to the low tidal volume and the huge gas exchange capability.

It is evident all these challenges require the work of strongly multidisciplinary teams, in which the role of chemical engineers is strongly advised. In turn, applications in the biochemical engineering may benefit of the technology transfer from the biomedical ground, on a very positive virtuous circle.

4.6 Conclusion

The history of blood oxygenator is a paradigm of how biomedical engineering is a strongly multidisciplinary field. Chemical engineering plays a key role in artificial organs development and still provides capital tools in challenging perspective (microfluidics). The chemical engineering principles have also evolved over time to meet the need of biomedical engineering and now many advancements in the biotechnology field exploit results from biomedical engineering applications. In this virtuous circle, it is possible that, in the near future, artificial wearable lungs will become a reality thanks also to a chemical engineering background in R&D teams.

List of Symbols

ARDS	Acute Respiratory Distress Syndrome
CFD	Computational Fluid Dynamics
ECLS	Extra-Corporeal Life Support
ECMO	Extra-Corporeal Membrane Oxygenator

References

1 Butt, W., and MacLaren, G. (2016) Extracorporeal membrane oxygenation 2016: an update. *F1000Research*, **5** (0), 4–9.

2 Bein, T., Weber, F., Philipp, A., et al. (2006) A new pumpless extracorporeal interventional lung assist in critical hypoxemia/hypercapnia. *Crit. Care Med.*, **34** (5), 1372–1377.

3 Kniazeva, T., Hsiao, J.C., Charest, J.L., and Borenstein, J.T. (2011) A microfluidic respiratory assist device with high gas permeance for artificial lung applications. *Biomed. Microdevices*, **13** (2), 315–323.

4 Potkay, J.A. (2014) The promise of microfluidic artificial lungs. *Lab Chip*, **14** (21), 4122–4138.

5 Potkay, J.A., Magnetta, M., Vinson, A., and Cmolik, B. (2011) Bio-inspired, efficient, artificial lung employing air as the ventilating gas. *Lab Chip*, **11** (17), 2901.

6 Wickramasinghe, S.R., and Han, B. (2005) Designing microporous hollow fibre blood oxygenators. *Chem. Eng. Res. Des.*, **83** (3), 256–267.

7 Fanelli, V., Ranieri, M. V., Mancebo, J., et al. (2016) Feasibility and safety of low-flow extracorporeal carbon dioxide removal to facilitate ultra-protective ventilation in patients with moderate acute respiratory distress syndrome. *Crit. Care*, **20** (1), 36.

8 Aaron, D., and Tsouris, C. (2005) Separation of CO_2 from flue gas: A review. *Separ. Sci. Technol.*, **40** (1–3), 321–348.

9 Iwahashi, H., Yuri, K., and Nosé, Y. (2004) Development of the oxygenator: Past, present, and future. *J. Artif. Organs*, 7 (3), 111–120.

10 Di Paola, L., Terrinoni, A.R., and Vitale, F. (2012) Extracorporeal membrane blood oxygenators: effect of membrane wetting on gas transfer and device performance. *Asia-Pacific J. Chem. Eng.*, 7 (S3), S348–S355.

11 Lim, M.W. (2006) The history of extracorporeal oxygenators. *Anaesthesia*, **61** (10), 984–995.

12 Wardrop, D., and Keeling, D. (2008) The story of the discovery of heparin and warfarin. *Br. J. Haematol.*, **141** (6), 757–763.

13 Gibbon, J.J. (1954) Application of a mechanical heart and lung apparatus to cardiac surgery. *Minn. Med.*, **37** (3), 171–185.

14 DeBakey, M. (1934) A simple continuous flow blood infusion instrument. *New Orleans Med. Surg. J.*, **87**, 386.

15 Clowes, G., Hopkins, A.L., and Neville, W. (1956) An artificial lung dependent upon diffusion of oxygen and carbon dioxide through plastic membrane. *J. Thorac. Surg.*, **32** (5), 630–637.

16 Kolff, W., Effler, D., Groves, L., et al. (1956) Disposable membrane oxygenator (heart-lung machine) and its use in experimental surgery. *Cleve. Clin. Q.*, **23** (2), 69–97.

17 Kolobow, T., Spragg, R., Pierce, J., and Zapol, W. (1971) Extended term (to 16 days) partial extracorporeal blood gas exchange with the spiral membrane lung in an unanesthetized lamb. *Trans. Am. Soc. Artif. Intern. Organs*, **17**, 350–354.

18 Suma, K., Tsuji, T., Takeuchi, Y., et al. (1981) Clinical performance of microporous polypropylene hollow-fiber oxygenator. *Ann. Thorac. Surg.*, **32** (6), 558–562.

19 Leverett, L.B., Hellums, J.D., Alfrey, C.P., and Lynch, E.C. (1972) Red blood cell damage by shear stress. *Biophys. J.*, **12** (3), 257–273.

20 Lequier, L., Horton, S., McMullan, D., and Bartlett, R.H. (2013) Extracorporeal membrane oxygenation circuitry. *Pediatr. Crit. care ...*, **14**, 1–10.

21 Jameel, S., Colah, S., and Klein, A.A. (2009) Recent advances in cardiopulmonary bypass techniques. *Contin. Educ. Anaesthesia, Crit. Care Pain*, **10** (1), 20–23.

22 Halaweish, I., Cole, A., Cooley, E., et al. (2015) Roller and centrifugal pumps. *ASAIO J.*, **61** (5), 496–501.

23 Mateják, M., Kulhánek, T., and Matoušek, S. (2015) Adair-based hemoglobin equilibrium with oxygen, carbon dioxide and hydrogen ion activity. *Scand. J. Clin. \& Lab. Investig.*, **75** (2), 113–120.

24 Tyuma, I. (1984) The Bohr effect and the Haldane effect in human hemoglobin. *Jpn J Physiol*, **34** (2), 205–216.

25 Changeux, J.-P. (2012) Allostery and the Monod-Wyman-Changeux model after 50 years. *Annu. Rev. Biophys.*, **41** (1), 103–133.

26 Monod, J., Wyman, J., and Changeux, J.P. (1965) On the nature of allosteric transitions: a plausible model. *J Mol Biol*, **12**, 88–118.

27 Bock, A., Field, H., and Adair, G. (1924) The oxygen and carbon dioxide dissociation curves of human blood. *J. Biol. Chem.*, **59**, 353–378.

28 Jensen, F.B. (2004) Red blood cell pH, the Bohr effect, and other oxygenation-linked phenomena in blood O2 and CO_2 transport. *Acta Physiol. Scand.*, **182**, 215–227.

29 Huestis, W.H., and Raftery, M.A. (1975) Conformation and cooperativity in hemoglobin. *Biochemistry*, **14** (9), 1886–1892.

30 Levantino, M., Spilotros, A., Cammarata, M., et al. (2012) The Monod-Wyman-Changeux allosteric model accounts for the quaternary transition dynamics in wild type and a recombinant mutant human hemoglobin. *PNAS*, **109** (37), 14894–14899.

31 Yuan, Y., Tam, M.F., Simplaceanu, V., and Ho, C. (2015) New look at hemoglobin allostery. *Chem. Rev.*, **115** (4), 1702–1724.

32 Di Cera, E., and Cera, E. Di (1998) Site-specific thermodynamics: understanding cooperativity in molecular recognition. *Chem. Rev.*, **98** (4), 1563–1592.

33 Goerke, A., Leung, J., and Wickramasinghe, S. (2002) Mass and momentum transfer in blood oxygenators. *Chem. Eng. Sci.*, **57** (11), 2035–2046.

34 Wickramasinghe, S.R., Goerke, A.R., Garcia, J.D., and Han, B. (2003) Designing blood oxygenators. *Ann. N. Y. Acad. Sci.*, **984**, 502–514.

35 Kawahito, S., Motomura, T., Glueck, J., and Nosé, Y. (2002) Development of a new hollow fiber silicone membrane oxygenator for ECMO: the recent progress. *Ann. Thorac. Cardiovasc. Surg.*, **8** (5), 268–274.

36 Ichikawa, S., and Nose, Y. (2002) Centrifugal blood pumps for various clinical needs. *Artif Organs*, **26** (11), 916–918.

37 Yano, T., Sekine, K., Mitoh, A., et al. (2003) An estimation method of hemolysis within an axial flow blood pump by computational fluid dynamics analysis. *Artif. Org.*, **27** (10), 920–925.

38 Vlasov, V., Karichev, Z., Muler, A., and Isaev, Y. (1988) Transport of carbon dioxide in membrane oxygenators. *Biomed. Eng.*, **21** (5), 168–174.

39 Kawahito, S., Maeda, T., Yoshikawa, M., et al. (2001) Blood trauma induced by clinically accepted oxygenators. *ASAIO J.*, **47** (5), 492–495.

40 Yeager, T., and Roy, S. (2017) Evolution of Gas Permeable Membranes for Extracorporeal Membrane Oxygenation. *Artif. Organs*, **0** (7).

41 Wickramasinghe, S.R., Kahr, C.M., and Han, B. (2002) Mass transfer in blood exygenators using blood analogue fluids. *Biotechnol. Prog.*, **18** (4), 867–873.

42 Zhang, J., Nolan, T.D.C., Zhang, T., et al. (2007) Characterization of membrane blood oxygenation devices using computational fluid dynamics. *J. Memb. Sci.*, **288** (1–2), 268–279.

43 Wickramasinghe, S.R., Garcia, J.D., and Han, B. (2002) Mass and momentum transfer in hollow fibre blood oxygenators. *J. Memb. Sci.*, **208** (1–2), 247–256.

44 Wickramasinghe, S.R., Semmens, M.J., and Cussler, E.L. (1992) Mass transfer in various hollow fiber geometries. *J. Membr. Sci.*, **69** (3), 235–250.

45 Arazawa, D.T., Oh, H. Il, Ye, S.H., et al. (2012) Immobilized carbonic anhydrase on hollow fiber membranes accelerates CO 2 removal from blood. *J. Membr. Sci.*, **403–404**, 25–31.

46 Kumar, P.S., Hogendoorn, J.A., Feron, P.H.M., and Versteeg, G.F. (2002) New absorption liquids for the removal of CO_2 from dilute gas streams using membrane contactors. *Chem. Eng. Sci.*, **57** (9), 1639–1651.

47 Wang, R., Zhang, H.Y., Feron, P.H.M., and Liang, D.T. (2005) Influence of membrane wetting on CO_2 capture in microporous hollow fiber membrane contactors. *Sep. Purif. Technol.*, **46** (1), 33–40.

48 Zhang, H.Y., Wang, R., Liang, D.T., and Tay, J.H. (2008) Theoretical and experimental studies of membrane wetting in the membrane gas-liquid contacting process for CO_2 absorption. *J. Membr. Sci.*, **308** (1–2), 162–170.

49 Kim, B.S., and Harriott, P. (1987) Critical entry pressure for liquids in hydrophobic membranes. *J. Colloid Interface Sci.*, **115** (1), 1–8.

50 Tuteja, A., Choi, W., Mabry, J.M., et al. (2008) Robust omniphobic surfaces. *Proc. Natl. Acad. Sci.*, **105** (47), 18200–18205.

51 Fair, J., Steinmeyer, D., Penney, W., and Crocker, B. (1997) Gas absorption and gas-liquid system design, in *Perry's Chemical Engineers' Handbook*, 7th Edn, MacGraw-Hill, New York.

52 Asai, S., Kimura, T., and Yoshizawa, H. (1987) Mass *Transfer in wetted-wall column with cocurrent laminar liquid-liquid flow*, **10** (2), 483–488.

53 Kantarci, N., Borak, F., and Ulgen, K.O. (2005) Bubble column reactors. *Process Biochem.*, **40** (7), 2263–2283.

54 Gabelman, A., and Hwang, S.-T. (1999) Hollow fiber membrane contactors. *J. Memb. Sci.*, **159** (1–2), 61–106.

55 Garcia-Ochoa, F., and Gomez, E. (2009) Bioreactor scale-up and oxygen transfer rate in microbial processes: An overview. *Biotech. Adv.*, **27** (2), 153–176.

56 Garcia-Ochoa, F., Gomez, E., Santos, V.E., and Merchuk, J.C. (2010) Oxygen uptake rate in microbial processes: An overview. *Biochem. Eng. J.*, **49** (3), 289–307.

57 Kawase, Y., Halard, B., and Moo-Young, M. (1992) Liquid-phase mass transfer coefficients in bioreactors. *Biotechnol. Bioeng.*, **39** (11), 1133–1140.

58 Kadic, E., and Heindel, T. (2014) Bubble column bioreactors, in *An Introduction to Bioreactor Hydrodynamics and Gas-Liquid Mass Transfer*. John Wiley & Sons, Inc., Hoboken, NJ: pp. 124–167.

59 Fujasová, M., Linek, V., and Moucha, T. (2007) Mass transfer correlations for multiple-impeller gas-liquid contactors. Analysis of the effect of axial

dispersion in gas and liquid phases on "local" kL a values measured by the dynamic pressure method in individual stages of the vessel. *Chem. Eng. Sci.*, **62** (6), 1650–1669.

60 Mansourizadeh, A., and Ismail, A.F. (2009) Hollow fiber gas-liquid membrane contactors for acid gas capture: A review. *J. Hazard. Mater.*, **171** (1–3), 38–53.

61 Wang, R., Li, D.F., and Liang, D.T. (2004) Modeling of CO_2 capture by three typical amine solutions in hollow fiber membrane contactors. *Chem. Eng. Process.*, **43** (7), 849–856.

62 Marhong, J.D., Munshi, L., Detsky, M., et al. (2015) Mechanical ventilation during extracorporeal life support (ECLS): a systematic review. *Intensive Care Med.*, **41** (6), 994–1003.

63 Squiers, J.J., Lima, B., and DiMaio, J.M. (2016) Contemporary extracorporeal membrane oxygenation therapy in adults: Fundamental principles and systematic review of the evidence. *J. Thorac. Cardiovasc. Surg.*, **152** (1), 20–32.

64 Tiruvoipati, R., Botha, J., and Peek, G. (2012) Effectiveness of extracorporeal membrane oxygenation when conventional ventilation fails: Valuable option or vague remedy? *J Crit. Care*, **27** (2), 192–198.

65 Khoshbin, E., Roberts, N., Harvey, C., et al. (2005) Poly-methyl pentene oxygenators have improved gas exchange capability and reduced transfusion requirements in adult extracorporeal membrane oxygenation. *ASAIO J.*, **51** (3), 281–287.

66 Zwischenberger, J.B., and Alpard, S.K. (2002) Artificial lungs: a new inspiration. *Perfusion*, **17** (4), 253–268.

67 Nolan, H., Wang, D., and Zwischenberger, J.B. (2011) Artificial lung basics. *Organogenesis*, **7** (1), 23–27.

5

Model Predictive Control for the Artificial Pancreas

M. Capocelli[1], L. De Santis[1], A. Maurizi[2], P. Pozzilli[2], and Vincenzo Piemonte[3]

[1] *Unit of Process Engineering, Department of Engineering, Università Campus Bio-Medico di Roma, Italy*
[2] *Unit of Endocrinology and Diabetes, Department of Medicine, Università Campus Bio-Medico di Roma, Italy*
[3] *Department of Engineering, University Campus Biomedico of Rome, Italy*

5.1 Introduction

Diabetes is one of the most common chronic diseases, affecting more than 400 million people in the world and resulting in 1.5 million deaths each year [1]. Achieving optimal glycemic control with intensive insulin therapy is the mainstay treatment for Type 1 diabetes (T1D) and is also required for many patients with Type 2 diabetes. Although many innovations have been made in the field of insulin manufacturing and delivery, many patients remain unable to achieve glycemic targets and good HbA1c levels [1]. Diabetes patients suffer from high morbidity and mortality rates due to complications that could be prevented with intensive treatment and novel technologies.

The CSII (Continuous Subcutaneous Insulin Infusion) was developed more than 30 years ago enabling patients to achieve physiologic replacement of insulin through the use of a mechanical delivery system [1–4]. To facilitate disease self-management and to improve patient quality of life, innovative CSII (Continuous Subcutaneous Insulin Infusion) devices, including both continuous glucose monitoring (CGM) and insulin delivery through insulin pumps, have been developed [1, 3]. The pump is connected to a thin plastic tube that has a small soft cannula placed under the skin and delivers a pre-programmed basal insulin dose to maintain blood glucose control between meals and overnight. Bolus doses are

Biomedical Engineering Challenges: A Chemical Engineering Insight, First Edition.
Edited by Vincenzo Piemonte, Angelo Basile, Taichi Ito, and Luigi Marrelli.

commonly given by patients before meals taking into account the pre-prandial blood glucose levels, carbohydrate intake at each meal, the insulin/carbohydrate ratio, and the predicted intensity of physical exercise.

Today, the Artificial Pancreas (AP), investigated and discussed since the 1970s [1–4], has become a hot area of translational research and industrial development, and is likely to be the most promising technological development in the treatment of diabetes [3–5]. The artificial pancreas (AP), schematically described in Figure 5.1, consists of a closed-loop control realized through a subcutaneous system combining a sensor for the CGM, a control algorithm, and an insulin infusion device.

The mathematical algorithm included in closed-loop control to maintain glucose concentration within the normal range must cope with [4]:

- The complex biological mechanism in glucose-insulin control regulations resulting in nonlinear dynamics,
- delays and inaccuracies in both glucose sensing and the insulin delivery device as the intrinsic delay between the insulin action (time needed for the occurrence of the insulin absorption peak after the injection) and the measuring of the interstitial glucose concentration,
- disturbances occurring in the system due to the meal ingestion or physical activities; the AP controller has to find a trade-off between slow-pace regulation (quasi-steady state) and drastic postprandial regulation; a slow response cannot provide a good attenuation of postprandial glucose peaks and an excessive responsiveness may result in system oscillations.

The first generation of AP are actually "hybrid closed-loop systems," due to the requirement of external intervention (usually for considering mealtime

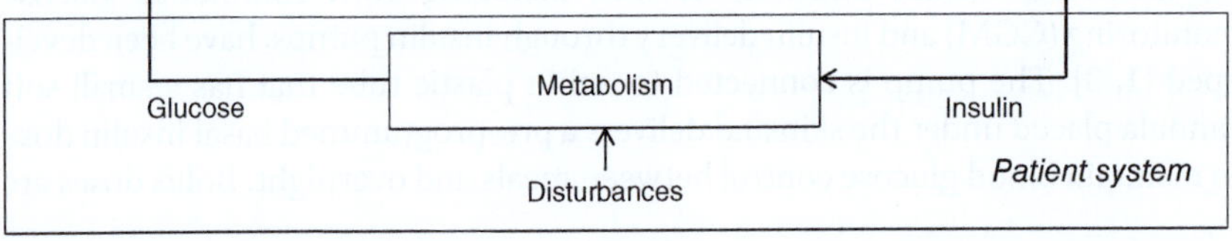

Figure 5.1 Descriptive scheme of the closed chain system for glucose-insulin control in an artificial pancreas (*Source:* V. Piemonte V., L. De Santis, M. Capocelli, A. Maurizi, P. Pozzilli. A Novel Three-Compartmental Model for Artificial Pancreas: Development and Validation. J Artificial Organs, 2017).

and exercise). Research effort is still needed to delivery fully closed-loop systems employing artificial intelligence. The class of model-predictive-control (MPC), using a quantitative mathematical model of the metabolic system with delays and constraints, is the most suitable approach for control system implementation and has overcome the difficulties encountered by standard proportional integral derivative (PID) controllers [3–9]. Figure 5.2 shows the strategy implemented by MPC. It predicts the future evolution of a system (glucose dynamics *G(t)* in the patient metabolic system) and optimizes the control signal *I(t)* to achieve the *G(t)* values ranging between defined set points. The MPC, acting like it might in a chess match, plans a sequence of control moves (changing in the insulin infusion rate) based on the history (glucose measurements) and predicted behavior. The first move is implemented and, after the observation of the biological response, the strategy is reassessed. This procedure, repeated at each sampling instant, offers several advantages [4–7]:

- it is customizable by using patient-specific model parameters obtained from the clinical data of each individual;
- it can solve the dilemma between postprandial and slow regulation using a predictive capacity that is tested beforehand and adjusted on real time analysis;
- it can be used for *in silico* trials with randomized process variables for the validation of the MPC model itself (or other equivalent models).

Since control strategy performances depend on model accuracy, model development and validation forms the most critical step in AP improvement.

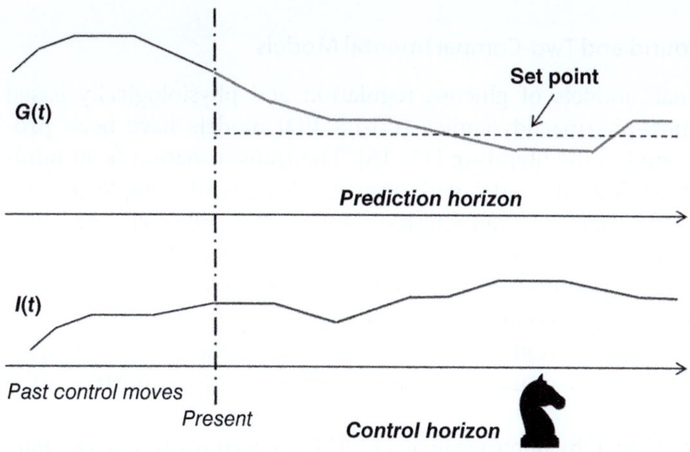

Figure 5.2 Basic concept of MPC during a chess game consisting of predicting and regulating glucose level. (*See insert for color representation of the figure.*)

In this chapter, we present an historical and technical review of MPC modeling including two three-compartment models as well as black block approaches. Moreover, we present the results of our recently published MPC [10] consisting of glucose-insulin interactions DDEs (delay differential equations) tested on actual clinical data. The model describes the glucose-insulin interactions in three compartments: the subcutaneous (SC) compartment, the plasma compartment (including the rapidly equilibrating tissues), and the interstitial fluid (ISF) compartment (including the slowly equilibrating tissues). The DDEs include six identifiable parameters able to account for inter-patient variability. The availability of realistic individual models is the basis for conducting an *in-silico* trial: the closed-loop control can be tuned individually and then tested on each virtual patient.

Although new models always perform better, the actual glucose regulation in the whole human body cannot be achieved through a mathematical model based on physiology, as it is far too sophisticated. At least 10 hormones participate in glucose regulation together with the important role played by the central nervous system (always neglected) [11]. A critical discussion on this theme was given by Teixeira and Malin [11]. The black-box approaches, mainly represented by the class of algorithms in Artificial Neural Networks (ANN), can be a valid alternative to phenomenological-based models since they present very low prediction errors, high robustness, "construction simplicity," as well as the availability of realizing self-training devices. The advantages and drawbacks of this class of model are also discussed in the following.

5.2 Phenomenological Models

5.2.1 Background and Two-Compartmental Models

Several "minimal" models of glucose regulation and physiologically based pharmacokinetics–pharmacodynamics (PB-PK-PD) models have been proposed and validated in the literature [12–15]. The transformation from minimal to phenomenological model with varying degrees of complexity and accuracy has been gradual and cannot precisely be historically collocated. The main difference between minimal and PB-PK-PD models is that the former assumes a large collection of tissues and organs as compartments, whereas the PB-PK-PD models treat each organ or tissue as a separate compartment (e.g., brain, liver, kidneys, etc.). A good compromise between the number of equations the model accuracy characterizes the different choices reported in the literature [10–18].

The "Minimal Model" by Bergman et al. [12, 13] has been realized to estimate insulin sensitivity and glucose effectiveness for clinical applications. Some successive modifications enabled improvement of model reliability in simulating individual glucose metabolism [16–18]. Bennet and Gourley [19] considered the delay of the insulin hepatic secretion; Sturis et al. proposed an ODE system with negative feedback and third-order delay [20] while the study by Chen at al. focused on the administration of glucose during meals [21, 22]. The model presented by Sturis et al. [20] has been formulated to understand the genesis and factors responsible for glucose-insulin system oscillations. The structure of the model consists of two compartments: one for insulin and one for glucose dynamics. Other variables were introduced to account for the delay between the two.

Hovorka et al. in their studies [23, 24] proposed a nonlinear model of the glucose-insulin regulation mechanism as an algorithm control for a closed-loop device to maintain the condition of normoglycemia. The model included a two-compartment structure to represent glucose: one including plasma and tissues, which are in rapid equilibrium with it and the other representing the interstitial and intracellular space that slowly comes into balance with the plasma.

Li et al. [25] introduced the presence of two explicit delays in the insulin-glucose system characterizing ultradian oscillations in insulin secretion. A study by Chen et al [22] focused on diabetic patients and simulated glucose administration to the body via carbohydrate ingestion during meals (as is usually in everyday life). The results of the study showed that the identified parameters are representative of a patient's silent conditions and that the model is capable of representing the dynamics of the glucose-insulin system. More recently, Wu et al. [6, 26] proposed a two-compartment of the glucose-insulin system with two explicit delays, with the aim of describing the oscillations in such a system related to the production of glucose from the liver and others related to secretion of insulin from the pancreas.

5.2.2 Three-Compartment Models

Insel et al. [27] and Cobelli et al. [28] were the first to show that a three-compartment model is required to describe basal glucose kinetics. Both models incorporated a priori physiological knowledge to achieve inter-patient specificity. Magni et al. reported an in silico trial, demonstrating that the three-compartmental MPC achieves satisfactory glycemic regulation in a virtual population of "type 1 diabetic patients" [5].

More recently, Piemonte et al. tested a three-compartment model that takes into account the main physiological factors playing a role in the blood glucose curve over time and the variability between patients (due to gender, age, weight, lifestyle, and other physiological factors) [10]. The model, extending the work of Wu at al. [26], includes the dynamics of insulin release and subcutaneous adsorption. The MPC structure, represented in Figure 5.3, meets the technology limits such as the delay caused by the absorption of sub-skin tissues and by the subcutaneous glucose sensor. The three compartments represented in

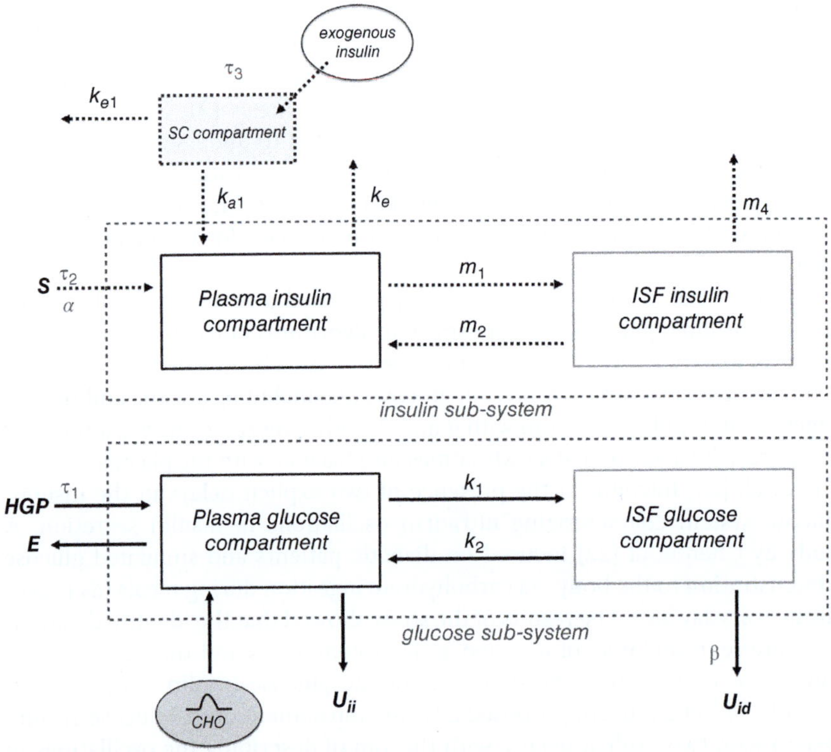

Figure 5.3 Block diagram of the model presented and tested by Piemonte et al. [10]. (*See insert for color representation of the figure.*)

Figure 5.3, are: (1) the plasma and rapidly equilibrating tissues, (2) the ISF compartment, representing the tissues slowly equilibrating with plasma, and (3) the SC compartment simulating the exogenous insulin dynamic in subcutaneous tissue. There are two glucose sources (the hepatic glucose production from the liver and the glucose intake from gut) and three mechanisms of glucose utilization in the model: insulin-independent, insulin-depending, and renal excretion. There are two sources of insulin in the system: the endogenous insulin secreted by the β-cells of the pancreas and the exogenous insulin injected in the subcutaneous tissue. The mathematical formulation of the glucose-insulin regulatory system consists of the DDEs that follow:

$$\dot{G}_P(t) = G_{in}(t) + HGP\left(I_p(t-\tau_1)\right) - U_{ii}\left(G_p(t)\right) - E\left(G_p(t)\right) - k_1 G_p(t) + k_2 G_i(t) \tag{5.1}$$

$$\dot{G}_i(t) = k_1 G_p(t) - k_2 G_i(t) - U_{ii}\left(G_p(t), I_i(t)\right) \tag{5.2}$$

$$\dot{I}_P(t)=k_{a1}I_{sc}(t)+\alpha\cdot S(G_p(t-\tau_2))-m_1I_p(t)+m_2I_i(t)-k_eI_p(t) \quad (5.3)$$

$$\dot{I}_i(t)=I_p(t)-m_2I_i(t)-m_4I_i(t) \quad (5.4)$$

$$\dot{I}_{sc}(t)=U(t-\tau_3)-k_{a1}I_{sc}(t)-k_{e1}I_{sc}(t) \quad (5.5)$$

where I_p and I_i are the plasma and ISF insulin in the insulin sub-system, I_{sc} is the insulin injected through the SC-compartment; G_p is the plasma glucose and G_i is the ISF glucose. S is insulin secreted by the endocrine pancreas; HGP stands for glucose hepatic production; U_{ii} is the insulin-independent glucose utilization; U_{id} is insulin-dependent glucose utilization; E is renal excretion; G_{in} is the glucose intake rate, and U is the exogenous insulin injected by the insulin pump. The mass concentration in compartments is calculated by assuming the distribution volumes of plasma glucose V_{gp}, ISF glucose V_{gi}, plasma insulin V_{ip}, and ISF insulin V_{ii}, fixed as proportional to body weight [10, 26].

Insulin-dependent glucose utilization assumes the shapes reported in Figure 5.4, depending on both insulin and glucose levels according to the following equation:

$$U_{id}(G_i,I_i) = \beta\cdot 0.01G_i / V_{gi}\cdot\left\{4+90\bigg/\left[1+\exp\left(-1.772\log\left[I_i\cdot\left(\frac{1}{V_{ii}}+\frac{0.03}{e}\right)\right]+7.76\right)\right]\right\} \quad (5.6)$$

Insulin-independent glucose utilization is mathematically formulated by Eq. 5.7. Figure 5.5 shows the effect of plasma glucose level on Uii.

$$U_{ii}(G_p)=72\left[1-\exp\left(-\frac{G_p}{144V_{gp}}\right)\right] \quad (5.7)$$

Hepatic glucose production increases with the decrease in plasma insulin concentration; the glucagon concentration is not included in this model. The function HGP, illustrated in Figure 5.6 versus the plasma insulin level, represents the effect of insulin on glucose production. High insulin concentrations completely inhibit glucose production, whereas lower insulin concentration increases the glucose output as reported by the following equation:

$$HGP(I_p)=170\bigg/\left[1+\exp\left(0.29\left(\frac{I_p}{V_{ip}}-17.5\right)\right)\right] \quad (5.8)$$

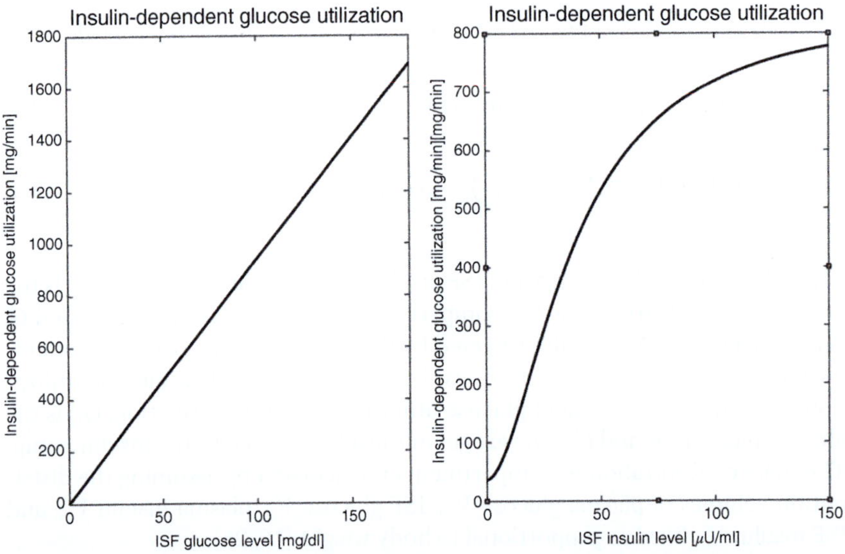

Figure 5.4 Insulin-dependent glucose utilization with ISF glucose level when $Ii(t) = 6\ \mu U/ml$ (left); change of insulin-dependent glucose utilization with ISF insulin level when $Gi(t) = 90\ mg/dl$ (right). (*See insert for color representation of the figure.*)

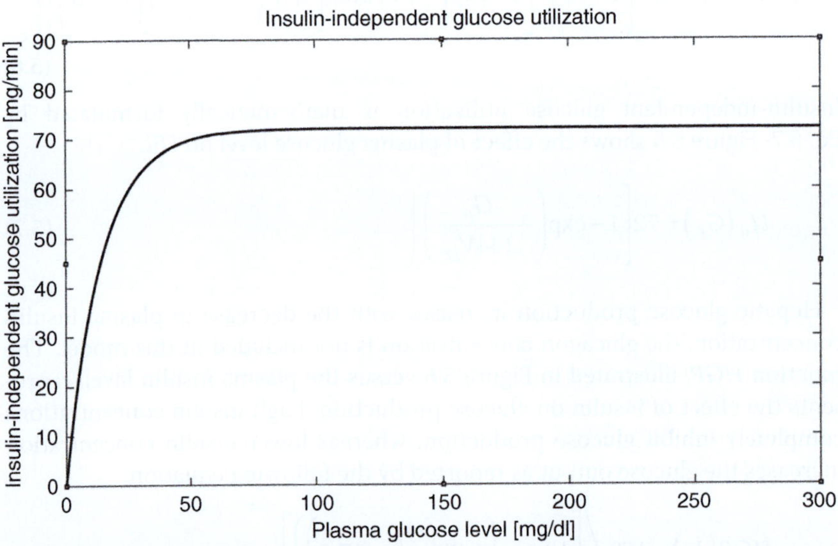

Figure 5.5 Effect of plasma glucose level on insulin-independent glucose utilization. (*See insert for color representation of the figure.*)

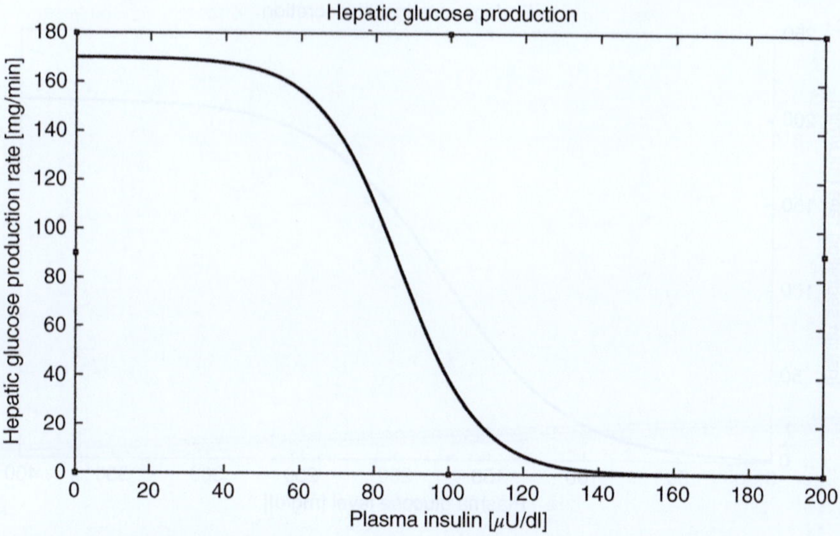

Figure 5.6 Change of hepatic glucose production with plasma insulin level. (*See insert for color representation of the figure.*)

Renal excretion occurs in the regulatory system when the plasma glucose level is higher than a threshold value as in Eq. 5.9. The parameter values can be found in the cited literature [10, 26].

$$E(G_p) = 0.0005\left[G_p(t) - k_{e2} \cdot BW\right] \tag{5.9}$$

The gut absorption profiles simulate the condition of the m-th meal ingestion G_m at time t_m. Glucose input G_{in} is obtained by integrating all meal intake G_m and is assumed to enter directly in to the plasma compartment. GM is obtained by the interpolation of experimental data. A parametric study of this equation is given by the work of Piemonte et al. [10].

$$G_{in}(t) = \sum_{\forall m \in M} G_m(t - t_m) \cdot \mathcal{U}(t - t_m) \tag{5.10}$$

The model includes insulin Lispro kinetics with bolus and continuous subcutaneous insulin infusion (CSII) methods of delivery. The insulin is absorbed in subcutaneous tissue with a time delay before its appearance in plasma. Insulin secretion in the bloodstream is stimulated by a high plasma glucose concentration as shown in Figure 5.7 and mathematically described by the following equation:

$$S(G_p) = 210 / \left[1 + \exp(5.21 - 0.003 G_p / V_{gp})\right] \tag{5.11}$$

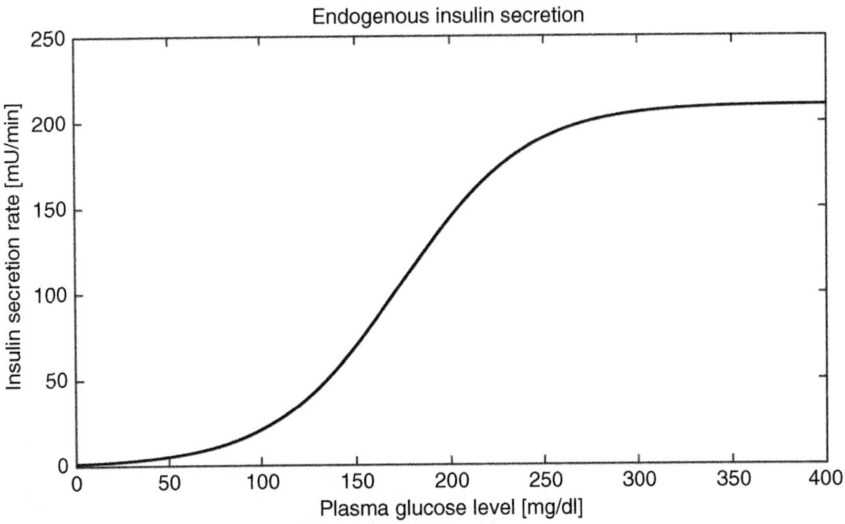

Figure 5.7 Insulin secretion rate with plasma glucose level. (*See insert for color representation of the figure.*)

This model has been tested by simulating clinical data, during a normal span of 4 days, from T1D patients attending the Endocrinology and Diabetes Department of University Campus Bio-Medico in Rome. The therapy pump allows setting a basal rate insulin all delivered throughout the day and night for normal body function without food and a bolus (or dose insulin on demand). Some model parameters assumed fixed values for each case study, both fixed and experimentally determined through the simulation of the clinical case studies), as reported and discussed in our recent work [10]. The identification of the six parameters in Table 5.1 (influencing the behavior of glucose regulation system) allowed us to tailor the simulation to a real clinical scenario (patient customization). Total time delay resulted in overall 100 min or more; the time-to-peak of plasma insulin concentration for short-acting insulin is 50–60 min followed by the delay in the insulin action (30–40 min) and the delay relative to the glucose diffusion from the plasma to the interstitial fluid of 10–20 min. Piemonte et al. discussed the values of calculated parameters in Table 5.1 from a physiological and pathological point of view confirming, through interpatient comparison, the actual clinical conditions of the patients [10].

Figure 5.8 depicts the values of $G_i(t)$, both simulated and experimentally evaluated, relative to one case study. The model output is represented by a continuous dark shaded line, the lighter shaded line represents the clinical data collected by the continuous glucose monitoring sensor; at the bottom of Figure 5.8, the solid curve is the injected insulin whereas the grey dashed curve

Table 5.1 Definition and explanation of the inter-patient adjustable parameters obtained by Piemonte et al. [10].

τ_1	Time delay of hepatic glucose production in minutes due to liver stimulation by plasma insulin for the production from glycogen stores. It presented strong inter-patient variability due to the complexity of the hepatic mechanisms.
τ_2	Time delay of pancreatic insulin production.
τ_3	Time delay of insulin subcutaneous delivery (from the insulin pump device to plasma); values showed a little variability due to the subcutaneous tissue resistance after a prolonged use.
k_{a1}	Transfer rate constant of exogenous insulin from subcutaneous tissue to plasma compartment; individual variable parameter based on subjects' clinical data (in the range 0–1 min^{-1}); observed values close to the upper limit for the studied cases.
α	Quantification of the residual capacity of pancreas to produce insulin in relation to high blood glucose level. The values are in the range 0–1.
β	Quantification of body sensibility to insulin action. Insulin acts on muscle and adipose cells stimulating both glucose uptake and metabolism. The values (in the range 0–1) were over 0.7 in agreement with Type 1 diabetics that do not present insulin resistance.

(*Source:* V. Piemonte V., L. De Santis, M. Capocelli, A. Maurizi, P. Pozzilli. A Novel Three-Compartmental Model for Artificial Pancreas: Development and Validation. J Artificial Organs, 2017).

represents the gut-absorption dynamic. In the cited paper, the good behavior of the proposed model has also been established by comparison with a previous two-compartment model [10, 26].

5.3 Black-Block Approach

The nonlinear nature of the problem, as well as the variety and variability of the disturbances (meal intake, physical activity, and stress), together with the inter-variability between subjects, reduces the robustness of the PB models in representing individual behavior. A purely mathematical black-box (BB) approach allows us to modulate the model on process input for predicting process variability without realizing a PB model based on actual phenomenology (with strong assumption and simplification) of the simulated process [29–32]. The artificial neural network (ANN) approach with adaptive connectionist structures is the most representative of the BB approaches and is able to overcome the issues and necessity of simplistic assumptions of PB models [31, 32].

In the field of process modeling and control, several control schemes can be realized starting from knowledge of the most important process parameters [32–34]. In the AP field, the predictive control neural models, inverse control

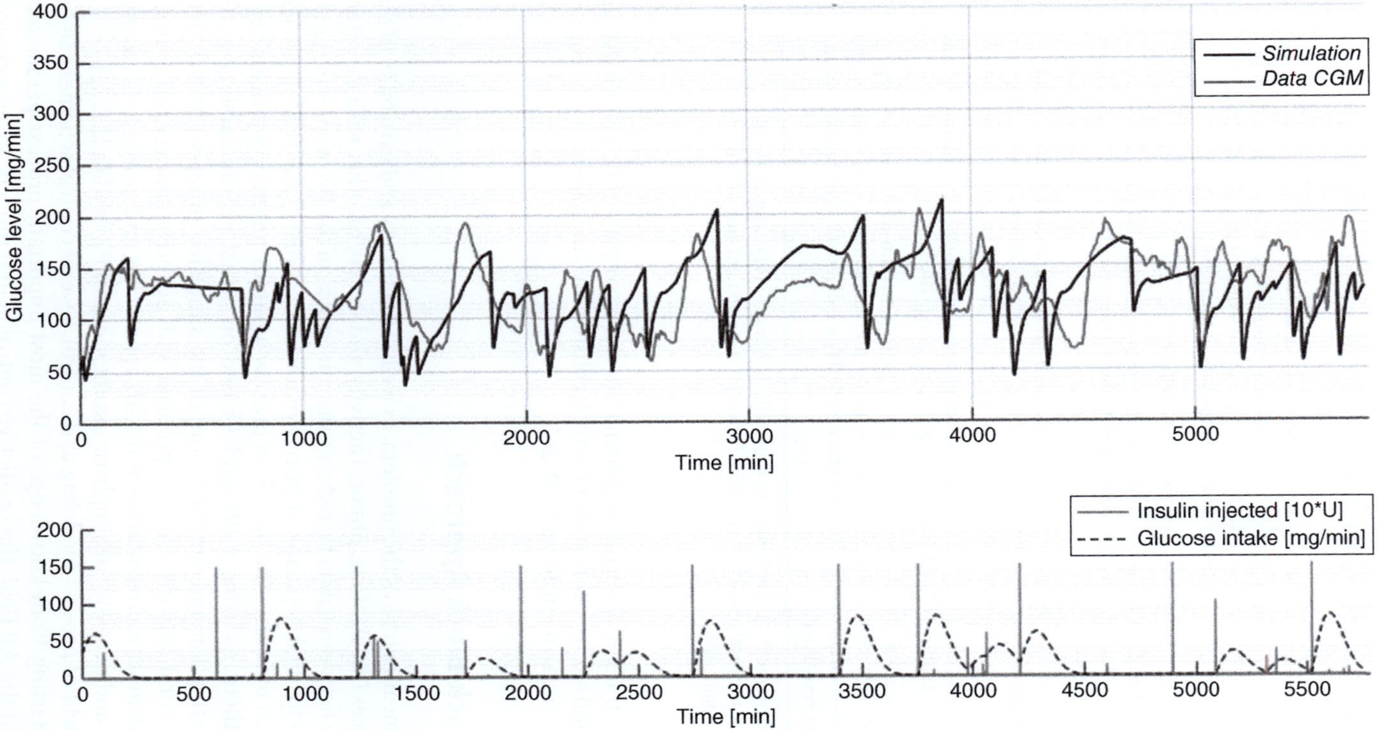

Figure 5.8 Glucose level profiles with the simulation and the insulin and glucose intake for one patient reporting in the validation study by Piemonte et al. [10] (*Source:* V. Piemonte V., L. De Santis, M. Capocelli, A. Maurizi, P. Pozzilli. A Novel Three-Compartmental Model for Artificial Pancreas: Development and Validation. J Artificial Organs, 2017). (*See insert for color representation of the figure.*)

models, and neural adaptive control models can be trained through clinical data, and employed to identify patient dynamics and to obtain reliable glucose–insulin models to be incorporated in the CGM [32].

The ANN approach is very interesting in the context of glucose prediction because of its natural ability to fit nonlinear functions, and the ease with which inputs belong to different domains (e.g., CGM, meal, and insulin). The ANN is a mathematical model consisting of processing elements (called neurons in analogy to the biological structure) connected by synapses in a network where each branch has a weight and each node has a bias and an activation function [30–32]. The relation between inputs and outputs is "learned" during the training phase thanks to a mathematical algorithm and the availability of a training and a validating data set of acceptable numerosity. Figure 5.9 shows how a simple feedforward neural network can be trained on the most important parameters regulating the glucose-insulin dynamics as the meal, physical activity, and physio-pathological parameters. In this simple case, the ANN training consists of the evaluation of the best values of weights and bias that allow reproduction of the desired output.

Up to now, the literature has not reported modeling results including all the variables showed in Figure 5.9. Moreover, the ANN has to be implemented in

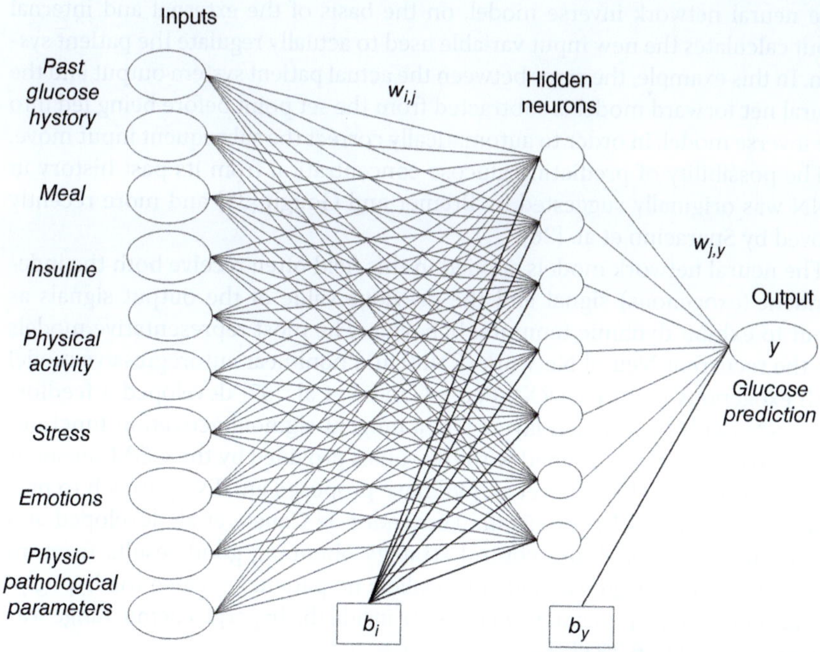

Figure 5.9 Conceptual scheme of an ideal glucose predictor able to exploit most of the inputs influencing glucose concentration. (*See insert for color representation of the figure.*)

both predicting the future trend (time t + k) and calculating the control move at time *t* (to modify the future system behavior). The complete implementation of ANN neural networks should also consider the utilization of inverse models to define the input in the control horizon. ANNs are highly suitable for incorporation in this multistep predictive control strategy approach predicting the outputs for some *k* time step into the future $y(t+k)$. Hussain individuated the normal cost function J to obtain the optimal sequence of control actions in a ANN-based predictive control:

$$J = \sum_{k=N_1}^{N_2} \left[e(t+k)\right]^2 + \sum_{i=1}^{N_3} \left[w_i \delta u(t+i)\right]^2 \tag{5.12}$$

Where N_1, N_2 define the prediction horizon (PH) is the control horizon, e is the error between target and prediction, w is the weighting factor, and δu the change in control sequences. The first term is a measure of the distance between network prediction and desired future trajectory and the second term penalizes excessive movement of the manipulated variables [30].

A simplified control methodology, extracted from [30] is reported in Figure 5.10. The overall strategy consists of placing the forward ANN model in parallel with the patient system to observe the model mismatches continuously. The neural network inverse model, on the basis of the external and internal input calculates the new input variable used to actually regulate the patient system. In this example, the error between the actual patient system output and the neural net forward model is subtracted from the set point before being fed into the inverse model, in order to automatically correct the subsequent input move.

The possibility of predicting glucose concentration from its past history in ANN was originally suggested by Bremer and Gough [35] and more recently proved by Sparacino et al. [36].

The neural network models applied in this field often receive both the independent (exogenous) signal and the previous value of the output signals as input to exhibit dynamic temporal behavior. The most representative models are the recurrent Neural Network (RNN) and nonlinear autoregressive model with exogenous input (NARX). Pérez-Gandia et al. [37] developed a feedforward NN (with two hidden layers and a tangent sigmoid activation function) for glucose prediction, trained with the values provided by the CGM sensor at three different PH. Pappada et al. [38] have proposed an NN approach to predict glycemia with a PH of 75 min. Fernandez de Canete et al. developed and tested an ANN model and control strategy obtaining good results from an experimental campaign carried out in silico: no patients reached the hypoglycemic zone and the percentage of patients inside the hyperglycaemic range was also very low (2.6–6.2%) [31].

The major criticisms of the models are: (1) ANN produces a loss of physiological and clinical information and does not increase the ability to understand

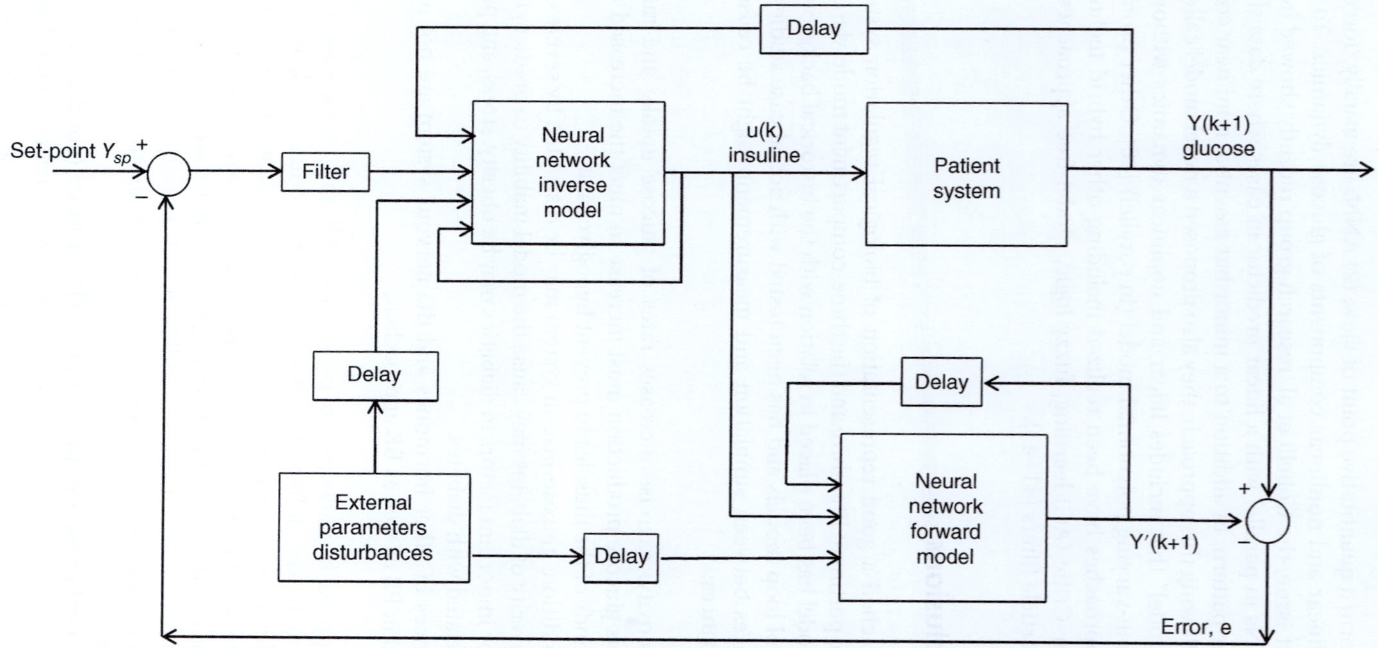

Figure 5.10 Control algorithm AP including ANN models (*Source:* adapted from [30] M.A. Hussain, Review of the applications of neural networks in chemical process control. Simulation and on-line implementations, Artif. Intell. Eng. 13 (1999) 55–68).

the therapy form a quantitative point of view, (2) ANNs are usually "forced" to model both linear and nonlinear components of glucose dynamics. To overcome this last issue, the Cobelli et al. research group recently showed how to couple the NNs in parallel with a linear predictor in cases where data show a marked linear pattern, in addition to a minor but essential nonlinear component [39]. Following this approach, they also proposed a novel model called the "jump NN model" that includes linear and nonlinear dynamics without the need for a time-varying polynomial model (in parallel) [40]. Other promising black-box approaches have been realized including other hybrid techniques such as Actor-Critic (AC) learning, fuzzy logic, model-free approaches, and Kalman or Particle filters [41–43].

5.4 Conclusions

MPC has reached a good representation of biological regulation system as proved by the pertinent literature and the three-compartment model described here. This model has been placed in relation with the historical background of closed-control loop models and has been tested with actual case studies [10]. The differences between simulation and measurement might be caused by other disturbances:

- physical activity: exercise increases rates of glucose uptake and rates of endogenous glucose production must increase to meet the increased metabolic demands of the muscles to prevent hypoglycemia.
- patient condition: glucose-insulin system may be disrupted by severe diabetes. The severity of diabetes may cause the model inability to predict glucose level. Other important factors in diabetics may be obesity, stress, and pathologies correlated with diabetes.
- the influences of other hormones and the nervous system are not usually considered in PB as well as BB approaches.

The PB approach allows us to personalize AP and to increase clinical knowledge to better define the therapy. On the other hand, in the BB approach, the absence of the intrinsic constraints caused by physiological meaning allows us to outperform the physiologically based (PB) MPC, to the detriment of the losses in information and external controllability. In this context, the realization of hybrid models, based on both physiological and artificial structures, could be a step forward in the realization of a reliable AP control-loop.

Moreover, further improvements to the AP hardware (not discussed in the present work) are needed to provide more accurate control systems. The control algorithm, however, is only one component in the closed-loop system. It is extremely important that the system is properly integrated with reliable sensors and pumps and an easy-to-use interface. In this framework, the choice of additional control variables could fall on the counter regulatory pancreatic hormone, the glucagon that is fast acting and shows a deficient secretion in diabetic individuals [11].

Nomenclature

Acronyms and Variables

AP	Artificial pancreas
B	Constant for the gut absorption equation
BW	Body weight
CGM	Continuous glucose monitoring
CHO	Carbohydrate size
CSII	Continuous subcutaneous insulin infusion
E	Error between simulated and monitored data
E	Renal excretion
I_{sc}	Exogenous insulin is injected through the SC-compartment
G_p	Plasma glucose concentration
G_i	ISF glucose concentration
G_{in}	Glucose intake rate
G_m	Glucose intake from meal
HGP	Hepatic glucose production
ISF	Interstitial fluid
Ii	Plasma insulin concentration in the insulin sub-system
Ip	ISF insulin concentration in the insulin sub-system
I_{sc}	Exogenous insulin is injected through the SC-compartment
k	Rate constant for the glucose gut absorption
k_1	Glucose transfer rate plasma-ISF compartment
k_2	Glucose ISF-plasma transfer rate
k_{a1}	Diffusion transfer rate of injected insulin to plasma compartment
k_e	Insulin elimination rate from plasma
k_{e1}	Subcutaneous local degradation elimination insulin, m-1
k_{e2}	Renal threshold of glucose
m	Meal
M	Domain of meal values
MPC	model-predictive-control
m_1	Insulin transfer rate to the ISF-plasma compartment
m_2	Insulin transfer rate from the ISF-plasma compartment
m_4	Insulin elimination rate from ISF carbon conversion
SC	Subcutaneous compartment
SSE	Sum of squares error

t Time
U Rate of insulin administration from insulin pump
U_{ii} Insulin-independent glucose utilization
U_{id} Insulin-dependent glucose utilization
V_{gp} Plasma glucose
V_{gi} Volume of the ISF glucose system
V_{ip} Plasma insulin
V_{ii}, ISF insulin

Greek Letters

α pancreatic insulin production
β insulin-dependent glucose utilization
τ_1 time delay of hepatic glucose production, min
τ_2 time delay of pancreatic insulin production, min
τ_3 time delay of the subcutaneous delivery from the insulin pump device to plasma, min

References

1 WHO Library Cataloguing-in-Publication Data Global report on diabetes. World Health Organization. ISBN 978 92 4 156525 7 World Health Organization 2016.
2 Bequette, B.W. A critical assessment of algorithm and challenges in the development of a closed-loop artificial pancreas. *Diabetes Technol Ther.* 1998;7:28–47.
3 Hovorka, R. Continuous glucose monitoring and closed-loop systems. *Diabet Med* 2005;23:1–12.
4 Cobelli, C., Renard, E., and Kovatchev, B., Artificial pancreas: past, present, future, *Diabetes*, 2011;60;2672–2682.
5 Magni, L., Raimondo, D.M., Bossi, L., Dalla Man, C., De Nicolao, G., Kovatchev, B. et al. Model predictive control of Type 1 diabetes: An in silico trial. *J. of Diabetes Science and Technology* 2007;1:804–811.
6 Wu, Z., Mathematical models with delays for glucose-insulin regulation and applications in artificial pancreas, 2013. PhD Thesis. Available online at: https://scholarbank.nus.edu.sg/handle/10635/37897 (accessed November 2017).
7 Li, J., Kuang, Y., and Mason C.C., Modeling the glucose-insulin regulatory system and ultradian insulin secretory oscillations with two explicit time delays, Journal of Theoretical Biology, 2006;722–735.

8 Mayne, D.Q., Rawlings, J.B., Rao, C.V., and Scokaert, P.O. Constrained model predictive control: stability and optimality. *Automatica* 2000;36:789–814.

9 Garcia, C.E., Prett, D.M., and Morari, M. Model predictive control theory and practice-a survey. *Automatica* 1989; 25(3):335–48.

10 Piemonte, V., Maurizi, A., Capocelli, M., De Santis, L. and Pozzilli, P. A Novel Three Compartmental Model for Artificial Pancreas: Development and Validation. *Artificial Organs*, 2017;41(12):E326–E336.

11 Teixeira, R.E. and Malin, S. The next generation of artificial pancreas control algorithms. M.S. *Journal of Diabetes Science and Technology* 2008;22: 105–112.

12 Bergman R.N., Ider, Y.Z. Bowden, C.R. and Cobelli, C. Quantitative estimation of insulin sensitivity." *American Journal of Physiology-Endocrinology and Metabolism*, 1979;236(6):E667.

13 Bergman, R. Minimal model: perspective from 2005. *Horm Res.* 2005;64:8–15.

14 Guyton, J.R., Foster, R.O., Soeldner, J.S. et al. A model of glucose-insulin homeostasis in man that incorporates the heterogeneous fast pool theory of pancreatic insulin release. *Diabetes.* 1978;27:1027–1042.

15 Dalla Man, C., Raimondo, D.M., Rizza, R.A., and Cobelli, C. GIM, simulation software of meal glucose–insulin model. *J Diabetes Sci Technol.* 2007;1:323.

16 Cobelli, C., Toolo, G., and Ferrannini. E. A model of glucose kinetics and their control by insulin, compartmental and noncompartmental approaches. *Mathematical Biosciences*, 1984;72(2):291–315.

17 Vicini, P., Caumo, A., and Cobelli, C. The hot IVGTT two-compartment minimal model: indexes of glucose effectiveness and insulin sensitivity. *American Journal of Physiology-Endocrinology and Metabolism*, 1997;273(5):E1024–E1032.

18 Toffolo, G. and Cobelli, C. The hot IVGTT two-compartment minimal model: an improved version," *American Journal of Physiology-Endocrinology and Metabolism*, 2003;284(2):E317–E321.

19 Bennett, D. and Gourley, S. Asymptotic properties of a delay differential equation model for the interaction of glucose with plasma and interstitial insulin, *Appl Math Comput* 2004:151:189–207.

20 Sturis, J. Computer model for mechanisms underlying ultradian oscillations of insulin and glucose, *Am J Physiol* 1991:260:E801–809.

21 Chen, C.-L. and Tsai, H.-W. Modeling the physiological glucose-insulin system on normal and diabetic subjects. *Computer Methods and Programs in Biomedicine*, 2010;97(2):130–140.

22 Chen, C.L., Tsai, H.W., and Wong, S.S., Modeling the physiological glucose-insulin dynamic system on diabetics, Journal of Theoretical Biology 2010:314–322.

23 Hovorka, R., Shojaee-Moradie, F., Carroll, P.V., Chassin, L.J., Gowrie, I.J., Jackson, N.C. et al. Partitioning glucose distribution/transport, disposal, and

endogenous production during IVGTT," *American Journal of Physiology-Endocrinology and Metabolism*, 2002;282:992–1007.

24 Hovorka, R., Canonico, V., Chassin, L.J., Haueter, U., Massi-Benedetti, M., Federici, M.O. et al., Nonlinear model predictive control of glucose concentration in subjects with type 1 diabetes, *Physiological Measurement*, 2004;25(4):905.

25 Li, J., Kuang, Y. and Mason, C.C. Modeling the glucose-insulin regulatory system and ultradian insulin secretory oscillations with two explicit time delays, *Journal of Theoretical Biology*, 2006;242(3):722–735.

26 Wu, Z., Chui, C.-K., Hong, G.-S. and Chang, S. Physiological analysis on oscillatory behavior of glucose-insulin regulation by model with delays, *Journal of Theoretical Biology* 2011;280:1–9.

27 Insel, P.A., Liljenquist, J.E., Tobin, J.D., Sherwin, R.S., Watkins, P. Andres, R., et al. Insulin control of glucose metabolism in man: a new kinetic analysis. *J Clin Invest* 1975:55:1057–1066.

28 Cobelli, C., Toffolo, G., and Ferrannini E. A model of glucose kinetics and their control by insulin, compartmental and noncompartmental approaches. *Math Biosci* 1984:72:291–315.

29 Thabit, H. and Hovorka, R, Coming of age: the artificial pancreas for type 1 diabetes. *Diabetologia* 2016:59:1795–1805.

30 Hussain, M.A. Review of the applications of neural networks in chemical process control. Simulation and on-line implementations, Artif. Intell. *Eng.* 1999;13:55–68.

31 Fernandez de Canete, J., Gonzalez-Perez, S. and Ramos-Diaz, J.C.. Artificial neural networks for closed loop control of in silico and ad hoc type 1 diabetes. *Computer Methods and Programs in Biomedicine* 2012;106:55–66.

32 Zecchin, C., Facchinetti, A., Sparacino, G., De Nicolao, G. and Cobelli, C. Neural network incorporating meal information improves accuracy of short-time prediction of glucose concentration, *IEEE Trans. Biomed. Eng.* 2012;59(6):1550–1560.

33 Capocelli, M., Prisciandaro, M., Lancia, A., and Musmarra, D. Application of ANN to hydrodynamic cavitation: Preliminary results on process efficiency evaluation. *Chemical Engineering Transactions*, 2014;36:199–204.

34 Capocelli, M., Prisciandaro, M., Lancia, A., and Musmarra, D. Factors influencing the ultrasonic degradation of emerging compounds: ANN analysis (2014) Chemical Engineering Transactions, 2014;39 (Special Issue):1777–1782.

35 Bremer, T. and Gough, D.A. Is blood glucose predictable from previous values? *A solicitation for data, Diabetes*, 1999;48:445–451.

36 Sparacino, G., Zanderigo, F., Corazza, S., Maran, A., Facchinetti, A. and Cobelli, C. Glucose concentration can be predicted ahead in time from continuous glucose monitoring sensor time-series. *IEEE Transactions on Biomedical Engineering*, 2007;54(5)May.

37 Pérez-Gandia, C., Facchinetti, A., Sparacino, G., Cobelli, C., Gomez, E., Rigla, M. et al. Artificial neural network algorithm for on-line glucose prediction from continuous glucose monitoring, *Diabetes Technol. Ther.*, 2010;12(1):81–88.

38 Pappada, S., Cameron, B., Rosman, P., Bourey, R., Papadimos, T., Oloruntu, W. et al. Neural network-based real-time prediction of glucose in patients with insulin-dependent diabetes, Diabetes Technol. *Ther.*, 2011;13(2)135–141.

39 Cobelli, C., Dalla Man, C., Sparacino, G., Magni, L, De Nicolao, G., Kovatchev, B., Diabetes: Models, signals, and control. *IEEE Rev Biomed Eng* 2009:2:54–96.

40 Zecchin, F. and Sparacino, C. Jump neural network for online short-time prediction of blood glucose from continuous monitoring sensors and meal information, *Computer Methods and Programs in Biomedicine* 2014,113:144–152.

41 Zhu, K.Y., Liu, W.D. and Xiao Y. Application of fuzzy logic control for regulation of glucose level of diabetic patient. In: Dua, S., Acharya, U., Dua, P. (eds). *Machine Learning in Healthcare Informatics. Intelligent Systems Reference Library*, Vol 56. Springer, Berlin, Heidelberg.

42 Wayne, B. Bequette Algorithms for a closed-loop artificial pancreas: the case for model predictive control. *Journal of Diabetes Science and Technology* 2013;7(6, November).

43 Daskalakia, E., Diemb, P., and Stavroula G. Mougiakakou: An Actor–Critic based controller for glucose regulation in type 1 diabetes. *Computer Methods Programs Biomed.* 2013;109(2):116–125.

6

Multiscale Synthetic Biology: From Molecules to Ecosystems

Luisa Di Paola[1] *and Alessandro Giuliani*[2]

[1] *Unit of Chemical-Physics Fundamentals in Chemical Engineering, Department of Engineering, Università Campus Bio-Medico di Roma, Italy*
[2] *Environment and Health Department, Istituto Superiore di Sanità, Rome, Italy*

6.1 Introduction: An Historical-Epistemological Perspective

The dream of a manmade living creature dates back to the dawn of civilization [1]: the Jewish legend of Golem and the late-romantic Mary Shelley's character of Frankenstein's Monster are just two of the most famous archetypes of synthetic (human) biology. The passage from myth and metaphor to science and technology has been continuous and, since the eighteenth century mechanic automata up until today, the basic view of synthetic biology research has basically been the same [2].

The emphasis shifted from the eighteenth-century "organism as a clockwork" metaphor to the nineteenth-century organism as a "thermal machine," ending up with the organism as a "computer" of the twentieth and twenty-first centuries (with a more marked emphasis on network systems these days with respect to the logical flux of information of the previous century). Notwithstanding that, both the fiction and technological ideas redounded about the idea of the existence of a "basic mechanism of life" that, albeit complex, could, be replicated in a laboratory (at least in principle).

Thus, it is not without interest to have a closer look at the concept of synthesis. Not all the projects of artificial life are properly "synthetic." We do not use the word *synthesis* (characterized by the Greek prefix "syn" pointing to the emergence of new features by the organic fusion of different elements) for cars or computer programs. On the other hand, we currently speak of organic

Biomedical Engineering Challenges: A Chemical Engineering Insight, First Edition.
Edited by Vincenzo Piemonte, Angelo Basile, Taichi Ito, and Luigi Marrelli.

synthesis, referring to the production of new organic molecules not present in nature and "synthesizer" is the name given machines devoted to the fusion of different sounds in electronic musical composition.

The chemistry-music pairing is far from being only contingent: Dimitri Ivanovic Mendeleev based the Periodic Table of Elements on harmonic rules. He was an excellent piano player, his colleague Borodin a great organic chemist and composer, and was a member of the Mighty Handful group who often played in Dimitri Ivanovic's home. The octave and the "simple ratios" were the basis for consonance corresponding to atomic element distribution along the Table and by valence rules to obtain consonant sounds and stable molecules, respectively. The presence of rules does not impose limits on the creation (hundreds of thousands of new organic molecules come out each year obeying carbon valence rules, billions of pieces of music stem from the same harmony rules). Organic synthesis builds upon "natural rules": the chemo-physical features of a brand new organic molecule can be predicted by its structural formula *before* its actual synthesis and the success of the synthesis is strictly dependent upon natural constraints as pH, temperature, catalysis, and so forth.

While an organic chemist is more similar to a chef (sooner or later he/she must put "manipulated ingredients" into an oven and wait for the physics to do its job), an engineer works in a very different way. The approach is no more synthetic: the engineer must only take into consideration global and largely aspecific physical properties (an artificial arm cannot be too heavy) and the environment acts in many cases as an obstacle (e.g., immune reaction to prosthetic elements) more than an "active driver" as in organic synthesis.

The work of a software engineer is even less dependent on natural constraints, while following the same local "action-reaction" if-then rules of mechanics: his/her creations live in the closed work of logic and the only "material" limit is the computing power that, by definition, has nothing to do with the meaning of the task the engineer wants to reach.

In Figure 6.1, we summarize different biology inspired artifacts along two hypothetical axes of synthesis and mechanics content.

In the following, we will mainly concentrate on bottom right part of the graph, the realm of the so-called "nano-machines" and of the (slightly) modified forms of life like OGMs. We will devote some attention to the top-left portion of the graph but focusing on the nature of biomimetic functional principles more than to strictly mechanical issues.

Before entering single applications, it is worth remembering a general principle unifying both the synthetic and the mechanic styles: in 1952, the Dutch electrical engineer Bernard Tellegen [3] developed a theorem whose relevance has largely been underestimated. Tellegen's theorem is a conservation principle (based on Kirchoff's laws of electrical circuits) of both potential and flux across a network. The flux does not need to be an electrical current and the same holds for the potential. Any system can be modeled by a set of nodes linked by

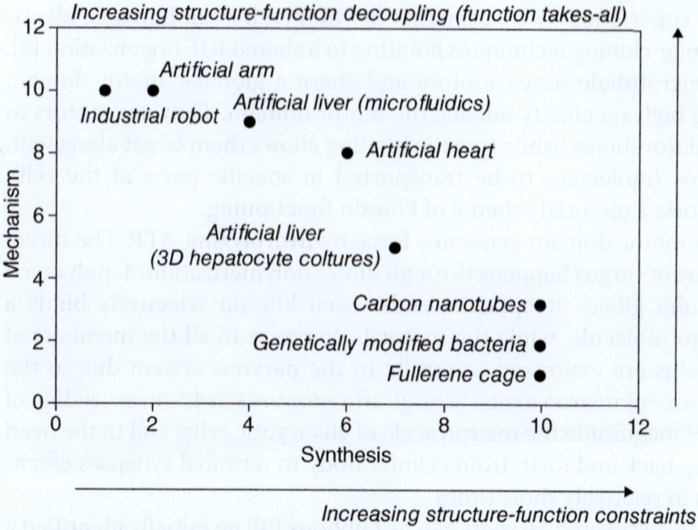

Figure 6.1 A synthetic biology sketch.

edges (i.e., being metabolites linked by chemical reactions transforming one into the other or mutually interacting persons in an office…) that have similar emerging properties independently of the physical nature of nodes and edges. As aptly stressed in [3], the theorem opens the way to "network thermodynamics," whose principles are strictly dependent from wiring architecture while largely independent of the constitutive laws governing the single elements. In the perspective of synthetic biology, the Tellegen theorem implies the presence of general "optimality principles" shared by more or less "artificial" devices located in every part of the Figure 6.1 graph.

6.2 Applications

6.2.1 Protein Synthetic Biology

The very basic elements of life machinery are proteins exerting a large number of functions by means of peculiar traits of their structure. The most sophisticated and efficient nano-machines are protein structures, working as motors or switches, typical engineering terminology for technological tools. These two functional classes of protein nano-machines cover an astounding number of functions in cells, from specialized transport to regulation.

Protein motors determine cell motility through a continuative use of energy from ATP hydrolysis [4]. Protein motors for intracellular transport belong to three super-families [5]: kinesins, dyneins, and myosins.

The kinesin superfamily (KIF) accounts for 45 members in humans, identified via molecular cloning techniques pointing to a shared KIF organization [6]. KIFs act as microtubule-based motors and share a globular motor domain while showing high specificity outside the motor domain. They use motors to move along microtubules, while specific binding allows them to get along with different cargos (molecules to be transported in specific parts of the cell). Figure 6.2 reports a pictorial scheme of kinesin functioning.

The kinesin motor domain generates force by hydrolyzing ATP. The directional transport of cargos happens through direct polymerization/depolymerization of tubulin gibers at opposite ends. Each kinesin selectively binds a preferred cargo molecule, while the motor is common to all the members of the KIF. Kinesins are expressed primarily in the nervous system due to the "out-of-scale" size of neuron axons (e.g., giraffe axons reach 4.5 m, exceeding of some orders of magnitude the micron scale of eukaryotic cells) and to the need of transporting back and forth from cellular body to terminal synapses essential molecules in relatively short times.

Dyneins were discovered 50 years ago by Gibbons [8], he initially identified a new protein as ATPase in *Tetrahymena pyriformis* and named after the dyne, the unit of force. Dyneins too are particularly abundant in the nervous system, where they drive intracellular transport along microtubules, as kinesins do. The similarity in function with kinesins arises from a strong similarity in the motor domain of the structure and in the core mechanism [9]. The dynein motor mechanism is far less well understood than that of myosin and kinesin, mainly due to their large size hindering conventional (crystallographic) structural investigation. Only recently, the motor domain of a dynein has been resolved, whereas its function in the cytoskeleton has been revealed by means of live-cell imaging [10] and electron microscopy [11].

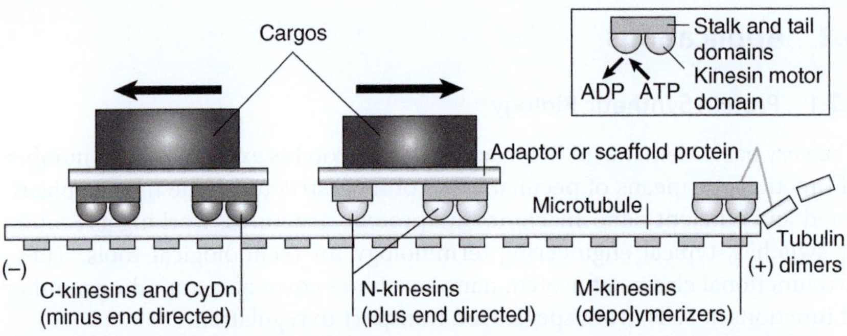

Figure 6.2 A sketch of kinesin functioning (*Source:* [7] N. Hirokawa, Y. Noda, Y. Tanaka, S. Niwa, Kinesin superfamily motor proteins and intracellular transport., *Nat. Rev. Mol. Cell Biol.* 10 (2009) 682–696. doi:10.1038/nrm2774). (*See insert for color representation of the figure.*)

Figure 6.3 shows a sketch of dynein bound to tubulin in microtubules. The structure, yet larger, resembles kinesin, witnessing the strong similarities between shape and function in these two super-families.

Dynein functions go beyond cargo transport along microtubules: they exert tension on cellular structures, helping cell division as well [13]; extracellular dyneins also operate as cargo transporters in cilia [14].

Myosins are motor proteins belonging to the actin family; they represent a wide family whose members are involved in many cellular pathways [15]. They act as mechano-enzymes, comprising actin- and ATP-binding sites. They are mainly involved in muscle contraction, with a direct use of energy through ATP hydrolysis. They are also involved in cytokinesis and cell

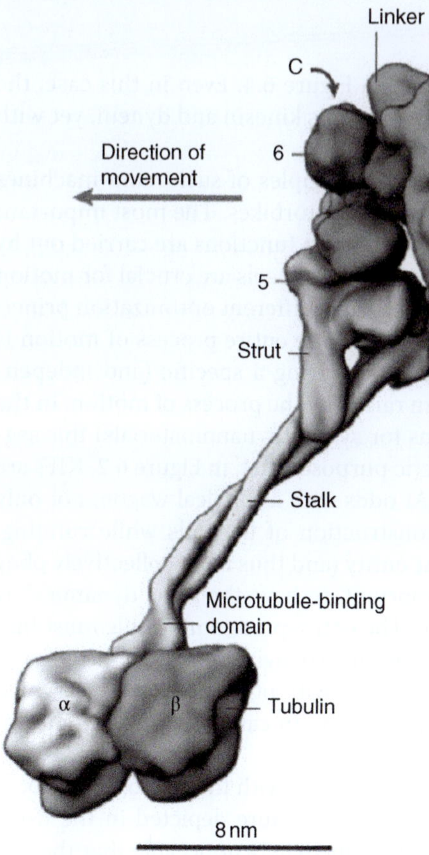

Figure 6.3 A 3D representation of a dynein bound to microtubule (*Source:* [12] A.J. Roberts, T. Kon, P.J. Knight, K. Sutoh, S.A. Burgess, Functions and mechanics of dynein motor proteins., *Nat. Rev. Mol. Cell Biol.* 14 (2013) 713–26. doi:10.1038/nrm3667). (*See insert for color representation of the figure.*)

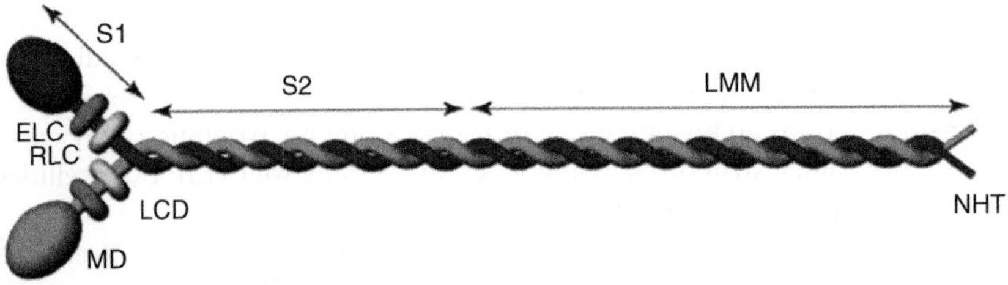

Figure 6.4 Myosin structure: the long tail (S2 + LMM) is a coiled coil, head S1 comprises the motor domain (MD) and the light chain domain (LCD) in turning including the essential light chain (ELC) and the regulatory LC (RLC) (*Source:* [16] M. Vicente-Manzanares, X. Ma, R.S. Adelstein, A.R. Horwitz, Non-muscle myosin II takes centre stage in cell adhesion and migration., *Nat. Rev. Mol. Cell Biol.* 10 (2009) 778–90. doi:10.1038/nrm2786). (*See insert for color representation of the figure.*)

division [16]. Myosin structure is shown in Figure 6.4. Even in this case, the structure resembles that of other motor proteins kinesin and dynein, yet with quite different mechanical functions.

It is worth comparing the "engineering" principles of such nano-machines with those adopted for machines like cars or motorbikes. The most important difference is that, in the case of cars, the different functions are carried out by rigidly separated parts of the artifact. Even if the wheels are crucial for motion and must be connected to the engine, they obey different optimization principles with respect to pistons or fuel-injectors. The entire process of motion is fragmented into dedicated parts each one offering a specific (and independently optimized) solution to a problem raised by the process of motion. In the case of proteins (but the same happens for synthetic nanomaterials) this segmentation is absent, it is only for didactic purposes that, in Figure 6.2, KIFs are depicted as wagons running on rails. At odds with a classical wagon, not only they are in charge of directing the construction of the rails while running, but they are part of the same molecular entity (and thus must collectively obey to the same constraints and, most important, to have the same dynamics) as the other parts of the protein molecule. The entire protein molecule must find its preferred configuration in its particular microenvironment conditions and must find an energetic favorable arrangement with other proteins with which it interacts. Looking at the actual form of a kinesin can help to get an idea of these concepts (Figure 6.5).

Figure 6.5 shows a domain of a kinesin molecule with the approximate location of different functionalities along the 3D structure depicted in the usual ribbon diagram by different shading. It is immediately noticeable that there is no clear border (and no macroscopic differences in structure) between different functionalities. If we compare this scheme with the scheme of a racing motorbike engine (Ducati 250 GT), we get an immediate idea of the drastic difference

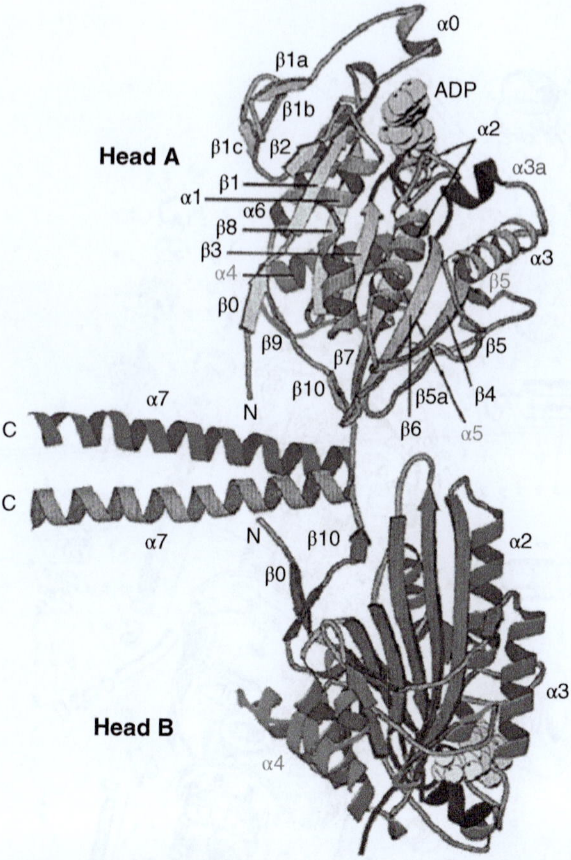

Figure 6.5 Ribbon structure of kinesin (*Source:* [17] F. Kozielski, S. Sack, A. Marx, M. Thormählen, E. Schönbrunn, V. Biou, A. Thompson, E.M. Mandelkow, E. Mandelkow, The crystal structure of dimeric kinesin and implications for microtubule-dependent motility, *Cell.* 91 (1997) 985–994. doi:10.1016/S0092-8674(00)80489-4). (*See insert for color representation of the figure.*)

between synthetic and mechanistic engineering, even in presence of a strong similarity to the desired function (regular and controlled motion) (Figure 6.6).

The elements of the engine interact while at the same time maintaining their independent and unique form (each one has a distinct label in Figure 6.6), there is no ambiguity as to the borders between different parts. The structure-function relation of kinesin is global and does not allow it to be factorized (bottom-right part of Figure 6.1), the global function of the engine dominates the engine construction, but the single parts must only obey to local constraints (top-left part of Figure 6.1).

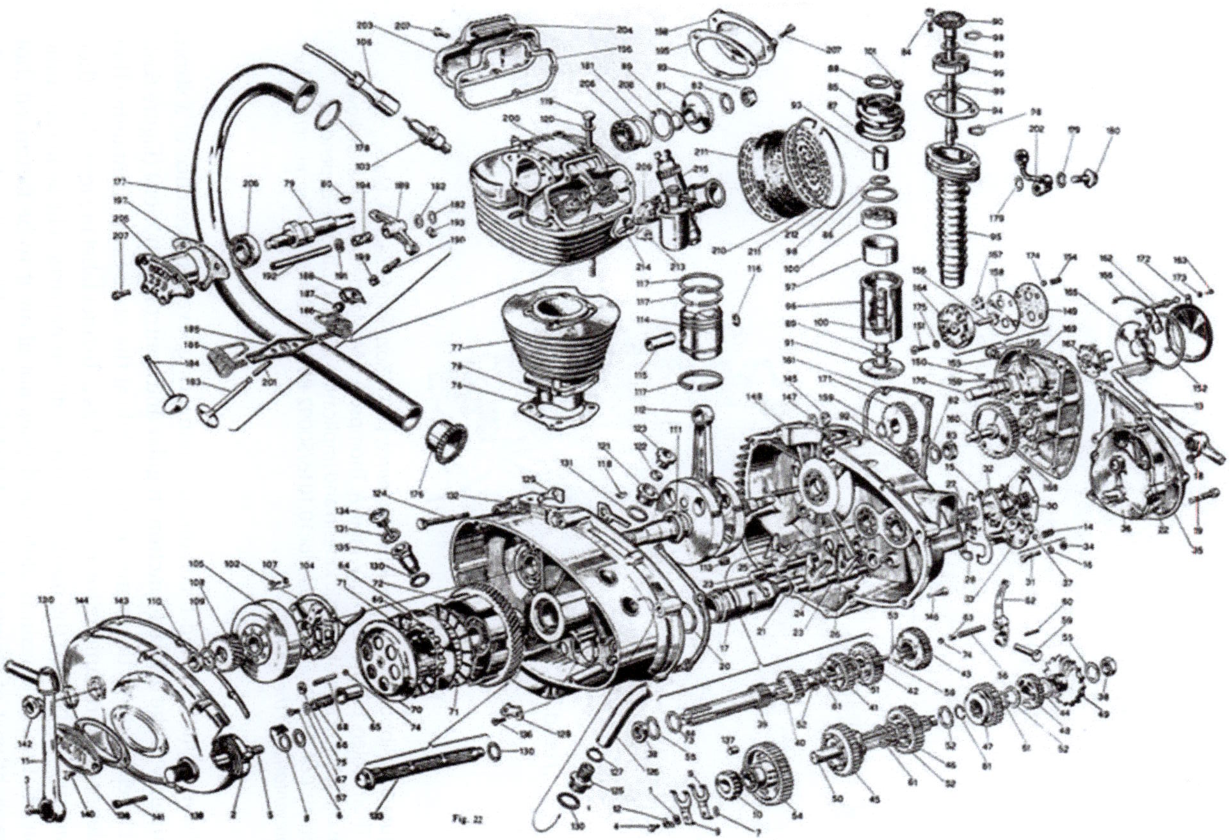

Figure 6.6 A racing motorbike blueprint (*Source:* Ducati 250 GTwww.ducatimeccanica.com/).

These differences force the synthetic engineer to a much softer and more sensible attitude with respect to the motorbike engine designer: if we want to adapt a KIF-type nano-machine to transport a drug across an axon to reach pre-synaptic space he cannot start from scratch. On the contrary, he must insert very small modifications (e.g., point mutations at specific sites of the gene codifying for the carrier protein) to an already existing KIF protein with the aim to modify its binding specificity so as to obtain a selective binding for the drug. In doing so, he only has approximate guidance in choosing the residue to be modified (in this case the formalization of protein structure as a contact network could be of use [18–21]). In any case, we must rely on a largely empirical approach, more similar to that adopted by a medicinal chemist, trying to modify a known drug lead to increase its activity than to a mechanical engineer. This said, the possibilities of making sensible modifications to protein molecules and complexes for pharmacology and/or diagnostic purposes are huge [22–25].

Protein switches clearly show the need for a different approach. Protein switches react to environmental (biochemical) input to process a given output. The switch is often based on a protein conformational change, which modifies protein affinity for a ligand (input). This regulatory activity is at the very basis of protein activity in natural systems and offers outstanding possibilities for engineered protein applications [26]. Regulation of protein binding depends on the activity of a chemical signal (effector) acting in the binding site (orthosteric) or in a different region (allosteric). Protein regulation is another face of protein adaptive character.

The allosterically regulated proteins are common protein switches, adapting their conformation to signals [27]. Conformational switching can be between ordered and disordered conformations or between ordered conformations. This latter case is generally applied when a large conformational changes is required (for instance, when a large fluorescence response is required [28]). On the other hand, transition between ordered conformations provides a very fine regulation of protein activity and represents a very challenging task in synthetic protein design [29]. There are few examples of engineered protein switches, all used as reporters, with a scarce application relevance [30]. In perspective, engineered protein switches may deeply change some fields, such as biotechnology and pharmacology, and for this reason, great research effort is devoted in this direction.

Synthetic biology can learn many lessons from far longer experience in protein engineering. Proteins are the smaller life elements that behave as systems and all life complexity in living systems arises from the adaptive nature of protein structure. The elusive relationship between protein composition, their function, and structure reveals the basic principle of life: the holistic, post-reductionist approach is the only path for life investigation and life lessons to synthetic biology must be interpreted only in this light. Citing Gierasch and Gershenson [31]:

> *This is extremely challenging: the old Mother Goose nursery rhyme about Humpty Dumpty pointed out that after Humpty "had a great fall; all the king's horses, and all the king's men, couldn't put Humpty together again." It is virtually impossible to reassemble an in vivo environment—that is, to put Humpty back together again. Instead, one can reassemble complexes and pathways and simply accept that this is an approximation of the cellular complexity. Optimally, we can take on the challenge and develop more and more powerful approaches to examine biochemical events in situ without disruption of the cellular complexity—that is, we can study Humpty before he falls.*

However, the message has been learned: the transition from modules construction to system is already the new wave of synthetic biology [32], approaching challenges in the complex system theory framework rather than specializing construction rules on single elements.

Protein Contact Networks (PCNs) represent a very talkative example of the systemic approach to protein engineering [33]. Complex networks formalism has been successfully applied to complex biological networks providing insightful perspectives and catching emerging features [34]. The Protein Contact Network paradigm is based on a translation of protein three-dimensional structure into a network, whose nodes are the protein residues while links are non-covalent intramolecular contacts between a pair of residues. It is evident these networks are geometrically constrained and this property influences the general network properties and provides to PCNs a typical fingerprint [20, 35]. Figure 6.7 reports a typical representation of protein structure as the PCN for the structure of the T-state in human hemoglobin.

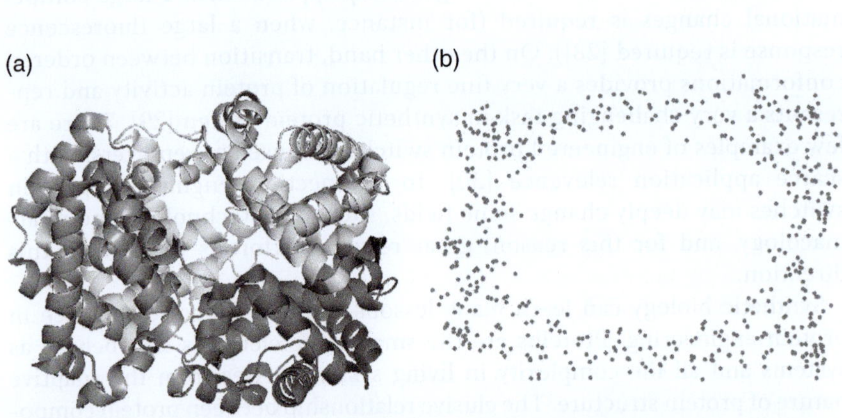

Figure 6.7 Protein contact networks. Panel (a) the ribbon representation of the T-state of human hemoglobin; Panel (b) the corresponding Protein Contact Network. (*See insert for color representation of the figure.*)

It leaps out that the PCN formalism strongly simplifies the whole structural information, yet provides insightful evidence of protein features and functions, from allostery [36] to folding [37]. Network properties describe both local and global features of protein structure and more generally describe the systemic nature of the protein structure. The simplest approach to network analysis is based on the node degree and degree distribution. The node degree in a network simply describes the number of links a node is involved in. In the perspective of PCN, high degree nodes are likely to correspond to residues largely contributing to protein stability. The degree distribution of network assigns the network model and provides a general insight on network general properties and dynamics [38].

The average shortest path, *asp* (or characteristic path length). is another peculiar trait of PCNs: the shortest path corresponds to the minimum number of links connecting two nodes. The average value over all node pairs is the average shortest path [39]. PCNs show small world character [40]: small world networks are those by a logarithmic scale of the average shortest path with the number of nodes and are characterized by complex networks where communication is the main function [41] (social networks, world wide web, biological networks). PCNs show a distinct small-world character, responsible for key processes such as folding [42] and allostery [43]. Based on the concept of the shortest path, the centrality metrics allow to assign a role to nodes in signal transmission. Closeness centrality assesses the property of a node to participate to small shortest paths, a kind of "communication highway" throughout the network [44]. High closeness nodes in technological networks are crucial to signal transmission. Analogously, high closeness nodes (residues) in PCNs participate in key processes, such as binding [45]. Betweenness centrality is based on the shortest path definition as well and computes the number of shortest paths the single node is involved in [46]. High betweenness nodes in PCN correspond to residue central in signal transmission throughout the protein structure, responsible for allosteric response.

Another distinct property of PCNs, shared also by smart technological devices [47], is modularity. The complexity of biological systems, in general, is grounded on modularity, and this feature has inspired the design of complex systems mimicking life [48]. In proteins, modularity is evident in the domain organization of the three-dimensional structure, responsible for primary regulatory and cooperative processes (binding [49] and folding [50]).

Clustering methodologies allow us to find modules (clusters) into networks; based on the clustering partition of a network in clusters (modules), Guimerà and Amaral proposed a method to classify nodes according to their topological role [51]. They defined two descriptors, the participation coefficient P and the intramodule connectivity z-score z, which outline, respectively, the proneness of nodes to establish links with nodes pertaining or not. These descriptors describe well the two roles of nodes (residues) in PCNs and in protein structures: nodes with high P and low z contribute to signal transmission throughout the protein structure rather than to stability; conversely, nodes with high z and low P contribute to stability but do not participate in signal transmission [52]. Eventually, in PCN, the P-z maps have a peculiar "dentist's chair" shape.

6.2.2 Tissue Engineering and Artificial Organs

Form has a crucial role in biology, one of the most striking effects of form is allowing metabolism. We know from physical chemistry that a tri-atomic collision in diffusive regime is almost impossible; notwithstanding that, biochemistry handbooks describe ordered reaction sequences encompassing dozens of ordered collisions in metabolic processes as phospholipid or amino-acid synthesis and energy metabolism. This is made possible by a highly organized mesh (cytoskeleton) shaping the cell and, through the action of cell-cell adhesion, making tissue an integrated whole [53]. In other words, metabolism is possible since chemical reactions do not happen in a diffusion regime but in a condensed state, where enzymes participating in the same biochemical pathway stay in close vicinity by means of the cytoskeleton mesh, so as to form integrated multi-enzymatic metabolic machines. This explains the otherwise unexplainable massive effect on gene expression of microgravity conditions [54] (gravity is an incredibly tiny force with respect to electromagnetic interactions and thus is unable *per se* to have molecular level effect). The same considerations hold true for the effect of physical viscosity and shear stress on development and cancer [55, 56].

Cell and tissue organization are thus not an unescapable consequence of metabolism and signaling but, due to the outlined constraints, the other way around. The tensegrity paradigm gives an elegant explanation to the role of form in biological regulation [57], while the simple consideration that pathologists still have cell form recognition at the microscope as a main diagnostic criterion gives us the sense of the relevance of form in biological regulation.

These considerations allow us to put in context the peculiar hybrid nature of tissue engineering application that can be located in widely different locations in the mechanics synthesis plane depicted in Figure 6.1: it all depends from where the peculiar shape of an engineered tissue comes from. Davies and Cachat [58] give a straightforward and clear situation of the gradient from mechanics to synthesis dominated in tissue engineering:

> *Classical tissue engineering is aimed mainly at producing anatomically and physiologically realistic replacements for normal human tissues. It is done either by encouraging cellular colonization of manufactured matrices or cellular recolonization of de-cellularized natural extracellular matrices from donor organs, or by allowing cells to self-organize into organs as they do during fetal life.*

"Encouraging cellular colonization of manufactured matrices" corresponds to the "mechanics" pole: the scaffold allowing for the generation of a correct tissue geometry is man made in a top-down albeit bio-mimetic manner (current 3D printing allowed huge progress along this research avenue [59]). This process is analogous to the growth of a vine plant driven by a pole, the geometry of tissue is forced by the manufactured matrix. Figure 6.8 reports a scheme of "top-down" design of tissue form by 3D bioprinting.

Going back to Davies and Cachat [58], the opposite (synthetic) pole is marked by the statement "allowing cells to self-organize into organs as they do during fetal life": this is the most ambitious avenue.

The intermediate step "cellular recolonization of de-cellularized natural extracellular matrices from donor organs" has to do with the transplant technology and builds upon the implant of extracellular matrix (that can be considered as a natural analogue of synthetic matrices) from a donor.

The simplest (and in many cases very effective) approach to tissue engineering and artificial organs production is the first one (top-down artificially imposed form) it is the strategy where the "classical engineering" content largely overcomes the "biological contribution." Pushing this approach to its limits we meet time-honored (and allowing for huge clinical advances) inventions like hemodialysis [61] or cardiopulmonary bypass [62], where the biology enters only as function to be simulated by an artificial device. In this approach, the biological structure/function relation is forgotten in the same way as an airplane does not fly like a bird. The limits of this approach stem from the impossibility of replacing many of the body parts and from the difficulty of creating an artificial-natural tissue interface (e.g., adverse immune reactions, bacterial infections).

The intermediate (transplant of extracellular matrix from a donor) approach while enlarging the reach of the previous approach, suffers of a lower reliability and does not eliminate adverse reactions.

The most truly synthetic approach could in principle solve the problems of the other two, guaranteeing the naturalness of the engineered tissues but the complexity of biological patterning mechanisms in both space and time still prevents its application. This is why basic research along this line is of crucial importance.

6.2.3 Biotechnology and Ecology Applications

Life is not only a target of synthetic biology, but also a tool in fields where the complexity and efficiency of living organisms is exploited in industrial productions [32]. Biotechnology is among the early application fields of synthetic biology and has also been a laboratory for testing the effect of genetic engineering on living organisms (e.g., plants). However, recent challenges emerged: the need for efficient and robust organisms in biofuel production and

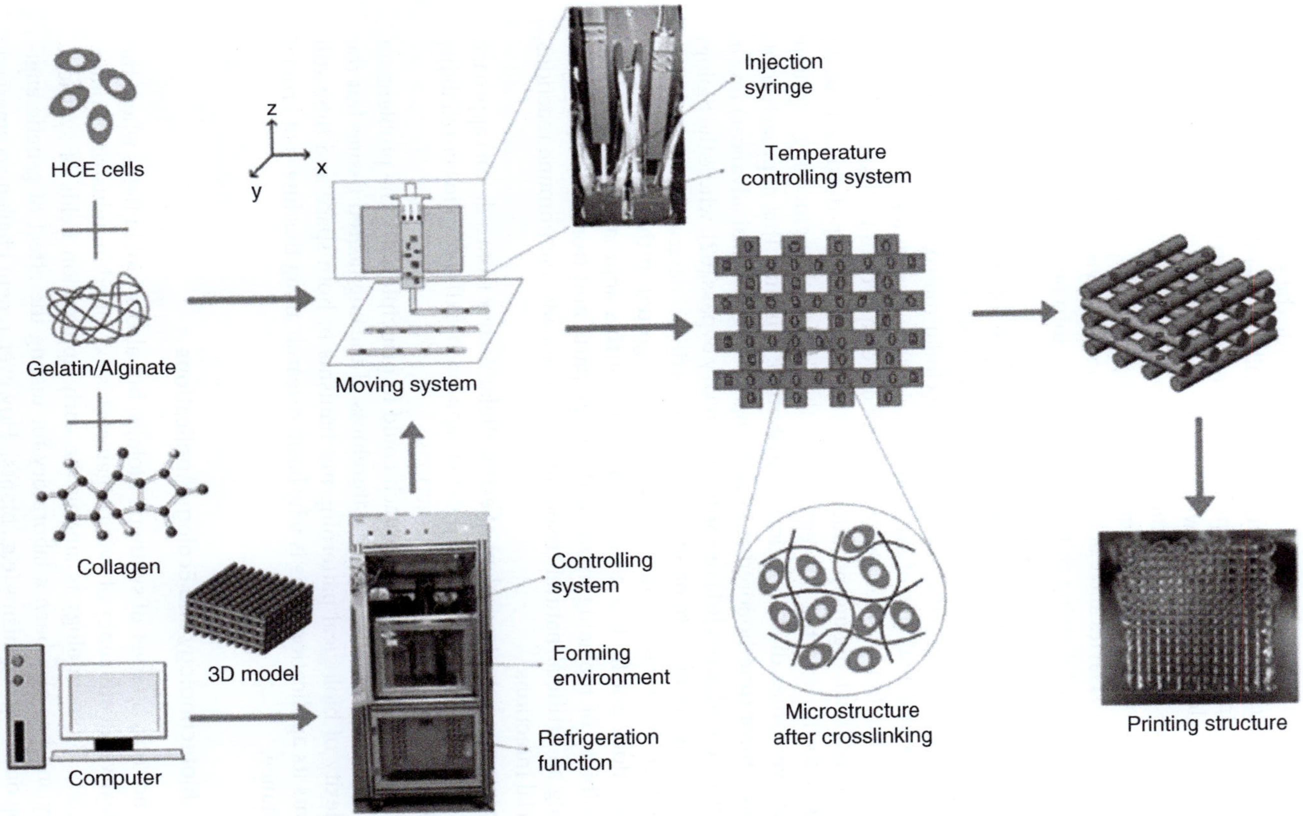

Figure 6.8 Construction of an organized scaffold for tissue engineering (*Source:* [60] Z. Wu, X. Su, Y. Xu, B. Kong, W. Sun, S. Mi, Bioprinting three-dimensional cell-laden tissue constructs with controllable degradation, *Sci. Rep.* 6 (2016) 24474. doi:10.1038/srep24474). (*See insert for color representation of the figure.*)

bioremediation [63–65] strongly pushed research effort toward the creation of purposed organisms. The reproduction of whole new organisms with specialized, optimized metabolic pathways (metabolic engineering) is the main tool in synthetic biology with widespread use in biotechnology [66]: this sort of application is not only useful to optimize bioenergy crops [67], but also allows us to modify microbial populations to convert pollutants in water or soil into valuable products (biofuels) [68, 69].

In recent years, the single strain paradigm has been converted into microbial consortia applications for more versatile biotechnological applications. It is evident the microbial ecosystem engineering poses new challenges, which again require a strong systemic background [70].

6.3 Conclusions

The viewpoint we adopted for this (very rough and partial) review on synthetic biology is the dialectic opposition between a mechanistic and a synthetic pole. This by no means must be considered the sole approach to this theme but, in our opinion, it allows us to put in context the fascinating relation between science and technology. The limits of a pure engineering approach to biological systems do not stem from the lack of "sufficiently tiny fingers" to manipulate the atoms in a molecule.

There is something of more fundamental: we can get a glimpse of this irreducible character of biology by the fascinating combinatorics of cell interactome (the set of all possible protein-protein interactions making metabolism possible) introduced by Tompa and Rose [71]. The authors, assuming the extremely minimalistic case of yeast, hosting only 4500 protein species (in humans this number is estimated at around 100,000), each present in a single copy, give an estimate of the total number of possible distinct patterns of interactions in the order of 10^{7200}. Shifting to the more realistic estimate of 3000 copies per cell for each protein, the number of possible interactomes is in the order of 10 elevated to 7.9. 10^{9}. This transfinite number tells us of an incredibly fine-tuned interactome wiring that does not allow us to generate a living cell from scratch. The selection of the correct interactome cannot happen by chance but following something similar to (still unknown) valence rules.

Organic chemists routinely manipulate matter by following both valence rules and reaction micro-environment constraints and their work follows a different style with respect to mechanical engineering. Notwithstanding that, chemists are able to create artifacts in a peculiar gray zone between the natural and the artificial: they cannot go against natural constraints and the molecules they synthetize are in some sense already taken into consideration as possible by nature. This is why synthetic chemicals can be recognized as proper binding partners by natural receptors. In their visionary paper [72], Hans Frauenfelder and Peter Wolynes located at the level of proteins the border between simple (where laws predominate) and complex (where boundary conditions rule) physical systems.

This is why we focused on protein nano-machines to give a general perspective on synthetic biology: the quest for a Periodic Table in biochemistry is in our opinion the most important scientific goal (dream?) to boost a sensible, effective (and potentially not harmful) synthetic biology.

Live evidence is the application of network theory to protein structure analysis, which not only drove many insightful findings about the structure-function relationship in proteins, but also inspired the design of technological networks. In the field of bioinspired design, the lesson comes back from nature to technology and nature instructs technology on the ground of efficiency and robustness.

List of Symbols

ATP	adenosine triphosphate
KIF	kinesin superfamily
PCN	protein contact networks

References

1 D. de S. Price, Automata in history: Automata and the origins of mechanism and mechanistic philosophy, *Technol. Cult.* 5 (1964) 9–23. www.jstor.org/stable/3101119.

2 J. Riskin, The Defecating Duck, or, the ambiguous origins of artificial life, *Crit. Inq.* 29 (2003) 599–633. doi:10.1086/377722.

3 D.C. Mikulecky, Network thermodynamics and complexity: A transition to relational systems theory, *Comput. Chem.* 25 (2001) 369–391. doi:10.1016/S0097-8485(01)00072-9.

4 R.D. Vale, R.A. Milligan, The way things move: Looking under the hood of molecular motor proteins, *Science* (80). 288 (2000) 88–95.

5 M. Schliwa, G. Woehlke, Molecular motors, *Nature.* 422 (2003) 759–65. doi:10.1038/nature01601.

6 N. Hirokawa, R. Takemura, Molecular motors and mechanisms of directional transport in neurons, *Nat. Rev. Neurosci.* 6 (2005) 201–214. doi:10.1038/nrn1624.

7 N. Hirokawa, Y. Noda, Y. Tanaka, S. Niwa, Kinesin superfamily motor proteins and intracellular transport., *Nat. Rev. Mol. Cell Biol.* 10 (2009) 682–96. doi:10.1038/nrm2774.

8 I.R. Gibbons, Studies on the protein components of cilia from tetrahymena pyriformis, *Proc. Natl. Acad. Sci. U. S. A.* 50 (1963) 1002–1010.

9 R.D. Vale, R.A. Milligan, *The way things move: looking under the hood of molecular motor proteins*, 288 (2000).
10 M.W. Ross, D.J. Mitchell, J.C. Cain, K.R. Blasier, K.K. Pfister, Live cell imaging of cytoplasmic dynein movement in transfected embryonic rat neurons, in: *Methods Cell Biol.*, 2016: pp. 253–267. doi:10.1016/bs.mcb.2015.06.020.
11 S.A. Burgess, M.L. Walker, H. Sakakibara, K. Oiwa, P.J. Knight, The structure of dynein-c by negative stain electron microscopy, in: *J. Struct. Biol.*, 2004: pp. 205–216. doi:10.1016/j.jsb.2003.10.005.
12 A.J. Roberts, T. Kon, P.J. Knight, K. Sutoh, S.A. Burgess, Functions and mechanics of dynein motor proteins, *Nat. Rev. Mol. Cell Biol.* 14 (2013) 713–26. doi:10.1038/nrm3667.
13 F.J. McNally, Mechanisms of spindle positioning, *J. Cell Biol.* 200 (2013) 131–140. doi:10.1083/jcb.201210007.
14 K.F. Lechtreck, IFT-cargo interactions and protein transport in cilia, *Trends Biochem. Sci.* 40 (2015) 765–778. doi:10.1016/j.tibs.2015.09.003.
15 M.A. Hartman, J.A. Spudich, The myosin superfamily at a glance, *J. Cell Sci.* 125 (2012) 1627–1632. doi:10.1242/jcs.094300.
16 M. Vicente-Manzanares, X. Ma, R.S. Adelstein, A.R. Horwitz, Non-muscle myosin II takes centre stage in cell adhesion and migration, *Nat. Rev. Mol. Cell Biol.* 10 (2009) 778–790. doi:10.1038/nrm2786.
17 F. Kozielski, S. Sack, A. Marx, M. Thormählen, E. Schönbrunn, V. Biou, et al., The crystal structure of dimeric kinesin and implications for microtubule-dependent motility, *Cell.* 91 (1997) 985–994. doi:10.1016/S0092–8674(00)80489–4.
18 A. Giuliani, L. Di Paola, Protein as networks: will contact maps hold the promise to represent the "structural-formula" of protein molecules?, *Curr. Protein Pept. Sci.* 17 (2016) 3.
19 L. Di Paola, M. De Ruvo, P. Paci, D. Santoni, A. Giuliani, Protein contact networks: an emerging paradigm in chemistry, *Chem. Rev.* 113 (2013) 1598–613. doi:10.1021/cr3002356.
20 D. Santoni, P. Paci, L. Di Paola, A. Giuliani, Are proteins just coiled cords? local and global analysis of contact maps reveals the backbone-dependent nature of proteins, *Curr. Protein Pept. Sci.* 17 (2016) 26–29. www.ncbi.nlm.nih.gov/pubmed/26412790.
21 L. Di Paola, P. Paci, D. Santoni, M. De Ruvo, A. Giuliani, Proteins as sponges: A statistical journey along protein structure organization principles, in: *J. Chem. Inf. Model.*, 2012: pp. 474–482. doi:10.1021/ci2005127.
22 D.E. Scott, A.R. Bayly, C. Abell, J. Skidmore, Small molecules, big targets: drug discovery faces the protein-protein interaction challenge, *Nat. Rev. Drug Discov.* 15 (2016) 533–550. doi:10.1038/nrd.2016.29.
23 B. Ma, R. Nussinov, Druggable orthosteric and allosteric hot spots to target protein-protein interactions, *Curr. Pharm. Des.* 20 (2014) 1293–1301. doi: 10.2174/13816128113199990073.

24 P. Csermely, T. Korcsmáros, H.J.M. Kiss, G. London, R. Nussinov, Structure and dynamics of molecular networks: A novel paradigm of drug discovery: A comprehensive review, *Pharmacol. Ther.* 138 (2013) 333–408. doi:10.1016/j.pharmthera.2013.01.016.

25 J.R. Wagner, C.T. Lee, J.D. Durrant, R.D. Malmstrom, V.A. Feher, R.E. Amaro, Emerging computational methods for the rational discovery of allosteric drugs, Chem. Rev. (2016) acs.chemrev.5b00631. doi:10.1021/acs.chemrev.5b00631.

26 M. Gerstein, N. Echols, Exploring the range of protein flexibility, from a structural proteomics perspective, *Curr. Opin. Chem. Biol.* 8 (2004) 14–19. doi:10.1016/j.cbpa.2003.12.006.

27 X.I. Ambroggio, B. Kuhlman, Design of protein conformational switches, *Curr. Opin. Struct. Biol.* 16 (2006) 525–530. doi:10.1016/j.sbi.2006.05.014.

28 C.I. Stains, J.R. Porter, A.T. Ooi, D.J. Segal, I. Ghosh, DNA sequence-enabled reassembly of the green fluorescent protein, *J. Am. Chem. Soc.* 127 (2005) 10782–10783. doi:10.1021/ja051969w.

29 M.J. Pandya, E. Cerasoli, A. Joseph, R.G. Stoneman, E. Waite, D.N. Woolfson, Sequence and structural duality: Designing peptides to adopt two stable conformations, *J. Am. Chem. Soc.* 126 (2004) 17016–17024. doi:10.1021/ja045568c.

30 V. Stein, K. Alexandrov, Synthetic protein switches: Design principles and applications, *Trends Biotechnol.* 33 (2015) 101–110. doi:10.1016/j.tibtech.2014.11.010.

31 L.M. Gierasch, A. Gershenson, Post-reductionist protein science, or putting Humpty Dumpty back together again, *Nat. Chem. Biol.* 5 (2009) 774–777. doi:10.1038/nchembio.241.

32 P.E.M. Purnick, R. Weiss, The second wave of synthetic biology: from modules to systems, *Nat. Rev. Mol. Cell Biol.* 10 (2009) 410–22. doi:10.1038/nrm2698.

33 L. Di Paola, M. De Ruvo, P. Paci, D. Santoni, A. Giuliani, Protein contact networks: an emerging paradigm in chemistry, *Chem. Rev.* 113 (2012) 1598–1613.

34 A.-L. Barabasi, Z.N.Z.N. Oltvai, A.-L. Barabási, Network biology: understanding the cell's functional organization, *Nat. Rev. Genet.* 5 (2004) 101–113. doi:10.1038/nrg1272.

35 L. Bartoli, P. Fariselli, R. Casadio, The effect of backbone on the small-world properties of protein contact maps, *Phys Biol.* 4 (2008) 1.

36 L. Di Paola, A. Giuliani, Protein contact network topology: A natural language for allostery, *Curr. Opin. Struct. Biol.* 31 (2015) 43–48. doi:10.1016/j.sbi.2015.03.001.

37 G. Bagler, S. Sinha, Assortative mixing in Protein Contact Networks and protein folding kinetics, *Bioinformatics.* 23 (2007) 1760.

38 J. Park, A.L. Barabàsi, Distribution of node characteristics in complex networks, *PNAS.* 104 (2007) 17916–17920.

39 R. Albert, A.L. Barabàsi, Statistical mechanics of complex networks, *Rev. Mod. Phys.* 74 (2002) 47.

40 A.R. Atilgan, P. Akan, C. Baysal, Small-world communication of residues and significance for protein dynamics, *Biophys J.* 86 (2004) 85.

41 D.J. Watts, S.H. Strogatz, Collective dynamics of "small-world" networks, *Nature.* 393 (1998) 440–442.

42 M. Vendruscolo, N. V Dokholyan, E. Paci, M. Karplus, Small-world view of the amino acids that play a key role in protein folding, *Phys. Rev. E.* 65 (2002) 61910.

43 N.R. Taylor, Small world network strategies for studying protein structures and binding, *Comput. Struct. Biotechnol. J.* 5 (2013) e201302006. doi:10.5936/csbj.201302006.

44 G. Sabidussi, The centrality index of a graph, *Psychometrika.* 31 (1966) 581–603. doi:10.1007/BF02289527.

45 G. Amitai, A. Shemesh, E. Sitbon, M. Shklar, D. Netanely, I. Venger, S. Pietrokovski, Network analysis of protein structures identifies functional residues, *J. Mol. Biol.* 344 (2004) 1135–1146. doi:10.1016/j.jmb.2004.10.055.

46 L.C. Freeman, A set of measures of centrality based on betweenness, *Sociometry.* 40 (1977) 35–41. doi:10.2307/3033543.

47 A. Kusiak, Integrated product and process design: A modularity perspective, *J. Eng. Des.* 13 (2002) 223–231. doi:10.1080/09544820110108926.

48 J. Pandremenos, E. Vasiliadis, G. Chryssolouris, Design architectures in biology, in: *Procedia CIRP*, 2012: pp. 448–452. doi:10.1016/j.procir.2012.07.077.

49 A. del Sol, M.J. Araúzo-Bravo, D. Amoros, R. Nussinov, Modular architecture of protein structures and allosteric communications: potential implications for signaling proteins and regulatory linkages, *Genome Biol.* 8 (2007) R92.

50 B. Vallat, C. Madrid-Aliste, A. Fiser, Modularity of protein folds as a tool for template-free modeling of structures, *PLoS Comput. Biol.* 11 (2015) 1–16. doi:10.1371/journal.pcbi.1004419.

51 R. Guimerà, M. Sales-Pardo, L.A.N. Amaral, Classes of complex networks defined by role-to-role connectivity profiles, *Nat Phys.* 3 (2006) 63–69.

52 A. Krishnan, J.P. Zbilut, M. Tomita, A. Giuliani, Proteins as networks: usefulness of graph theory in protein science, *Curr Protein Pept Sc.* 9 (2008) 28–38.

53 D.E. Conway, M.T. Breckenridge, E. Hinde, E. Gratton, C.S. Chen, M.A. Schwartz, Fluid shear stress on endothelial cells modulates mechanical tension across VE-cadherin and PECAM-1, *Curr. Biol.* 23 (2013) 1024–1030. doi:10.1016/j.cub.2013.04.049.

54 T.G. Hammond, E. Benes, K.C. O'Reilly, D.A. Wolf, R.M. Linnehan, A. Taher, et al., Mechanical culture conditions effect gene expression: gravity-induced changes on the space shuttle, *Physiol Genomics.* 3 (2000) 163–173. www.ncbi.nlm.nih.gov/pubmed/11015612.

55 S. Huang, D.E. Ingber, The structural and mechanical complexity of cell-growth control, *Nat. Cell Biol.* 1 (1999) E131–8. doi:10.1038/13043.

56 D.T. Butcher, T. Alliston, V.M. Weaver, A tense situation: forcing tumour progression, *Nat. Rev. Cancer.* 9 (2009) 108–22. doi:10.1038/nrc2544.

57 D.E. Ingber, Tensegrity: the architectural basis of cellular mechanotransduction, *Annu. Rev. Physiol.* 59 (1997) 575–99. doi:10.1146/annurev.physiol.59.1.575.

58 J.A. Davies, E. Cachat, Synthetic biology meets tissue engineering, *Biochem. Soc. Trans.* 44 (2016) 696–701. doi:10.1042/BST20150289.

59 S. V Murphy, A. Atala, 3D bioprinting of tissues and organs, *Nat. Biotechnol.* 32 (2014) 773–785. doi:10.1038/nbt.2958.

60 Z. Wu, X. Su, Y. Xu, B. Kong, W. Sun, S. Mi, Bioprinting three-dimensional cell-laden tissue constructs with controllable degradation, *Sci. Rep.* 6 (2016) 24474. doi:10.1038/srep24474.

61 A.R. Nissenson, R.N. Fine, *Handbook of Dialysis Therapy* (2008). Elsevier. Philadelphia, PA.

62 L.H. Cohn, Fifty years of open-heart surgery, *Circulation.* 107 (2003) 2168–2170. doi:10.1161/01.CIR.0000071746.50876.E2.

63 H.H. Huang, D. Camsund, P. Lindblad, T. Heidorn, Design and characterization of molecular tools for a Synthetic Biology approach towards developing cyanobacterial biotechnology, *Nucleic Acids Res.* 38 (2010) 2577–2593. doi:10.1093/nar/gkq164.

64 S.W. Seo, J. Yang, B.E. Min, S. Jang, J.H. Lim, H.G. Lim, et al., Synthetic biology: Tools to design microbes for the production of chemicals and fuels, *Biotechnol. Adv.* 31 (2013) 811–817. doi:10.1016/j.biotechadv.2013.03.012.

65 J.S. Singh, P.C. Abhilash, H.B. Singh, R.P. Singh, D.P. Singh, Genetically engineered bacteria: An emerging tool for environmental remediation and future research perspectives, *Gene.* 480 (2011) 1–9. doi:10.1016/j.gene.2011.03.001.

66 S.A. Lynch, R.T. Gill, Synthetic biology: New strategies for directing design, *Metab. Eng.* 14 (2012) 205–211. doi:10.1016/j.ymben.2011.12.007.

67 P.M. Shih, Y. Liang, D. Loqué, Biotechnology and synthetic biology approaches for metabolic engineering of bioenergy crops, Plant J. (2016). doi:10.1111/tpj.13176.

68 M. Urgun-Demirtas, B. Stark, K. Pagilla, Use of genetically engineered microorganisms (GEMs) for the bioremediation of contaminants, *Crit. Rev. Biotechnol.* 26 (2006) 145–164. doi:10.1080/07388550600842794.

69 A.A. Raghoebarsing, A. Pol, K.T. van de Pas-Schoonen, A.J.P. Smolders, K.F. Ettwig, W.I.C. Rijpstra, et al., A microbial consortium couples anaerobic methane oxidation to denitrification, *Nature.* 440 (2006) 918–21. doi:10.1038/nature04617.

70 A.E. Escalante, M. Rebolleda-Gómez, M. Benítez, M. Travisano, Ecological perspectives on synthetic biology: Insights from microbial population biology, *Front. Microbiol.* 6 (2015) 1–10. doi:10.3389/fmicb.2015.00143.

71 P. Tompa, G.D. Rose, The Levinthal paradox of the interactome, *Protein Sci.* 20 (2011) 2074–2079. doi:10.1002/pro.747.

72 P.G. Wolynes, Biomolecules: Where the physics of complexity and simplicity meet, *Phys. Today.* 47 (1994) 58–64.

7

Chemical Reaction Engineering Methodologies for Biomedical Imaging Analysis

Masahiro Kawahara

Department of Chemistry and Biotechnology, Graduate School of Engineering, The University of Tokyo, Japan

7.1 Introduction

In vivo imaging has become a powerful and indispensable technology that aids modern medicine. Ultrasonography, X-ray computed tomography, and magnetic resonance imaging (MRI) are three major imaging modalities that are routinely used in clinics. These modalities can identify lesions in organs based on the differences of physical properties. More recently, molecular imaging techniques have been developed to detect lesions based on biochemical properties. These techniques are based on the modalities such as MRI, nuclear modalities (positron emission tomography (PET) and single photon emission computed tomography (SPECT)), and fluorescence imaging. In such techniques, specific biochemical probes are designed to report biochemical states such as oxygen concentration, accumulation of specific metabolites, and expression of specific disease-marker proteins. Such functional identification provides specific information to aid therapeutic directions.

This chapter focuses on the aspects of chemical reaction engineering in the molecular imaging techniques (MRI, PET/SPECT, and fluorescence imaging) and summarizes their advances, capability, and significance in the biomedical field.

7.2 Magnetic Resonance Imaging (MRI)

MRI enables tomographic imaging of the whole body with the use of magnetism and frequency modulation (FM)-band electromagnetic waves. Using MRI, soft tissues can be three-dimensionally visualized in high resolution even if they are

Biomedical Engineering Challenges: A Chemical Engineering Insight, First Edition.
Edited by Vincenzo Piemonte, Angelo Basile, Taichi Ito, and Luigi Marrelli.

located deep in the body. MRI is especially useful for visualizing tissues surrounded by bones, which cannot be attained by X-ray computed tomography that clearly visualizes bones (Logothetis, 2008; Poldrack and Farah, 2015). Because MRI does not use ionizing radiation, MRI is completely non-invasive, thereby allowing repetitive imaging of each patient. Moreover, MRI enables functional imaging such as blood flow and metabolic information.

7.2.1 ^{1}H-MRI

The principle of MRI is based on nuclear magnetic resonance (NMR), in which ^{1}H of water located in magnetic fields is excited by absorbing the energy of electromagnetic waves with specific frequencies. The use of gradient magnetic fields endows resonance frequencies with positional information, which enables non-invasive three-dimensional imaging (Lauterbur, 1973). The longitudinal (T1) and transverse (T2) relaxation times of ^{1}H are often used for MRI analyses in clinics. T1-weighted images exhibit higher signals in tissues with shorter T1 times and provide morphological information. On the contrary, T2-weighted images exhibit higher signals in tissues with longer T2 times and indicate areas with higher water content. The lesions of tissues can be identified by the fact that lesions are of low T1 and high T2 signals in most cases.

To achieve more functional imaging, many types of contrast agents have been developed (Boros et al., 2015). One of the principles that are clinically used already is based on the fact that magnetic materials promote relaxation of their surrounding ^{1}H. The substances that prominently shorten T1 times can be detected as higher signals in T1-weighted images. Such substances are called positive contrast agents and include paramagnetic metal complexes with gadolinium (Gd) or manganese (Mn) ions. Gd-chelate complexes such as Gd-1,4,7,10-tetraazacyclododecane-1,4,7,10-tetraacetic acid (DOTA) and Gd-diethylenetriaminepentaacetic acid (DTPA) are used in clinics because Gd ions compete with Ca ions *in vivo*, which exhibit high toxicity. Gd-DOTA and Gd-DTPA are also chemically modified to attain higher sensitivity (Datta and Raymond, 2009; Hermann et al., 2008). Generally, linear polymers are chemically conjugated with the Gd-chelate complexes (Villaraza et al., 2010). Polyethylene glycol conjugates have lower toxicity and higher water solubility with better elimination from the body (Desser et al., 1994). Poly-L-lysine conjugates have lower toxicity and longer blood retention time due to strong interaction with serum albumin (Sieving et al., 1990). Diphenylcyclohexyl conjugates reversibly interact with serum albumin and lead to blood vessel visualization (Goyen et al., 2005; Lauffer et al., 1996). Several types of nanocarriers including dendrimer conjugates (Kojima et al., 2011; Lim et al., 2012), liposomes (Bulte et al., 1999; Kono et al., 2011), and block copolymers (Kaida et al., 2010; Mi et al., 2015) are designed to attain higher sensitivity and longer blood retention time, as well as functionalities including enhanced permeability and retention in solid tumors, active targeting through appropriate modifications, and therapeutic drug delivery.

On the other hand, superparamagnetic iron oxide (SPIO) prominently shortens T2 times and thereby lowers signals in T2-weighted images. This negative contrast agent is a nanoparticle and distorts the surrounding magnetic field area whose volume is 10–100 times as much as the actual volume, which leads to high sensitivity. Among SPIOs with various particle sizes ranging from 10 nm to 1 μm, small-size SPIOs are resistant to entrapment by reticuloendothelial systems (Deserno et al., 2004). SPIO is chemically modified with polyethylene glycol or used as a core of block copolymer-based vesicles to attain longer blood retention time and enhanced tumor accumulation (Kokuryo et al., 2013; Kumagai et al., 2010). Another interesting application is to label specific cells with SPIO by introducing a gene encoding a SPIO-forming protein. In fact, ferritin- or MagA-introduced cells can be detected in T2-weighted images even after transplantation *in vivo* (Aung et al., 2009; Cohen et al., 2007; Goldhawk et al., 2009; Naumova et al., 2010; Rohani et al., 2014).

7.2.2 ^{19}F-MRI

The previously mentioned MRI observes water protons and therefore cannot detect specific molecules of interest. One solution to this limitation is to use nuclei other than ^{1}H. Recently, ^{19}F-MRI contrast agents have been developed. ^{19}F-NMR has the highest detection sensitivity (~80% of ^{1}H NMR) in multi-nuclear NMR. Since cellular endogenous ^{19}F concentration is low, ^{19}F-MRI allows quantitative detection of ^{19}F-MRI contrast agents in high signal-to-noise ratios (Chen et al., 2010). Therefore, the development of stimuli-responsive ^{19}F-MRI contrast agents allows detailed analysis of biochemical reactions *in vivo*.

One example is an imaging technique for caspase activities (Mizukami et al., 2008). An oligopeptide substrate for caspase is flanked by a Gd-chelate complex and a ^{19}F substituent, which shortens the ^{19}F relaxation time due to paramagnetic relaxation enhancement effects from the Gd-chelate complex. Cleavage of the agent with caspase releases ^{19}F from the effects by the Gd-chelate complex, which elongates the ^{19}F relaxation time and increases the ^{19}F-MRI signals. This principle is also applied for gene expression imaging using a hairpin DNA linker that dissociates in the presence of its target nucleic acid sequence (Sakamoto et al., 2011).

Another example is based on the phenomenon that the transverse relaxation time (T2) in NMR is shortened in solid states compared to liquid states. A self-assembled chemical probe in which a ligand for an enzyme is connected to a ^{19}F-containing group is designed to mimic solid states (Takaoka et al., 2009). While this probe is NMR-silent, it recovers the ^{19}F-NMR signals in the presence of the enzyme because ligand binding leads to disassembly of the probe, which mimics liquid states. Similarly, enzyme activity is probed by a silica nanoparticle and a ^{19}F-containing group that are connected by a substrate peptide of the enzyme (Tanaka et al., 2012). The ^{19}F-NMR signals are recovered upon cleavage of the probe at the substrate peptide. This principle can also be applied to a pH-responsive contractile nanogel with a ^{19}F-containing group (Oishi et al., 2007). Swelling of the nanogel recovers the ^{19}F-NMR signals.

7.2.3 MRI using Magnetization Transfer

Recently, the magnetization transfer technique was eagerly applied to biomedical imaging. Multiple resonance NMR is a method of shifting magnetic coherence between nuclei with different Larmor frequencies, which can selectively detect a chemical species of interest. A method to investigate a metabolic process has been reported using ^{13}C- and ^{2}D-labeled glucose, leading to detection of a specific proton on the glucose (Doura et al., 2012).

Chemical exchange saturation transfer (CEST) enables indirect detection of endogenous molecules with amide or hydroxyl groups that have water-exchanging protons (Sherry and Woods, 2008). In CEST, radio-wave pulses that specifically inhibit NMR signals of such protons are irradiated, and the resultant protons exchange with bulk water protons, which leads to negative contrasting effects. CEST has been applied to amide proton transfer imaging to detect distribution of peptides/proteins (Bai et al., 2016), myo-inositol imaging (Haris et al., 2011), glycogen imaging (glycoCEST) (van Zijl et al., 2007), and glycosaminoglycan imaging (gagCEST) (Ling et al., 2008) to detect various tissue defects. The use of paramagnetic metal chelates allows frequency shifts of CEST signals, which facilitates the sensitivity (PARACEST) (Hancu et al., 2010; Zhang et al., 2003).

In hyperpolarization, the distribution of energy levels of nuclear spins is shifted from the thermal equilibrium state, which dramatically enhances sensitivity (Viale and Aime, 2010). One of the methods to attain hyperpolarization is dynamic nuclear polarization (Comment, 2016). The mixture composed of a target compound and an unpaired electrons-containing compound is irradiated by a microwave corresponding to the electron spin resonance in a strong magnetic field at extremely low temperatures, which leads to hyperpolarization of nuclear spins with the sensitivity magnified more than 10^5-fold.^{13}C-labeled pyruvate, which has a relatively long relaxation time, can be applied to metabolic analysis and tumor imaging, in which the conversion process from pyruvate to lactate is analyzed based on the difference of their ^{13}C chemical shifts (Golman et al., 2006). An enzyme inhibitor is conjugated with a cage-like compound incorporating ^{129}Xe, and the resultant probe detects interaction between an enzyme and its inhibitor based on the difference of the ^{129}Xe chemical shifts (Chambers et al., 2009). Since hyperpolarized states are relaxed rapidly at room temperature, it is still challenging to design hyperpolarized probes with sufficiently longer relaxation times for sensitive detection. To circumvent this issue, a recent trend is to firstly find out platform structures to secure long relaxation times, and then add a sensing moiety on the structures to probe a specific biological response (Nishihara et al., 2016; Nonaka et al., 2013, 2017).

7.3 Positron Emission Tomography (PET) and Single Photon Emission Computed Tomography (SPECT)

PET and SPECT allow quantitative imaging of γ-rays emitted from radioisotope-labeled chemical agents that are administered in the body (Mariani et al., 2010; Vaquero and Kinahan, 2015). The PET radioisotopes include ^{11}C, ^{13}N, ^{15}O, and ^{18}F, which emit two γ-rays with 511 keV when positrons produced by β^+ decay collide with electrons. Meanwhile, SPECT radioisotopes include ^{67}Ga, ^{99m}Tc, ^{111}In, and ^{123}I, which emit single γ-rays by electron capture and isomeric transition. The half-lives of the PET radioisotopes range from 2 to 110 min, whereas those of the SPECT radioisotopes range from 6 to 78 h. PET imaging detects two γ-rays with a pair of detectors by coincidence counting, which allows high resolution and quantativity. The PET radioisotopes are produced by cyclotrons that shoot accelerated protons into nuclei. Since the half-lives of the PET radioisotopes are short, it is challenging to rapidly synthesize final sterile compounds for immediate medical use. To achieve this, hospitals should be equipped with the following three items; cyclotrons, hot cells with sterility and sufficient radiation-shielding capability, and automated chemical synthesizers. In contrast, SPECT radioisotopes are commercially available owing to their relatively long half-lives, which is advantageous compared to PET radioisotopes in terms of versatility.

7.3.1 PET

Because ^{11}C, ^{13}N, and ^{15}O are the radioisotopes for the elements composed of organic compounds, many metabolites *per se* can be candidates for radio-labeling. Furthermore, the radio-labeled compounds retain native chemical characteristics and facilitate accurate biochemical analyses *in vivo*. However, the half-lives of these radioisotopes are very short (<20 min), which limits their applications. ^{18}F has a similar atomic size compared to ^{1}H and therefore can replace ^{1}H of the probes of interest. Since the half-life of ^{18}F is relatively long (~110 min), radio-labeled compounds can be distributed from cyclotron-equipped manufacturing centers to nearby hospitals. In this case, hospitals need PET detectors only and no expensive equipment.

Nowadays, ^{18}F-fluoro-2-deoxyglucose (FDG) is widely used for PET cancer screening in clinics. This screening detects faster-proliferating cancer cells that necessitate glucose as an energy source. While the ^{18}F-FDG PET contributes to early-stage cancer screening, ^{18}F-FDG also accumulates at inflammatory sites and normal brain tissues, which causes high background signals (Kubota et al., 1994;

Sugiyama et al., 2004). Because the protein metabolism of normal brain tissues is relatively low, amino acids are considered as the superior imaging agents. L-type amino acid transporter 1 (LAT1) is expressed and activated in higher levels in cancer cells that require amino acids for sufficient protein synthesis and active proliferation (Yanagida et al., 2001). To utilize this fact, the amino acid derivatives such as L-methyl-^{11}C-methionine, *O*-^{18}F-fluoromethyl-L-tyrosine, and their D-isomers are applied to specifically detect several types of cancer *in vivo* without unspecific detection of inflammatory sites (Burger et al., 2014; Ishiwata et al., 2005; Takeda et al., 1984; Tsukada et al., 2006).

Another promising approach is based on DNA synthesis, which is a direct measure of cell proliferation activity. ^{18}F-3'-fluoro-3'-deoxy-L-thymidine (FLT) is readily incorporated into cells and phosphorylated by thymidine kinase 1 (TK1), which is expressed during the S-phase of the cell cycle (Shields et al., 1998; Wagner et al., 2003). Since the resultant monophosphate is impermeable across the plasma membrane, it is accumulated in actively proliferating cells. The efficacy of ^{18}F-FLT for cancer imaging has already been demonstrated in clinical trials (Chen et al., 2005; Muzi et al., 2006).

Lipid metabolism is another important target for PET imaging. One of the target molecules is acetate. When administered *in vivo*, ^{11}C-acetate is converted into ^{11}C-acetyl-coenzyme A, integrated into the tricarboxylic acid (TCA) cycle, and finally cleared as ^{11}C-CO_2. In normal tissues, the initial accumulation of ^{11}C reflects blood flow, whereas the clearance of ^{11}C reflects oxygen metabolism (Hussain et al., 2009). In proliferating cancerous tissues, the ^{11}C-acetyl-coenzyme A is actively utilized in fatty acid synthesis, leading to higher accumulation levels (Yoshimoto et al., 2001). ^{11}C-acetate-PET is shown to be useful for imaging prostate cancer, renal cell carcinoma, hepatocellular carcinoma, and so on (Ho et al., 2003; Oyama et al., 2002, 2009; Tsuchida et al., 2008). Another target is choline, which is a precursor of acetylcholine. When administered *in vivo*, methyl-^{11}C-choline is phosphorylated to produce ^{11}C-phosphatidylcholine, which is then used for membrane phospholipid synthesis (Hara et al., 1997). The accumulation of ^{11}C-phosphatidylcholine reflects choline kinase activity, which is higher in actively proliferating cancer cells. Methyl-^{11}C-choline-PET is applied to detect a variety of cancer types, including brain, lung, liver, and bladder cancers (de Jong et al., 2002; Shinoura et al., 1997; Treglia et al., 2012). The ^{18}F derivatives of acetate and choline, namely ^{18}F-fluoroacetate and ^{18}F-fluoromethylcholine, respectively, are also used as long-lived alternatives (Kwee et al., 2007; Ponde et al., 2007).

Besides metabolite labeling, hypoxia PET imaging techniques using ^{18}F-fluoromisonidazole (FMISO) and ^{64}Cu-diacetyl-bis (N^4-methylthiosemicarbazone) (ATSM) have also been demonstrated (Jiang et al., 2014; Yoshii et al., 2010). Because cancer stem cells are adapted to hypoxia conditions with a non-proliferative phenotype, hypoxia PET imaging is an important means to identify sites including cancer stem cells.

7.3.2 SPECT

There are many SPECT imaging probes for detection of pathogenic states, especially cancer. Two of the imaging probes are ^{67}Ga-citrate and ^{201}Tl-chloride, which are accumulated at general cancerous sites (Edwards and Hayes, 1969; Salvatore et al., 1976). The mechanism for the accumulation of ^{67}Ga-citrate is believed to be due to the exchange reaction with transferrin, which is then incorporated by transferrin receptors expressed on cancer cells. ^{201}Tl-chloride, originally developed as a myocardial blood flow indicator, is accumulated at cancerous sites with high blood flow and Na^+-K^+ ATPase activities, which suggests the importance of both enough supply and active transport of ^{201}Tl-chloride in the mechanism of the ^{201}Tl-chloride uptake. Similarly, myocardial blood flow indicators ^{99m}Tc-hexakis methoxy isobutyl isonitrile (^{99m}Tc-MIBI) and ^{99m}Tc-tetrofosmin are used for cancer imaging (Fonti et al., 2015; Romanidis et al., 2014).

Other functional SPECT probes to detect more specific cellular states have also been developed. A noradrenaline receptor substrate ^{123}I-meta-iodobenzylguanidine (MIBG) allows imaging of neuronal functions as well as cancerous pheochromocytoma and neuroblastoma (Bomanji et al., 1987; Koopmans et al., 2009). A somatostatin analog ^{99m}Tc/^{111}In-octreotide enables imaging of endocrine tumors expressing a somatostatin receptor (Graham and Menda, 2011; Krenning et al., 1992). A glucagon-like peptide-1 (GLP-1) agonist ^{99m}Tc/^{111}In-exendin-4 visualizes insulinoma expressing a GLP-1 receptor (Christ et al., 2013), whereas an integrin-binding tripeptide ^{99m}Tc/^{111}In-RGD visualizes angiogenesis and cancer metastasis (Gaertner et al., 2012). ^{99m}Tc/^{111}In-prostate-specific membrane antigen (PSMA) inhibitors allow imaging of prostate cancers (Barrett et al., 2013), whereas a phosphatidylserine-binding protein ^{99m}Tc/^{111}In-annexin V allows imaging of cells undergoing early-stage apoptosis (Mochizuki et al., 2003). ^{99m}Tc-methylene diphosphonate (MDP) and its hydroxymethylene derivative are accumulated at high bone-forming activities, reflect bone metabolism, and are useful for diagnosis of metastatic bone cancer (Fogelman et al., 1979). Furthermore, ^{123}I/^{99m}Tc/^{111}In-antibodies allow targeted imaging of cells expressing their specific antigens (Kaur et al., 2012). Due to their high specificity, antibody-based probes hold great promise for visualizing various types of cancer.

Most of these functional probes are composed of peptides, which are labeled with ^{123}I, ^{111}In, or ^{99m}Tc. The ^{123}I-labeling using NaI is directly conducted to tyrosine residues on the peptides, or indirectly conducted to lysine residues on the peptides using the Bolton–Hunter reagent or more recently deiodination-resistant N-succinimidyl 3-iodobenzoate (SIB), which is advantageous for safety *in vivo* (Vaidyanathan and Zalutsky, 2006). The other two radioisotopes are firstly complexed with bifunctional chelating agents with peptide-binding moieties. For example, ^{111}In has been complexed with diethylenetriaminepentaacetic acid (DTPA), 1-(4-isothiocyanatobenzyl) ethylenediaminetetraacetic acid (SCN-Bz-EDTA), 2-(4-isothiocyanatobenzyl)-1,4,7,10-tetraazacyclododecane-1,4,7,10-tetraacetic acid (SCN-Bz-DOTA), and so on (Hnatowich et al., 1983; Vervoort et al., 2010), whereas ^{99m}Tc has been complexed with protected diaminedithiol (N_2S_2), hydrazinonicotinamide (HYNIC), 4'-aminomethyl-N,N'-trimethylenedibenzohydroxamamide (H_2N-$C_3(BHam)_2$), and so on (Liu and Edwards, 1999; Xu et al., 1999).

7.4 Fluorescence Imaging

Fluorescence is the emission of light with longer wavelengths when fluorophores are excited by light with specific wavelengths. Clinically, fluorescence imaging is applied to a lesser extent than other modalities mainly due to the poor tissue permeability of light. However, recent progress in endoscope technology has opened up new possibilities for fluorescence imaging in future medicine (Mitsunaga et al., 2013; Mizushima et al., 2016; Weissleder and Nahrendorf, 2015). Fluorescent molecules are categorized into fluorescent proteins and small organic fluorophores, both of which contain the extended structures of conjugated double bonds (Dean and Palmer, 2014). These fluorescent molecules are often linked to specific proteins to trace their subcellular localization in living cells. Although fluorescent proteins may affect physiological functions of the tethered proteins, the fusion proteins can be genetically encoded and easily expressed in living cells via gene transfection/transduction. Small organic fluorophores only minimally affect functions of the tethered proteins due to their small sizes. Taking advantage of fluorescent molecules, various biomedical applications from basic to clinic are demonstrated.

7.4.1 Fluorescent Proteins

Fluorescent proteins are widely used nowadays as a research tool to mark specific cells or proteins of interest. The milestone of fluorescent proteins has been the discovery of green fluorescent protein (GFP) from *Aequorea Victoria* (Shimomura et al., 1962). The fluorophore of GFP is formed by post-translational autocyclization/dehydration/oxidation and does not need other cofactors. It provides a technically easy platform to fluorescently label a protein of interest by genetic engineering. Various GFP mutants with different fluorescence properties such as blue fluorescent protein (BFP), cyan fluorescent protein (CFP), and yellow fluorescent protein (YFP) have been developed to enable multi-color imaging (Heim et al., 1994; Ormo et al., 1996; Tomosugi et al., 2009). Fluorescent proteins derived from species other than *Aequorea Victoria* are also available (Lin et al., 2009; Shcherbo et al., 2010). For example, DsRed is isolated from *Zoanthus* and its variants (mRFP, mCherry, tdTomato, mBanana, and so on) have also been developed (Campbell et al., 2002; Matz et al., 1999; Shaner et al., 2004; Wang et al., 2004).

Intriguingly, protein engineering approaches generate fluorescent protein variants that are irreversibly photo-activatable (Patterson and Lippincott-Schwartz, 2002; Subach et al., 2009), irreversibly photo-convertible (Ando et al., 2002; Chudakov et al., 2007), and have reversibly photo-switchable properties (Ando et al., 2004; Grotjohann et al., 2011).

Fluorescent proteins are utilized for various analyses to probe physiological states of proteins. Genetic fusion of a fluorescent protein with a protein of interest visualizes the physiological state and subcellular localization of the fusion protein even at single-molecule resolutions (Hibino et al., 2009; Iino et al., 2001). Protein–protein interactions are visualized by fusing two proteins of interest with different fluorescent proteins that can be a pair of the donor and acceptor in fluorescence resonance energy transfer (FRET) (Jares-Erijman and Jovin, 2003). FRET is usually analyzed by the ratio of fluorescence emission intensities from the donor and acceptor when the donor is excited. Since the FRET efficiency is affected by the distance and orientation of the donor and acceptor, the linker between the protein of interest and fluorophore usually needs to be optimized (Ibraheem et al., 2011; Komatsu et al., 2011a). The use of luciferase instead of the donor fluorophore allows bioluminescence resonance energy transfer (BRET) to the acceptor fluorophore in the presence of luciferin, which dispenses with fluorescence excitation that may cause background fluorescence, photobleaching, and phototoxicity (De et al., 2013; Xia and Rao, 2009). In bimolecular fluorescence complementation (BiFC), a fluorescent protein is divided into two fragments to which proteins of interest are fused. The interaction between the proteins of interest reconstitutes the fluorescent protein (Hu et al., 2002). This approach is generalized as the protein fragment complementation assay (PCA) and includes the use of enzymes that can catalytically convert substrates into fluorescent or luminescent products (Shekhawat and Ghosh, 2011; Wehr and Rossner, 2016).

An intriguing approach using FRET analysis is to visualize signal transduction based on intramolecular FRET biosensors (Aoki et al., 2013; Kiyokawa et al., 2006; Mochizuki et al., 2001). One example of this approach utilizes a single chimeric protein composed of a donor fluorescent protein, a substrate peptide for a kinase of interest in signal transduction, the binding domain of the phosphorylated substrate peptide, and an acceptor protein. Activation of the kinase of interest leads to phosphorylation of the substrate peptide of the chimeric protein, intramolecular interaction of the binding domain to form a hairpin structure, and approximation of the donor and acceptor fluorescent proteins to cause FRET. This approach is used to visualize kinase activities of various signaling molecules in the Ras/mitogen-activated protein kinase (MAPK) pathway (Kamioka et al., 2012).

Another important utility of fluorescent proteins is to report specific gene expression. This allows marking of the desired cells of interest, such as a specific lineage of cells in development and malignant cells in cancer. A facile design of multi-color reporter gene expressions enables cell cycle imaging named Fucci, in which each phase of the cell cycle in single cells are marked in different colors (Abe et al., 2013; Bajar et al., 2016; Sakaue-Sawano et al., 2008). While fluorescent proteins require fluorescence excitation that is harmful for cells in long-term tracking, a fusion protein composed of a luciferase (donor) and a fluorescent protein (acceptor) does not require fluorescence excitation but emits fluorescence in the presence of the luciferase substrate as a result of BRET (Hoshino et al., 2007; Saito et al., 2012). Improvements in brightness and fluorescence emission properties have recently been achieved, which facilitate multi-color imaging of different gene expressions in single cells (Suzuki et al., 2016).

7.4.2 Small Organic Fluorophores

So far, a variety of chemical probes are designed for labeling specific proteins of interest. In most cases, tagged proteins need to be exogenously expressed in the cells. A protein of interest is genetically fused with a tetracysteine-containing short peptide tag, which is labeled with As-containing fluorogenic probes FlAsH and ReAsH in cytoplasmic reducing conditions (Martin et al., 2005). These probes non-covalently bind to the tag with high affinities, and dramatically increase their fluorescence intensity upon binding. A non-covalent coiled-coil interaction is used for specific labeling of a membrane protein of interest, in which an exogenously prepared fluorescence-labeled peptide binds to the coiled-coil partner within the fused membrane protein (Yano and Matsuzaki, 2011). A tag–metal chelate interaction such as His tag–Ni-nitrilotriacetic acid (NTA)-dye is also used for membrane protein labeling (Fessenden, 2009). Such non-covalent modification methods, even with high-affinity binding properties, are difficult to apply to proteins whose expression levels are low (Dean and Palmer, 2014). This difficulty is caused by fluorescence signals derived from non-specific binding of the probes.

Therefore, recent progress in protein fluorescence labeling is prominent in covalent modification methods. His- or flag-tagged membrane proteins are labeled by fluorophore-conjugated metal chelates, which receive a nucleophilic attack from the side chains within the tag to yield covalent bonds between the tagged proteins and the fluorophore (Nonaka et al., 2009; Uchinomiya et al., 2009). Membrane proteins tagged with a substrate peptide of *Staphylococcus aureus* sortase A are covalently labeled using another fluorophore-conjugated substrate peptide and the sortase A enzyme, both of which are exogenously added (Hirota et al., 2010). Similarly, lipoic acid ligase and biotin ligase are used for enzymatic covalent fluorophore labeling of the cognate small substrate peptide tags (Howarth et al., 2005; Liu et al., 2012; Uttamapinant et al., 2010). Halo-tag (Los et al., 2008), SNAP tag (Keppler et al., 2003), CLIP-tag (Gautier et al., 2008), and TMP-tag (Gallagher et al., 2009) have been developed by obtaining hints from naturally occurring covalent bond formation of haloalkane dehalogenase, O^6-alkylguanine transferase, and dihydrofolate reductase. Although their sizes are larger than the aforementioned small peptide tags, their substrates are small organic compounds, and the cell-permeable fluorophore-conjugated substrates are easily available. Intriguingly, the SNAP tag substrate is engineered to fluoresce only upon binding to the SNAP tag, which enables wash-free fluorogenic labeling and real-time imaging of the SNAP-tagged target protein in living cells (Komatsu et al., 2011b; Sun et al., 2011).

Although the previously mentioned methods are useful for understanding the behavior of exogenously overexpressed tagged proteins, it is also important to develop methods to label non-tagged endogenous proteins. A photoaffinity labeling method allows light-induced protein labeling using its ligand that is conjugated with a fluorophore and a photoactive group (Chen et al., 2011; Dejda et al., 2011). This method leads to the protein that is covalently modified with both fluorophore and ligand. Another intriguing method also utilizes a ligand for protein labeling, but finally releases the ligand from the fluorophore-conjugated protein. In the principle of this method named ligand-directed tosyl chemistry, the ligand is modified with an electrophilic tosyl ester group and a fluorophore (Tsukiji et al., 2009). Upon binding of the modified ligand to its target protein, a proximal nucleophilic amino acid residue (mainly Glu, Tyr, and His) attacks a tosyl ester group to yield a fluorophore-conjugated protein and a phenylsulfonated ligand, the latter of which is dissociated from the former. In the principle of a similar method named affinity-guided dimethylaminopyridine (DMAP) chemistry, a DMAP-modified ligand is used to catalyze nucleophilic attack of a proximal amino acid residue (mainly Lys) on a fluorophore-conjugated acyl donor, resulting in a fluorophore-conjugated protein (Tamura and Hamachi, 2015; Wang et al., 2011). Since DMAP is a catalyst for acylation, the reaction rate is enhanced by incorporating multiple DMAP moieties of the modified ligand. Using these ligand-directed chemical methods, several cytoplasmic and cell-surface proteins are successfully labeled. These chemistry-based labeling methods enable selective and chemically minimal labeling of endogenous proteins, which is promising for visual analysis of protein dynamics in living cells.

Another important application of small organic fluorophores is to detect specific molecules or enzyme activities of interest based on the alteration of their fluorescence properties. The first milestone of biomedically useful small organic fluorophores was the development of a fluorescent calcium indicator Fura-2 (Grynkiewicz et al., 1985). This calcium indicator has a strong fluorescence intensity, and its excitation spectrum is sufficiently blue-shifted upon Ca^{2+} binding. This feature allows ratiometric fluorescence detection in two different wavelengths, which facilitates Ca^{2+} concentration measurement irrespective of fluorescent indicator concentration, cellular thickness, and light intensity. The key success of this indicator lies in the improvement of the Ca^{2+}

recognition moiety. The well-known Ca^{2+}-selective ligand ethyleneglycol-tetraacetic acid (EGTA) is converted into 1,2-bis (o-aminophenoxy) ethane-N,N,N',N'-tetraacetic acid (BAPTA), which has an aromatic backbone with low pK_a, leading to the fine-tuned detection of Ca^{2+} at neutral pH. This BAPTA ligand is also used in the next-generation fluorescein-based Ca^{2+} indicator Fluo-3, whose fluorescence is turned on in the presence of Ca^{2+}(Kao et al., 1989; Minta et al., 1989).

The fluorescence turn-on feature is greatly useful for biomedical imaging analyses, because it enables wash-free labeling of specific molecules with high signal-to-noise ratios, which facilitates spatiotemporal molecular imaging in high contrasts. Although fluorescence imaging has a major drawback that it is impossible to detect probes existing deep in the body due to poor tissue permeability of light, it is the superior modality that can relatively easily endow the probes with a turn-on nature. Historically, the rational design of such fluorescence turn-on probes has been difficult because the fluorescence properties including the fluorescence quantum yields are difficult to predict in advance (de Silva et al., 1997). However, a detailed analysis revealed that the fluorescence turn-on of Fluo-3 is attained by photoinduced electron transfer within the fluorescein moiety (Miura et al., 2003). Intriguingly, because the xanthene and benzene moieties of fluorescein are orthogonal, electrons between these moieties are decoupled. Therefore, the fluorescence property is solely controlled by the intramolecular photoinduced electron transfer between these moieties, which can be logically designed by modifying the electron-donating property of the benzene moiety (Tanaka et al., 2001; Urano, 2012; Urano et al., 2005). When the energy level of highest-occupied molecular orbital (HOMO) of the benzene moiety exceeds a threshold, the fluorescence is almost quenched by the fast photoinduced electron transfer process. In contrast, when the energy level of HOMO of the benzene moiety is lower than the threshold, the fluorescence emission process dominates the slow photoinduced electron transfer process. This principle is also applied to other fluorophores such as rhodamine, BODIPY, and cyanine. Based on this principle, a series of useful probes for acidic pH (Urano et al., 2009), singlet oxygen (Tanaka et al., 2001), nitric oxide (Izumi et al., 2009), highly reactive oxygen species (Setsukinai et al., 2003), peroxynitrite (Ueno et al., 2006), glutathione S-transferase (Fujikawa et al., 2008), and β-galactosidase (Urano et al., 2005) have also been developed.

Another principle that enables turn-on fluorescent probes is intramolecular spirocyclization (Urano, 2012). This is based on rhodamine, which has strong fluorescence irrespective of pH. When the carboxyl group of the benzene moiety of rhodamine is substituted by the hydroxymethyl group, the xanthene moiety of rhodamine is nucleophilically attacked by the hydroxymethyl group in strongly basic conditions, leading to a spirocyclized product. This product does not bear the fluorescent xanthene moiety with the extended conjugated double bonds any more, but instead bear non-fluorescent, three mutually unconjugated benzene rings. Using this principle, the carboxyl group of tetramethylrhodamine or its silyl derivative is substituted by the mercaptomethyl group, and the resultant spirocyclized non-fluorescent product specifically reacts with ClO^- to yield the oxidized methylsulfonate group and fluorescence-recovered xanthene group (Kenmoku et al., 2007; Koide et al., 2011). Hydroxymethyldiethylrhodol (HMDER) that is alkylated at the phenolic hydroxyl group is spirocyclized and non-fluorescent, whereas it recovers fluorescence upon removal of the alkyl group by hydrolysis. The HMDER-based approach is used to create fluorescent probes to detect β-galactosidase activity (Kamiya et al., 2011). Similarly, hydroxymethyl rhodamine green (HMRG) is used instead of HMDER as a platform molecule for designing a probe to detect γ-glutamyl transpeptidase activity that is enhanced in several types of cancer. This probe (gGlu-HMRG) has proven to be useful to rapidly visualize even small-size cancerous sites *in vivo* (Sakabe et al., 2013; Urano et al., 2011), which has not been attained by other modalities due to their low signal contrasts.

7.5 Conclusion

This chapter has introduced chemical reaction engineering in three imaging modalities (MRI, PET/SPECT imaging, and fluorescence imaging). The deep knowledge and understanding on chemical reaction engineering have realized imaging of specific molecules or activities *in vivo*. Numerous studies other than those introduced in this chapter are going on all over the world, and the progresses are so fast. Moreover, new molecular imaging modalities including photoacoustic imaging (Valluru and Willmann, 2016; Wang et al., 2016) and Raman spectroscopic imaging (Cheng and Xie, 2015; Kong et al., 2015) have recently emerged. There exist a variety of compounds which are the signatures of the physiological states of the body. As it is ultimately essential to detect them *in vivo* for realizing effective therapies for various diseases, many types of probes will be developed. Thus, chemical reaction engineering will be increasingly important for biomedical imaging in the future.

List of Abbreviations

BAPTA	1,2-bis (o-aminophenoxy) ethane-N,N,N',N'-tetraacetic acid
BRET	bioluminescence resonance energy transfer
CEST	chemical exchange saturation transfer
DMAP	dimethylaminopyridine
DOTA	1,4,7,10-tetraazacyclododecane-1,4,7,10-tetraacetic acid
DTPA	diethylenetriaminepentaacetic acid

FDG	fluoro-2-deoxyglucose
FLT	3'-fluoro-3'-deoxy-L-thymidine
FRET	fluorescence resonance energy transfer
GFP	green fluorescent protein
HMDER	hydroxymethyl diethylrhodol
HMRG	hydroxymethyl rhodamine green
HOMO	highest-occupied molecular orbital
MRI	magnetic resonance imaging
NMR	nuclear magnetic resonance
PET	positron emission tomography
SPECT	single photon emission computed tomography
SPIO	superparamagnetic iron oxide
T1	longitudinal relaxation time
T2	transverse relaxation time

References

Abe, T., Sakaue-Sawano, A., Kiyonari, H., Shioi, G., Inoue, K., Horiuchi, T., et al., Visualization of cell cycle in mouse embryos with Fucci2 reporter directed by Rosa26 promoter, *Development*, 2013; **140**:237–246.

Ando, R., Hama, H., Yamamoto-Hino, M., Mizuno, H., Miyawaki, A., An optical marker based on the UV-induced green-to-red photoconversion of a fluorescent protein, *Proc Natl Acad Sci U S A*, 2002; **99**:12651–12656.

Ando, R., Mizuno, H., Miyawaki, A., Regulated fast nucleocytoplasmic shuttling observed by reversible protein highlighting, *Science*, 2004; **306**:1370–1373.

Aoki, K., Kamioka, Y., Matsuda, M., Fluorescence resonance energy transfer imaging of cell signaling from in vitro to in vivo: basis of biosensor construction, live imaging, and image processing, *Dev Growth Differ*, 2013; **55**:515–522.

Aung, W., Hasegawa, S., Koshikawa-Yano, M., Obata, T., Ikehira, H., Furukawa, T., et al., Visualization of in vivo electroporation-mediated transgene expression in experimental tumors by optical and magnetic resonance imaging, *Gene Ther*, 2009; **16**:830–839.

Bai, Y., Lin, Y., Zhang, W., Kong, L., Wang, L., Zuo, P., et al., Noninvasive amide proton transfer magnetic resonance imaging in evaluating the grading and cellularity of gliomas, *Oncotarget*, 2016; **2016**:13970.

Bajar, B.T., Lam, A.J., Badiee, R.K., Oh, Y.H., Chu, J., Zhou, X.X., et al., Fluorescent indicators for simultaneous reporting of all four cell cycle phases, *Nat Methods*, 2016; **13**:993–996.

Barrett, J.A., Coleman, R.E., Goldsmith, S.J., Vallabhajosula, S., Petry, N.A., Cho, S., et al., First-in-man evaluation of 2 high-affinity PSMA-avid small molecules for imaging prostate cancer, *J Nucl Med*, 2013; **54**:380–387.

Bomanji, J., Levison, D.A., Zuzarte, J., Britton, K.E., Imaging of carcinoid tumors with iodine-123 metaiodobenzylguanidine, *J Nucl Med*, 1987; **28**:1907–1910.

Boros, E., Gale, E.M., Caravan, P., MR imaging probes: design and applications, *Dalton Trans*, 2015; **44**:4804–4818.

Bulte, J.W., de Cuyper, M., Despres, D., Frank, J.A., Short- vs. long-circulating magnetoliposomes as bone marrow-seeking MR contrast agents, *J Magn Reson Imaging*, 1999; **9**:329–335.

Burger, I.A., Zitzmann-Kolbe, S., Pruim, J., Friebe, M., Graham, K., Stephens, A., et al., First clinical results of (D)-18F-Fluoromethyltyrosine (BAY 86–9596) PET/CT in patients with non-small cell lung cancer and head and neck squamous cell carcinoma, *J Nucl Med*, 2014; **55**:1778–1785.

Campbell, R.E., Tour, O., Palmer, A.E., Steinbach, P.A., Baird, G.S., Zacharias, D.A., Tsien, R.Y., A monomeric red fluorescent protein, *Proc Natl Acad Sci U S A*, 2002; **99**:7877–7882.

Chambers, J.M., Hill, P.A., Aaron, J.A., Han, Z., Christianson, D.W., Kuzma, N.N., Dmochowski, I.J., Cryptophane xenon-129 nuclear magnetic resonance biosensors targeting human carbonic anhydrase, *J Am Chem Soc*, 2009; **131**:563–569.

Chen, J., Lanza, G.M., Wickline, S.A., Quantitative magnetic resonance fluorine imaging: today and tomorrow, *Wiley Interdiscip Rev Nanomed Nanobiotechnol*, 2010; **2**:431–440.

Chen, W., Cloughesy, T., Kamdar, N., Satyamurthy, N., Bergsneider, M., Liau, L., et al., Imaging proliferation in brain tumors with 18F-FLT PET: comparison with 18F-FDG, *J Nucl Med*, 2005; **46**:945–952.

Chen, Y., Falck, J.R., Manthati, V.L., Jat, J.L., Campbell, W.B., 20-Iodo-14,15-epoxyeicosa-8(Z)-enoyl-3-azidophenylsulfonamide: photoaffinity labeling of a 14,15-epoxyeicosatrienoic acid receptor, *Biochemistry*, 2011; **50**:3840–3848.

Cheng, J.X., Xie, X.S., Vibrational spectroscopic imaging of living systems: An emerging platform for biology and medicine, *Science*, 2015; **350**:aaa8870.

Christ, E., Wild, D., Ederer, S., Behe, M., Nicolas, G., Caplin, M.E., et al., Glucagon-like peptide-1 receptor imaging for the localisation of insulinomas: a prospective multicentre imaging study, *Lancet Diabetes Endocrinol*, 2013; **1**:115–122.

Chudakov, D.M., Lukyanov, S., Lukyanov, K.A., Tracking intracellular protein movements using photoswitchable fluorescent proteins PS-CFP2 and Dendra2, *Nat Protoc*, 2007; **2**:2024–2032.

Cohen, B., Ziv, K., Plaks, V., Israely, T., Kalchenko, V., Harmelin, A., et al., MRI detection of transcriptional regulation of gene expression in transgenic mice, *Nat Med*, 2007; **13**:498–503.

Comment, A., Dissolution DNP for in vivo preclinical studies, *J Magn Reson*, 2016; **264**:39–48.

Datta, A., Raymond, K.N., Gd-hydroxypyridinone (HOPO)-based high-relaxivity magnetic resonance imaging (MRI) contrast agents, *Acc Chem Res*, 2009; **42**:938–947.

De, A., Jasani, A., Arora, R., Gambhir, S.S., Evolution of BRET biosensors from live cell to tissue-scale in vivo imaging, *Front Endocrinol (Lausanne)*, 2013; **4**:131.

de Jong, I.J., Pruim, J., Elsinga, P.H., Jongen, M.M., Mensink, H.J., Vaalburg, W., Visualisation of bladder cancer using (11)C-choline PET: first clinical experience, *Eur J Nucl Med Mol Imaging*, 2002; **29**:1283–1288.

de Silva, A.P., Gunaratne, H.Q., Gunnlaugsson, T., Huxley, A.J., McCoy, C.P., Rademacher, J.T., Rice, T.E., Signaling recognition events with fluorescent sensors and switches, *Chem Rev*, 1997; **97**:1515–1566.

Dean, K.M., Palmer, A.E., Advances in fluorescence labeling strategies for dynamic cellular imaging, *Nat Chem Biol*, 2014; **10**:512–523.

Dejda, A., Bourgault, S., Doan, N.D., Letourneau, M., Couvineau, A., Vaudry, H., et al., Identification by photoaffinity labeling of the extracellular N-terminal domain of PAC1 receptor as the major binding site for PACAP, *Biochimie*, 2011; **93**:669–677.

Deserno, W.M., Harisinghani, M.G., Taupitz, M., Jager, G.J., Witjes, J.A., Mulders, P.F., et al., Urinary bladder cancer: preoperative nodal staging with ferumoxtran-10-enhanced MR imaging, *Radiology*, 2004; **233**:449–456.

Desser, T.S., Rubin, D.L., Muller, H.H., Qing, F., Khodor, S., Zanazzi, G., et al., Dynamics of tumor imaging with Gd-DTPA-polyethylene glycol polymers: dependence on molecular weight, *J Magn Reson Imaging*, 1994; **4**:467–472.

Doura, T., Nonaka, H., Sando, S., Atom arrangement strategy for designing a turn-on 1H magnetic resonance probe: a dual activatable probe for multimodal detection of hypochlorite, *Chem Commun (Camb)*, 2012; **48**:1565–1567.

Edwards, C.L., Hayes, R.L., Tumor scanning with 67Ga citrate, *J Nucl Med*, 1969; **10**:103–105.

Fessenden, J.D., Forster resonance energy transfer measurements of ryanodine receptor type 1 structure using a novel site-specific labeling method, *PLoS One*, 2009; **4**:e7338.

Fogelman, I., Citrin, D.L., McKillop, J.H., Turner, J.G., Bessent, R.G., Greig, W.R., A clinical comparison of Tc-99m HEDP and Tc-99m MDP in the detection of bone metastases: concise communication, *J Nucl Med*, 1979; **20**:98–101.

Fonti, R., Pace, L., Cerchione, C., Catalano, L., Salvatore, B., De Luca, S., et al., 18F-FDG PET/CT, 99mTc-MIBI, and MRI in the prediction of outcome of patients with multiple myeloma: a comparative study, *Clin Nucl Med*, 2015; **40**:303–308.

Fujikawa, Y., Urano, Y., Komatsu, T., Hanaoka, K., Kojima, H., Terai, T., et al., Design and synthesis of highly sensitive fluorogenic substrates for glutathione S-transferase and application for activity imaging in living cells, *J Am Chem Soc*, 2008; **130**:14533–14543.

Gaertner, F.C., Kessler, H., Wester, H.J., Schwaiger, M., Beer, A.J., Radiolabelled RGD peptides for imaging and therapy, *Eur J Nucl Med Mol Imaging*, 2012; **39 Suppl** 1:S126–138.

Gallagher, S.S., Sable, J.E., Sheetz, M.P., Cornish, V.W., An in vivo covalent TMP-tag based on proximity-induced reactivity, *ACS Chem Biol*, 2009; **4**:547–556.

Gautier, A., Juillerat, A., Heinis, C., Correa, I.R., Jr., Kindermann, M., Beaufils, F., et al., An engineered protein tag for multiprotein labeling in living cells, *Chem Biol*, 2008; **15**:128–136.

Goldhawk, D.E., Lemaire, C., McCreary, C.R., McGirr, R., Dhanvantari, S., Thompson, R.T., et al., Magnetic resonance imaging of cells overexpressing MagA, an endogenous contrast agent for live cell imaging, *Mol Imaging*, 2009; **8**:129–139.

Golman, K., in 't Zandt, R., Thaning, M., Real-time metabolic imaging, *Proc Natl Acad Sci U S A*, 2006; **103**:11270–11275.

Goyen, M., Edelman, M., Perreault, P., O'Riordan, E., Bertoni, H., Taylor, J., et al., MR angiography of aortoiliac occlusive disease: a phase III study of the safety and effectiveness of the blood-pool contrast agent MS-325, *Radiology*, 2005; **236**:825–833.

Graham, M.M., Menda, Y., Radiopeptide imaging and therapy in the United States, *J Nucl Med*, 2011; **52 Suppl 2**:56S–63S.

Grotjohann, T., Testa, I., Leutenegger, M., Bock, H., Urban, N.T., Lavoie-Cardinal, F., et al., Diffraction-unlimited all-optical imaging and writing with a photochromic GFP, *Nature*, 2011; **478**:204–208.

Grynkiewicz, G., Poenie, M., Tsien, R.Y., A new generation of Ca2+ indicators with greatly improved fluorescence properties, *J Biol Chem*, 1985; **260**:3440–3450.

Hancu, I., Dixon, W.T., Woods, M., Vinogradov, E., Sherry, A.D., Lenkinski, R.E., CEST and PARACEST MR contrast agents, *Acta Radiol*, 2010; **51**:910–923.

Hara, T., Kosaka, N., Shinoura, N., Kondo, T., PET imaging of brain tumor with [methyl-11C]choline, *J Nucl Med*, 1997; **38**:842–847.

Haris, M., Cai, K., Singh, A., Hariharan, H., Reddy, R., In vivo mapping of brain myo-inositol, *Neuroimage*, 2011; **54**:2079–2085.

Heim, R., Prasher, D.C., Tsien, R.Y., Wavelength mutations and posttranslational autoxidation of green fluorescent protein, *Proc Natl Acad Sci U S A*, 1994; **91**:12501–12504.

Hermann, P., Kotek, J., Kubicek, V., Lukes, I., Gadolinium(III) complexes as MRI contrast agents: ligand design and properties of the complexes, *Dalton Trans*, 2008:3027–3047.

Hibino, K., Hiroshima, M., Takahashi, M., Sako, Y., Single-molecule imaging of fluorescent proteins expressed in living cells, *Methods Mol Biol*, 2009; **544**:451–460.

Hirota, N., Yasuda, D., Hashidate, T., Yamamoto, T., Yamaguchi, S., Nagamune, T., et al., Amino acid residues critical for endoplasmic reticulum export and trafficking of platelet-activating factor receptor, *J Biol Chem*, 2010; **285**:5931–5940.

Hnatowich, D.J., Layne, W.W., Childs, R.L., Lanteigne, D., Davis, M.A., Griffin, T.W., Doherty, P.W., Radioactive labeling of antibody: a simple and efficient method, *Science*, 1983; **220**:613–615.

Ho, C.L., Yu, S.C., Yeung, D.W., 11C-acetate PET imaging in hepatocellular carcinoma and other liver masses, *J Nucl Med*, 2003; **44**:213–221.

Hoshino, H., Nakajima, Y., Ohmiya, Y., Luciferase-YFP fusion tag with enhanced emission for single-cell luminescence imaging, *Nat Methods*, 2007; **4**:637–639.

Howarth, M., Takao, K., Hayashi, Y., Ting, A.Y., Targeting quantum dots to surface proteins in living cells with biotin ligase, *Proc Natl Acad Sci U S A*, 2005; **102**:7583–7588.

Hu, C.D., Chinenov, Y., Kerppola, T.K., Visualization of interactions among bZIP and Rel family proteins in living cells using bimolecular fluorescence complementation, *Mol Cell*, 2002; **9**:789–798.

Hussain, R., Kudo, T., Tsujikawa, T., Kobayashi, M., Fujibayashi, Y., Okazawa, H., Validation of the calculation of the clearance rate constant (k(mono)) of [(11)C] acetate using parametric k(mono) image for myocardial oxidative metabolism, *Nucl Med Biol*, 2009; **36**:877–882.

Ibraheem, A., Yap, H., Ding, Y., Campbell, R.E., A bacteria colony-based screen for optimal linker combinations in genetically encoded biosensors, *BMC Biotechnol*, 2011; **11**:105.

Iino, R., Koyama, I., Kusumi, A., Single molecule imaging of green fluorescent proteins in living cells: E-cadherin forms oligomers on the free cell surface, *Biophys J*, 2001; **80**:2667–2677.

Ishiwata, K., Tsukada, H., Kubota, K., Nariai, T., Harada, N., Kawamura, K., et al., Preclinical and clinical evaluation of O-[11C]methyl-L-tyrosine for tumor imaging by positron emission tomography, *Nucl Med Biol*, 2005; **32**:253–262.

Izumi, S., Urano, Y., Hanaoka, K., Terai, T., Nagano, T., A simple and effective strategy to increase the sensitivity of fluorescence probes in living cells, *J Am Chem Soc*, 2009; **131**:10189–10200.

Jares-Erijman, E.A., Jovin, T.M., FRET imaging, *Nat Biotechnol*, 2003; **21**:1387–1395.

Jiang, L., Tu, Y., Shi, H., Cheng, Z., PET probes beyond (18)F-FDG, *J Biomed Res*, 2014; **28**:435–446.

Kaida, S., Cabral, H., Kumagai, M., Kishimura, A., Terada, Y., Sekino, M., et al., Visible drug delivery by supramolecular nanocarriers directing to single-platformed diagnosis and therapy of pancreatic tumor model, *Cancer Res*, 2010; **70**:7031–7041.

Kamioka, Y., Sumiyama, K., Mizuno, R., Sakai, Y., Hirata, E., Kiyokawa, E., Matsuda, M., Live imaging of protein kinase activities in transgenic mice expressing FRET biosensors, *Cell Struct Funct*, 2012; **37**:65–73.

Kamiya, M., Asanuma, D., Kuranaga, E., Takeishi, A., Sakabe, M., Miura, M., et al., beta-Galactosidase fluorescence probe with improved cellular accumulation based on a spirocyclized rhodol scaffold, *J Am Chem Soc*, 2011; **133**: 12960–12963.

Kao, J.P., Harootunian, A.T., Tsien, R.Y., Photochemically generated cytosolic calcium pulses and their detection by fluo-3, *J Biol Chem*, 1989; **264**: 8179–8184.

Kaur, S., Venktaraman, G., Jain, M., Senapati, S., Garg, P.K., Batra, S.K., Recent trends in antibody-based oncologic imaging, *Cancer Lett*, 2012; **315**:97–111.

Kenmoku, S., Urano, Y., Kojima, H., Nagano, T., Development of a highly specific rhodamine-based fluorescence probe for hypochlorous acid and its application to real-time imaging of phagocytosis, *J Am Chem Soc*, 2007; **129**:7313–7318.

Keppler, A., Gendreizig, S., Gronemeyer, T., Pick, H., Vogel, H., Johnsson, K., A general method for the covalent labeling of fusion proteins with small molecules in vivo, *Nat Biotechnol*, 2003; **21**:86–89.

Kiyokawa, E., Hara, S., Nakamura, T., Matsuda, M., Fluorescence (Forster) resonance energy transfer imaging of oncogene activity in living cells, *Cancer Sci*, 2006; **97**:8–15.

Koide, Y., Urano, Y., Hanaoka, K., Terai, T., Nagano, T., Development of an Si-rhodamine-based far-red to near-infrared fluorescence probe selective for hypochlorous acid and its applications for biological imaging, *J Am Chem Soc*, 2011; **133**:5680–5682.

Kojima, C., Turkbey, B., Ogawa, M., Bernardo, M., Regino, C.A., Bryant, L.H., Jr., et al., Dendrimer-based MRI contrast agents: the effects of PEGylation on relaxivity and pharmacokinetics, *Nanomedicine*, 2011; **7**:1001–1008.

Kokuryo, D., Anraku, Y., Kishimura, A., Tanaka, S., Kano, M.R., Kershaw, J., et al., SPIO-PICsome: development of a highly sensitive and stealth-capable MRI nano-agent for tumor detection using SPIO-loaded unilamellar polyion complex vesicles (PICsomes), *J Control Release*, 2013; **169**:220–227.

Komatsu, N., Aoki, K., Yamada, M., Yukinaga, H., Fujita, Y., Kamioka, Y., Matsuda, M., Development of an optimized backbone of FRET biosensors for kinases and GTPases, *Mol Biol Cell*, 2011a; **22**:4647–4656.

Komatsu, T., Johnsson, K., Okuno, H., Bito, H., Inoue, T., Nagano, T., Urano, Y., Real-time measurements of protein dynamics using fluorescence activation-coupled protein labeling method, *J Am Chem Soc*, 2011b; **133**:6745–6751.

Kong, K., Kendall, C., Stone, N., Notingher, I., Raman spectroscopy for medical diagnostics – From in-vitro biofluid assays to in-vivo cancer detection, *Adv Drug Deliv Rev*, 2015; **89**:121–134.

Kono, K., Nakashima, S., Kokuryo, D., Aoki, I., Shimomoto, H., Aoshima, S., et al., Multi-functional liposomes having temperature-triggered release and magnetic resonance imaging for tumor-specific chemotherapy, *Biomaterials*, 2011; **32**:1387–1395.

Koopmans, K.P., Neels, O.N., Kema, I.P., Elsinga, P.H., Links, T.P., de Vries, E.G., Jager, P.L., Molecular imaging in neuroendocrine tumors: molecular uptake mechanisms and clinical results, *Crit Rev Oncol Hematol*, 2009; **71**:199–213.

Krenning, E.P., Bakker, W.H., Kooij, P.P., Breeman, W.A., Oei, H.Y., de Jong, M., et al., Somatostatin receptor scintigraphy with indium-111-DTPA-D-Phe-1-octreotide in man: metabolism, dosimetry and comparison with iodine-123-Tyr-3-octreotide, *J Nucl Med*, 1992; **33**:652–658.

Kubota, R., Kubota, K., Yamada, S., Tada, M., Ido, T., Tamahashi, N., Microautoradiographic study for the differentiation of intratumoral macrophages, granulation tissues and cancer cells by the dynamics of fluorine-18-fluorodeoxyglucose uptake, *J Nucl Med*, 1994; **35**:104–112.

Kumagai, M., Sarma, T.K., Cabral, H., Kaida, S., Sekino, M., Herlambang, N., et al., Enhanced in vivo magnetic resonance imaging of tumors by PEGylated iron-oxide-gold core-shell nanoparticles with prolonged blood circulation properties, *Macromol Rapid Commun*, 2010; **31**:1521–1528.

Kwee, S.A., DeGrado, T.R., Talbot, J.N., Gutman, F., Coel, M.N., Cancer imaging with fluorine-18-labeled choline derivatives, *Semin Nucl Med*, 2007; **37**:420–428.

Lauffer, R.B., Parmelee, D.J., Ouellet, H.S., Dolan, R.P., Sajiki, H., Scott, D.M., et al., MS-325: a small-molecule vascular imaging agent for magnetic resonance imaging, *Acad Radiol*, 1996; **3 Suppl 2**:S356–358.

Lauterbur, P.C., Image formation by induced local interactions: examples employing nuclear magnetic resonance, *Nature*, 1973; **242**:190–191.

Lim, J., Turkbey, B., Bernardo, M., Bryant, L.H., Jr., Garzoni, M., Pavan, G.M., et al., Gadolinium MRI contrast agents based on triazine dendrimers: relaxivity and in vivo pharmacokinetics, *Bioconjug Chem*, 2012; **23**:2291–2299.

Lin, M.Z., McKeown, M.R., Ng, H.L., Aguilera, T.A., Shaner, N.C., Campbell, R.E., et al., Autofluorescent proteins with excitation in the optical window for intravital imaging in mammals, *Chem Biol*, 2009; **16**:1169–1179.

Ling, W., Regatte, R.R., Navon, G., Jerschow, A., Assessment of glycosaminoglycan concentration in vivo by chemical exchange-dependent saturation transfer (gagCEST), *Proc Natl Acad Sci U S A*, 2008; **105**:2266–2270.

Liu, D.S., Phipps, W.S., Loh, K.H., Howarth, M., Ting, A.Y., Quantum dot targeting with lipoic acid ligase and HaloTag for single-molecule imaging on living cells, *ACS Nano*, 2012; **6**:11080–11087.

Liu, S., Edwards, D.S., 99mTc-labeled small peptides as diagnostic radiopharmaceuticals, *Chem Rev*, 1999; **99**:2235–2268.

Logothetis, N.K., What we can do and what we cannot do with fMRI, *Nature*, 2008; **453**:869–878.

Los, G.V., Encell, L.P., McDougall, M.G., Hartzell, D.D., Karassina, N., Zimprich, C., et al., HaloTag: a novel protein labeling technology for cell imaging and protein analysis, *ACS Chem Biol*, 2008; **3**:373–382.

Mariani, G., Bruselli, L., Kuwert, T., Kim, E.E., Flotats, A., Israel, O., et al., A review on the clinical uses of SPECT/CT, *Eur J Nucl Med Mol Imaging*, 2010; **37**:1959–1985.

Martin, B.R., Giepmans, B.N., Adams, S.R., Tsien, R.Y., Mammalian cell-based optimization of the biarsenical-binding tetracysteine motif for improved fluorescence and affinity, *Nat Biotechnol*, 2005; **23**:1308–1314.

Matz, M.V., Fradkov, A.F., Labas, Y.A., Savitsky, A.P., Zaraisky, A.G., Markelov, M.L., Lukyanov, S.A., Fluorescent proteins from nonbioluminescent Anthozoa species, *Nat Biotechnol*, 1999; **17**:969–973.

Mi, P., Dewi, N., Yanagie, H., Kokuryo, D., Suzuki, M., Sakurai, Y., et al., Hybrid calcium phosphate-polymeric micelles incorporating gadolinium chelates for imaging-guided gadolinium neutron capture tumor therapy, *ACS Nano*, 2015; **9**:5913–5921.

Minta, A., Kao, J.P., Tsien, R.Y., Fluorescent indicators for cytosolic calcium based on rhodamine and fluorescein chromophores, *J Biol Chem*, 1989; **264**:8171–8178.

Mitsunaga, M., Kosaka, N., Choyke, P.L., Young, M.R., Dextras, C.R., Saud, S.M., et al., Fluorescence endoscopic detection of murine colitis-associated colon cancer by topically applied enzymatically rapid-activatable probe, *Gut*, 2013; **62**:1179–1186.

Miura, T., Urano, Y., Tanaka, K., Nagano, T., Ohkubo, K., Fukuzumi, S., Rational design principle for modulating fluorescence properties of fluorescein-based probes by photoinduced electron transfer, *J Am Chem Soc*, 2003; **125**:8666–8671.

Mizukami, S., Takikawa, R., Sugihara, F., Hori, Y., Tochio, H., Walchli, M., et al., Paramagnetic relaxation-based 19f MRI probe to detect protease activity, *J Am Chem Soc*, 2008; **130**:794–795.

Mizushima, T., Ohnishi, S., Shimizu, Y., Hatanaka, Y., Hatanaka, K.C., Hosono, H., et al., Fluorescent imaging of superficial head and neck squamous cell carcinoma using a gamma-glutamyltranspeptidase-activated targeting agent: a pilot study, *BMC Cancer*, 2016; **16**:411.

Mochizuki, N., Yamashita, S., Kurokawa, K., Ohba, Y., Nagai, T., Miyawaki, A., Matsuda, M., Spatio-temporal images of growth-factor-induced activation of Ras and Rap1, *Nature*, 2001; **411**:1065–1068.

Mochizuki, T., Kuge, Y., Zhao, S., Tsukamoto, E., Hosokawa, M., Strauss, H.W., et al., Detection of apoptotic tumor response in vivo after a single dose of chemotherapy with 99mTc-annexin V, *J Nucl Med*, 2003; **44**:92–97.

Muzi, M., Spence, A.M., O'Sullivan, F., Mankoff, D.A., Wells, J.M., Grierson, J.R., et al., Kinetic analysis of 3'-deoxy-3'-18F-fluorothymidine in patients with gliomas, *J Nucl Med*, 2006; **47**:1612–1621.

Naumova, A.V., Reinecke, H., Yarnykh, V., Deem, J., Yuan, C., Murry, C.E., Ferritin overexpression for noninvasive magnetic resonance imaging-based tracking of stem cells transplanted into the heart, *Mol Imaging*, 2010; **9**:201–210.

Nishihara, T., Yoshihara, H.A., Nonaka, H., Takakusagi, Y., Hyodo, F., Ichikawa, K., et al., Direct monitoring of gamma-glutamyl transpeptidase activity in vivo using a hyperpolarized (13) C-labeled molecular probe, *Angew Chem Int Ed Engl*, 2016; **55**:10626–10629.

Nonaka, H., Fujishima, S.H., Uchinomiya, S.H., Ojida, A., Hamachi, I., FLAG-tag selective covalent protein labeling via a binding-induced acyl-transfer reaction, *Bioorg Med Chem Lett*, 2009; **19**:6696–6699.

Nonaka, H., Hata, R., Doura, T., Nishihara, T., Kumagai, K., Akakabe, M., et al., A platform for designing hyperpolarized magnetic resonance chemical probes, *Nat Commun*, 2013; **4**:2411.

Nonaka, H., Hirano, M., Imakura, Y., Takakusagi, Y., Ichikawa, K., Sando, S., Design of a 15N molecular unit to achieve long retention of hyperpolarized spin state, *Sci Rep*, 2017; **7**:40104.

Oishi, M., Sumitani, S., Nagasaki, Y., On-off regulation of 19F magnetic resonance signals based on pH-sensitive PEGylated nanogels for potential tumor-specific smart 19F MRI probes, *Bioconjug Chem*, 2007; **18**:1379–1382.

Ormo, M., Cubitt, A.B., Kallio, K., Gross, L.A., Tsien, R.Y., Remington, S.J., Crystal structure of the *Aequorea victoria* green fluorescent protein, *Science*, 1996; **273**:1392–1395.

Oyama, N., Akino, H., Kanamaru, H., Suzuki, Y., Muramoto, S., Yonekura, Y., et al., 11C-acetate PET imaging of prostate cancer, *J Nucl Med*, 2002; **43**:181–186.

Oyama, N., Okazawa, H., Kusukawa, N., Kaneda, T., Miwa, Y., Akino, H., et al., 11C-Acetate PET imaging for renal cell carcinoma, *Eur J Nucl Med Mol Imaging*, 2009; **36**:422–427.

Patterson, G.H., Lippincott-Schwartz, J., A photoactivatable GFP for selective photolabeling of proteins and cells, *Science*, 2002; **297**:1873–1877.

Poldrack, R.A., Farah, M.J., Progress and challenges in probing the human brain, *Nature*, 2015; **526**:371–379.

Ponde, D.E., Dence, C.S., Oyama, N., Kim, J., Tai, Y.C., Laforest, R., et al., 18F-fluoroacetate: a potential acetate analog for prostate tumor imaging – in vivo evaluation of 18F-fluoroacetate versus 11C-acetate, *J Nucl Med*, 2007; **48**:420–428.

Rohani, R., Figueredo, R., Bureau, Y., Koropatnick, J., Foster, P., Thompson, R.T., et al., Imaging tumor growth non-invasively using expression of MagA or modified ferritin subunits to augment intracellular contrast for repetitive MRI, *Mol Imaging Biol*, 2014; **16**:63–73.

Romanidis, K., Karathanos, E., Nagorni, E.A., Giatromanolaki, A., Sibridis, E., Zissimopoulos, A., et al., Parathyroid adenoma detected with 99mTc-tetrofosmin dual-phase scintigraphy: a case report, *BMC Res Notes*, 2014; **7**:335.

Saito, K., Chang, Y.F., Horikawa, K., Hatsugai, N., Higuchi, Y., Hashida, M., et al., Luminescent proteins for high-speed single-cell and whole-body imaging, *Nat Commun*, 2012; **3**:1262.

Sakabe, M., Asanuma, D., Kamiya, M., Iwatate, R.J., Hanaoka, K., Terai, T., et al., Rational design of highly sensitive fluorescence probes for protease and glycosidase based on precisely controlled spirocyclization, *J Am Chem Soc*, 2013; **135**:409–414.

Sakamoto, T., Shimizu, Y.K., Sasaki, J., Hayakawa, H., Fujimoto, K., Signal turn-on probe for nucleic acid detection based on (19)F nuclear magnetic resonance, *Bioorg Med Chem Lett*, 2011; **21**:303–306.

Sakaue-Sawano, A., Kurokawa, H., Morimura, T., Hanyu, A., Hama, H., Osawa, H., et al., Visualizing spatiotemporal dynamics of multicellular cell-cycle progression, *Cell*, 2008; **132**:487–498.

Salvatore, M., Carratu, L., Porta, E., Thallium-201 as a positive indicator for lung neoplasms: preliminary experiments, *Radiology*, 1976; **121**:487–488.

Setsukinai, K., Urano, Y., Kakinuma, K., Majima, H.J., Nagano, T., Development of novel fluorescence probes that can reliably detect reactive oxygen species and distinguish specific species, *J Biol Chem*, 2003; **278**:3170–3175.

Shaner, N.C., Campbell, R.E., Steinbach, P.A., Giepmans, B.N., Palmer, A.E., Tsien, R.Y., Improved monomeric red, orange and yellow fluorescent proteins derived from *Discosoma* sp. red fluorescent protein, *Nat Biotechnol*, 2004; **22**:1567–1572.

Shcherbo, D., Shemiakina, II, Ryabova, A.V., Luker, K.E., Schmidt, B.T., Souslova, E.A., et al., Near-infrared fluorescent proteins, *Nat Methods*, 2010; **7**:827–829.

Shekhawat, S.S., Ghosh, I., Split-protein systems: beyond binary protein-protein interactions, *Curr Opin Chem Biol*, 2011; **15**:789–797.

Sherry, A.D., Woods, M., Chemical exchange saturation transfer contrast agents for magnetic resonance imaging, *Annu Rev Biomed Eng*, 2008; **10**:391–411.

Shields, A.F., Grierson, J.R., Dohmen, B.M., Machulla, H.J., Stayanoff, J.C., Lawhorn-Crews, J.M., et al., Imaging proliferation in vivo with [F-18]FLT and positron emission tomography, *Nat Med*, 1998; **4**:1334–1336.

Shimomura, O., Johnson, F.H., Saiga, Y., Extraction, purification and properties of aequorin, a bioluminescent protein from the luminous hydromedusan, *Aequorea, J Cell Comp Physiol*, 1962; **59**:223–239.

Shinoura, N., Nishijima, M., Hara, T., Haisa, T., Yamamoto, H., Fujii, K., et al., Brain tumors: detection with C-11 choline PET, *Radiology*, 1997; **202**:497–503.

Sieving, P.F., Watson, A.D., Rocklage, S.M., Preparation and characterization of paramagnetic polychelates and their protein conjugates, *Bioconjug Chem*, 1990; **1**:65–71.

Subach, F.V., Patterson, G.H., Manley, S., Gillette, J.M., Lippincott-Schwartz, J., Verkhusha, V.V., Photoactivatable mCherry for high-resolution two-color fluorescence microscopy, *Nat Methods*, 2009; **6**:153–159.

Sugiyama, M., Sakahara, H., Sato, K., Harada, N., Fukumoto, D., Kakiuchi, T., et al., Evaluation of 3'-deoxy-3'-18F-fluorothymidine for monitoring tumor response to radiotherapy and photodynamic therapy in mice, *J Nucl Med*, 2004; **45**:1754–1758.

Sun, X., Zhang, A., Baker, B., Sun, L., Howard, A., Buswell, J., et al., Development of SNAP-tag fluorogenic probes for wash-free fluorescence imaging, *Chembiochem*, 2011; **12**:2217–2226.

Suzuki, K., Kimura, T., Shinoda, H., Bai, G., Daniels, M.J., Arai, Y., et al., Five colour variants of bright luminescent protein for real-time multicolour bioimaging, *Nat Commun*, 2016; **7**:13718.

Takaoka, Y., Sakamoto, T., Tsukiji, S., Narazaki, M., Matsuda, T., Tochio, H., et al., Self-assembling nanoprobes that display off/on 19F nuclear magnetic resonance signals for protein detection and imaging, *Nat Chem*, 2009; **1**:557–561.

Takeda, A., Goto, R., Tamemasa, O., Chaney, J.E., Digenis, G.A., Biological evaluation of radiolabeled D-methionine as a parent compound in potential nuclear imaging, *Radioisotopes*, 1984; **33**:213–217.

Tamura, T., Hamachi, I., Labeling proteins by affinity-guided DMAP chemistry, *Methods Mol Biol*, 2015; **1266**:229–242.

Tanaka, K., Kitamura, N., Chujo, Y., Heavy metal-free 19F NMR probes for quantitative measurements of glutathione reductase activity using silica nanoparticles as a signal quencher, *Bioorg Med Chem*, 2012; **20**:96–100.

Tanaka, K., Miura, T., Umezawa, N., Urano, Y., Kikuchi, K., Higuchi, T., Nagano, T., Rational design of fluorescein-based fluorescence probes. Mechanism-based design of a maximum fluorescence probe for singlet oxygen, *J Am Chem Soc*, 2001; **123**:2530–2536.

Tomosugi, W., Matsuda, T., Tani, T., Nemoto, T., Kotera, I., Saito, K., et al., An ultramarine fluorescent protein with increased photostability and pH insensitivity, *Nat Methods*, 2009; **6**:351–353.

Treglia, G., Giovannini, E., Di Franco, D., Calcagni, M.L., Rufini, V., Picchio, M., Giordano, A., The role of positron emission tomography using carbon-11 and fluorine-18 choline in tumors other than prostate cancer: a systematic review, *Ann Nucl Med*, 2012; **26**:451–461.

Tsuchida, T., Takeuchi, H., Okazawa, H., Tsujikawa, T., Fujibayashi, Y., Grading of brain glioma with 1–11C-acetate PET: comparison with 18F-FDG PET, *Nucl Med Biol*, 2008; **35**:171–176.

Tsukada, H., Sato, K., Fukumoto, D., Nishiyama, S., Harada, N., Kakiuchi, T., Evaluation of D-isomers of O-11C-methyl tyrosine and O-18F-fluoromethyl tyrosine as tumor-imaging agents in tumor-bearing mice: comparison with L- and D-11C-methionine, *J Nucl Med*, 2006; **47**:679–688.

Tsukiji, S., Miyagawa, M., Takaoka, Y., Tamura, T., Hamachi, I., Ligand-directed tosyl chemistry for protein labeling in vivo, *Nat Chem Biol*, 2009; 5:341–343.

Uchinomiya, S.H., Nonaka, H., Fujishima, S.H., Tsukiji, S., Ojida, A., Hamachi, I., Site-specific covalent labeling of His-tag fused proteins with a reactive Ni(II)-NTA probe, *Chem Commun (Camb)*, 2009; 5880–5882.

Ueno, T., Urano, Y., Kojima, H., Nagano, T., Mechanism-based molecular design of highly selective fluorescence probes for nitrative stress, *J Am Chem Soc*, 2006; **128**:10640–10641.

Urano, Y., Novel live imaging techniques of cellular functions and in vivo tumors based on precise design of small molecule-based 'activatable' fluorescence probes, *Curr Opin Chem Biol*, 2012; **16**:602–608.

Urano, Y., Asanuma, D., Hama, Y., Koyama, Y., Barrett, T., Kamiya, M., et al., Selective molecular imaging of viable cancer cells with pH-activatable fluorescence probes, *Nat Med*, 2009; **15**:104–109.

Urano, Y., Kamiya, M., Kanda, K., Ueno, T., Hirose, K., Nagano, T., Evolution of fluorescein as a platform for finely tunable fluorescence probes, *J Am Chem Soc*, 2005; **127**:4888–4894.

Urano, Y., Sakabe, M., Kosaka, N., Ogawa, M., Mitsunaga, M., Asanuma, D., et al., Rapid cancer detection by topically spraying a gamma-glutamyltranspeptidase-activated fluorescent probe, *Sci Transl Med*, 2011; **3**:110ra119.

Uttamapinant, C., White, K.A., Baruah, H., Thompson, S., Fernandez-Suarez, M., Puthenveetil, S., et al., A fluorophore ligase for site-specific protein labeling inside living cells, *Proc Natl Acad Sci U S A*, 2010; **107**:10914–10919.

Vaidyanathan, G., Zalutsky, M.R., Preparation of N-succinimidyl 3-[*I] iodobenzoate: an agent for the indirect radioiodination of proteins, *Nat Protoc*, 2006; **1**:707–713.

Valluru, K.S., Willmann, J.K., Clinical photoacoustic imaging of cancer, *Ultrasonography*, 2016; **35**:267–280.

van Zijl, P.C., Jones, C.K., Ren, J., Malloy, C.R., Sherry, A.D., MRI detection of glycogen in vivo by using chemical exchange saturation transfer imaging (glycoCEST), *Proc Natl Acad Sci U S A*, 2007; **104**:4359–4364.

Vaquero, J.J., Kinahan, P., Positron emission tomography: current challenges and opportunities for technological advances in clinical and preclinical imaging systems, *Annu Rev Biomed Eng*, 2015; **17**:385–414.

Vervoort, L., Burvenich, I., Staelens, S., Dumolyn, C., Waegemans, E., Van Steenkiste, M., et al., Preclinical evaluation of monoclonal antibody 14C5 for targeting pancreatic cancer, *Cancer Biother Radiopharm*, 2010; **25**:193–205.

Viale, A., Aime, S., Current concepts on hyperpolarized molecules in MRI, *Curr Opin Chem Biol*, 2010; **14**:90–96.

Villaraza, A.J., Bumb, A., Brechbiel, M.W., Macromolecules, dendrimers, and nanomaterials in magnetic resonance imaging: the interplay between size, function, and pharmacokinetics, *Chem Rev*, 2010; **110**:2921–2959.

Wagner, M., Seitz, U., Buck, A., Neumaier, B., Schultheiss, S., Bangerter, M., et al., 3′-[18F]fluoro-3′-deoxythymidine ([18F]-FLT) as positron emission tomography tracer for imaging proliferation in a murine B-cell lymphoma model and in the human disease, *Cancer Res*, 2003; **63**:2681–2687.

Wang, H., Koshi, Y., Minato, D., Nonaka, H., Kiyonaka, S., Mori, Y., et al., Chemical cell-surface receptor engineering using affinity-guided, multivalent organocatalysts, *J Am Chem Soc*, 2011; **133**:12220–12228.

Wang, L., Jackson, W.C., Steinbach, P.A., Tsien, R.Y., Evolution of new nonantibody proteins via iterative somatic hypermutation, *Proc Natl Acad Sci U S A*, 2004; **101**:16745–16749.

Wang, S., Lin, J., Wang, T., Chen, X., Huang, P., Recent advances in photoacoustic imaging for deep-tissue biomedical applications, *Theranostics*, 2016; **6**:2394–2413.

Wehr, M.C., Rossner, M.J., Split protein biosensor assays in molecular pharmacological studies, *Drug Discov Today*, 2016; **21**:415–429.

Weissleder, R., Nahrendorf, M., Advancing biomedical imaging, *Proc Natl Acad Sci U S A*, 2015; **112**:14424–14428.

Xia, Z., Rao, J., Biosensing and imaging based on bioluminescence resonance energy transfer, *Curr Opin Biotechnol*, 2009; **20**:37–44.

Xu, L.C., Nakayama, M., Harada, K., Kuniyasu, A., Nakayama, H., Tomiguchi, S., et al., Bis(hydroxamamide)-based bifunctional chelating agent for 99mTc labeling of polypeptides, *Bioconjug Chem*, 1999; **10**:9–17.

Yanagida, O., Kanai, Y., Chairoungdua, A., Kim, D.K., Segawa, H., Nii, T., et al., Human L-type amino acid transporter 1 (LAT1): characterization of function and expression in tumor cell lines, *Biochim Biophys Acta*, 2001; **1514**:291–302.

Yano, Y., Matsuzaki, K., Fluorescence ratiometric detection of ligand-induced receptor internalization using extracellular coiled-coil tag-probe labeling, *FEBS Lett*, 2011; **585**:2385–2388.

Yoshii, Y., Furukawa, T., Kiyono, Y., Watanabe, R., Waki, A., Mori, T., et al., Copper-64-diacetyl-bis (N4-methylthiosemicarbazone) accumulates in rich regions of CD133+ highly tumorigenic cells in mouse colon carcinoma, *Nucl Med Biol*, 2010; **37**:395–404.

Yoshimoto, M., Waki, A., Yonekura, Y., Sadato, N., Murata, T., Omata, N., et al., Characterization of acetate metabolism in tumor cells in relation to cell proliferation: acetate metabolism in tumor cells, *Nucl Med Biol*, 2001; **28**:117–122.

Zhang, S., Merritt, M., Woessner, D.E., Lenkinski, R.E., Sherry, A.D., PARACEST agents: modulating MRI contrast via water proton exchange, *Acc Chem Res*, 2003; **36**:783–790.

8

Noninvasive and Label-Free Characterization of Cells for Tissue Engineering Purposes

Shunsuke Tomita

Biomedical Research Institute, National Institute of Advanced Industrial Science and Technology, and DAILAB, Tsukuba, Ibaraki, Japan

8.1 Introduction

In order to generate sufficient cell quantities for tissue engineering in the context of cell-based therapy, disease modeling, and drug discovery, *ex vivo* culture expansion of cells is usually inevitable. As long-term *ex vivo* cultivation may impose various deleterious effects on the cells, every step in the process of tissue engineering must be continuously evaluated. Conventional techniques for the characterization of cultured cells rely on genetic activity profiles, or the expression of cell-surface and intracellular markers. For instance, fluorophore-conjugated antibodies that are specific to cell-surface markers associated with target events are often used to evaluate the cells at the single or population level. However, these techniques usually involve invasive cell processing, such as staining or the extraction of specific molecules, which demands perturbation or, in some cases, sacrifice of tissue engineering constructs. Moreover, reliable surface markers are difficult to establish; uni- or bivariate approaches are insufficient for the cell assessment if the abundance of target cellular components is susceptible to unrelated cellular events. Therefore, new objective assessment approaches that are noninvasive and label-free are required in order to correctly characterize cultured cells; such approaches would reduce the cost and time necessary to monitor the cell state, and to screen the effects of stimuli on the cell behavior.

In recent years, "signature"-based cell profiling has emerged as an alternative approach, in which cultured cells are characterized based on cell

Biomedical Engineering Challenges: A Chemical Engineering Insight, First Edition.
Edited by Vincenzo Piemonte, Angelo Basile, Taichi Ito, and Luigi Marrelli.

signatures that are acquired using various analytical techniques. These signatures are rich in information on cell phenotypes, and a statistical investigation using multivariate analyses enables the characterization of cultured cells. So far, cellular signatures have been generated noninvasively by several different approaches in a label-free manner: (1) *vibrational spectroscopic features*; infrared (IR) and Raman spectroscopy are most widely used for high-throughput nondestructive signature-based cell profiling, which provides a spectral signature that reflects specific information on proteins, lipids, nucleic acids, and polysaccharides; these techniques can be used to detect biologically relevant molecules both when they are isolated or when they are incorporated in higher-order structures; (2) *morphological features*; phase-contrast microscopy is typically used to extract various morphological features of single cells (e.g., area, perimeter, length, and inner radius) or cell populations (e.g., orientation and colony shape); (3) *secreted molecule features*; small molecules or proteins that are consumed by or secreted from cultured cells reflect the function and state of the cells. The detection of a multitude of small molecules in cell culture media by mass spectrometry, or the acquisition of response patterns using cross-reactive molecules are used to generate cellular signatures. The initial purpose of this chapter is to provide a brief overview of multivariate analyses as key statistical techniques that are commonly used in signature-based cell profiling. Subsequently, the aforementioned three approaches for noninvasive and label-free characterization of cultured cells are presented using selected examples.

8.2 Multivariate Analyses

Cellular signatures can be regarded as multidimensional data containing hidden data structures that are difficult to assess via a simple visual inspection of the raw data. Multivariate analyses have therefore been essential to the understanding of the underlying changes and differences in cellular signatures associated with, for example, diseases, biological processes, or the influence of external stimuli. Multivariate analyses are usually employed for two purposes: (1) exploratory analysis and (2) sample classification. The exploratory analysis provides an understanding of specific differences related to each class in the data set. The most frequently used methods for exploratory analysis include principal component analysis (PCA) and hierarchical clustering analysis (HCA). For the classification of samples, linear discriminant analysis (LDA) is commonly used, which employs the identity of samples in a training data set to generate a model that allows the classification of unknown samples. A brief description of selected multivariate analyses is given in the following sections using representative case studies.

8.2.1 Principal Component Analysis (PCA)

PCA is an unsupervised exploratory technique aimed at describing variations present in the data set by reducing the dimensionality of the data, while retaining most of the variation within the data set. PCA uses a multidimensional data set such as spectra or morphological parameters for each sample and generates a set of orthogonal eigenvectors, the so-called principal components (PCs) that contain the coefficients of a linear combination. The degree of variance included in the individual PCs decreases according to their numerical order, that is, the first PC comprises the highest, while the second PC comprises the second highest degree of variance, and so on. Thus, the PCA concentrates the information within the data set into a lower dimensional space and ranks the new dimensions in order of importance. Each signature is plotted as one point in two- or three-dimensional linearly transformed PC space, using two or three PCs of choice. Usually, the first two to three PCs are used, as they contain the most variance, thus ensuring optimal visualization of the data spread [1]. For example, Figure 8.1(a) shows signatures of three different samples described by ten-dimensional data.

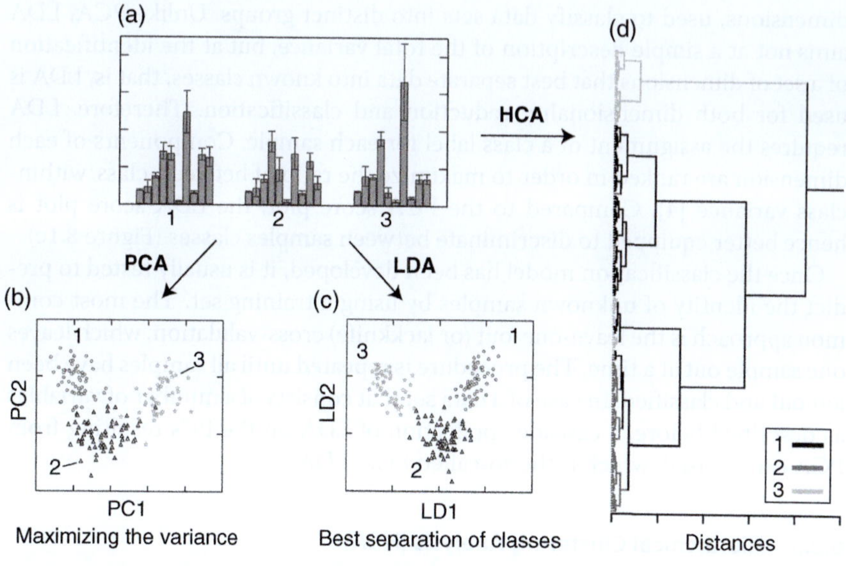

Figure 8.1 Data outputs analyzed by representative multivariate analyses. (a) Signatures of three different samples (**1–3**). In the context of, for example, spectroscopy, each variable may correspond to the intensity at each wavenumber. (b) PCA score plot. (c) LDA score plot. (d) HCA dendrogram. The datasets were reproduced with permission from the UCI machine learning repository [5]. (*Source:* Data from Lichman, M. UCI Machine Learning Repository [http://archive.ics.uci.edu/ml]. Irvine, CA: University of California, School of Information and Computer Science.)

The PCA score plot based on the first two PCs shows that the PCA identifies the directions along which the data exhibits the largest spread (Figure 8.1b). Therefore, PCA provides a straightforward method to visualize the variability of the sample set; that is, it shows the degree of similarity between the samples [2].

In the context of spectroscopy, the intensity at each wavenumber is generally used as input variables for a classification analysis. However, a prior step is usually required in order to reduce the number of variables and avoid the so-called "curse of dimensionality,"[3] which may manifest as an overfit that can lead to poor performance of classifiers when tested with independent data. It has been recommended that the number of spectra should be 5–10 times higher than the number of variables in the data set [3]. Therefore, the number of variables is often reduced by PCA before the following classification analysis, such as LDA (*vide infra*). LDA analyses typically use the first 10 PCs, as they capture in most cases >99% of the variance of the original data [4]. In contrast, PCs >10 mainly contain noise and are thus infrequently included in classification analyses.

8.2.2 Linear Discriminant Analysis (LDA)

LDA is a supervised classification technique that generates a set of orthogonal dimensions, used to classify data sets into distinct groups. Unlike PCA, LDA aims not at a simple description of the total variance, but at the identification of a set of dimensions that best separate data into known classes, that is, LDA is used for both dimensionality reduction and classification. Therefore, LDA requires the assignment of a class label for each sample. Components of each dimension are ranked in order to maximize the ratio of between-class/within-class variance [4]. Compared to the PCA score plot, the LDA score plot is hence better equipped to discriminate between samples classes (Figure 8.1c).

Once the classification model has been developed, it is usually tested to predict the identity of unknown samples by using a training set. The most common approach is the leave-one-out (or Jackknife) cross-validation, which leaves one sample out at a time. The procedure is repeated until all samples have been left out and classified. In case of a data set that consists of a number of variables as described before, a cascade application of LDA on the PCs resulting from PCA can be used, which is the so-called PCA-LDA.

8.2.3 Hierarchical Clustering Analysis (HCA)

HCA is an unsupervised exploratory technique, which determines clusters on the basis of the Euclidean distances between elements of a data set. In its simplest form, nearest-neighbor points are initially paired into one cluster, before these clusters are paired with another nearest neighbor or cluster, until all points and clusters are connected to each other. The minimum variance method (Ward's algorithm) is the most common clustering criterion used in

HCA (Figure 8.1d). The resulting dendrograms show connectivity and some measure of the distance between each of the pairs. In Figure 8.1d, many of the data points of Sample 3 could be successfully clustered, while a part was taken into the set of Sample 1 and 2. In the HCA, the connectivity pattern affords information on relationship similarity, while the distance furnishes information on the magnitude of similarity [2].

8.2.4 Other Multivariate Analyses

Several other models for exploratory analysis and classification of samples have been reported. Among these, partial least-squares (PLS) is another commonly used technique, which constructs a set of linear combinations of variables, but uses the data classes in the construction; PLS is hence a supervised technique [6]. PLS aims to find a sequence of new variables that are maximally correlated with a numerical representation of the data classes, while being orthogonal to each other. Support vector machine (SVM) has recently been recognized as a reliable method for data sets with several distinct classes that cannot be easily separated by linear boundaries [7].

8.3 Vibrational Spectroscopic Features

One of several high-throughput nondestructive techniques for the evaluation of single cells is vibrational spectroscopy, which generates spectra with hundreds of variables (e.g., the intensities at each wavenumber). The cellular sample is thus represented by very large data sets.

Cellular characteristics have been studied predominantly using two techniques: (1) Fourier transform infrared (FTIR) and (2) Raman spectroscopy [8, 9]. FTIR spectroscopy is based on mid-IR (1800–900 cm^{-1}) light that interacts with cell samples and measures the vibrational modes of the functional groups of biomolecules present in the cells. Although the vibrational motions of individual chemical bonds for all types molecules fall in the 4000–400 cm^{-1} range, the fundamental vibrational modes of biological structures are contained within the mid-IR range (Figure 8.2a) [9, 10].

Raman spectroscopy detects a wide range of chemical bonds, and is, in contrast to FTIR spectroscopy, not restricted to those with a dipole moment. The two techniques are therefore complementary, as weak bands in the FTIR spectrum often correspond to strong bands in the Raman spectrum (Figure 8.2b) [11]. The Raman effect is an inelastic light scattering process of photons following the interaction between a monochromatic light source and the molecules constituting the cell sample. During this interaction, energy of the photons is transferred to the molecules in the form of vibrational energy (Stokes scattering), that is, the energy loss of the scattered photon corresponds to the

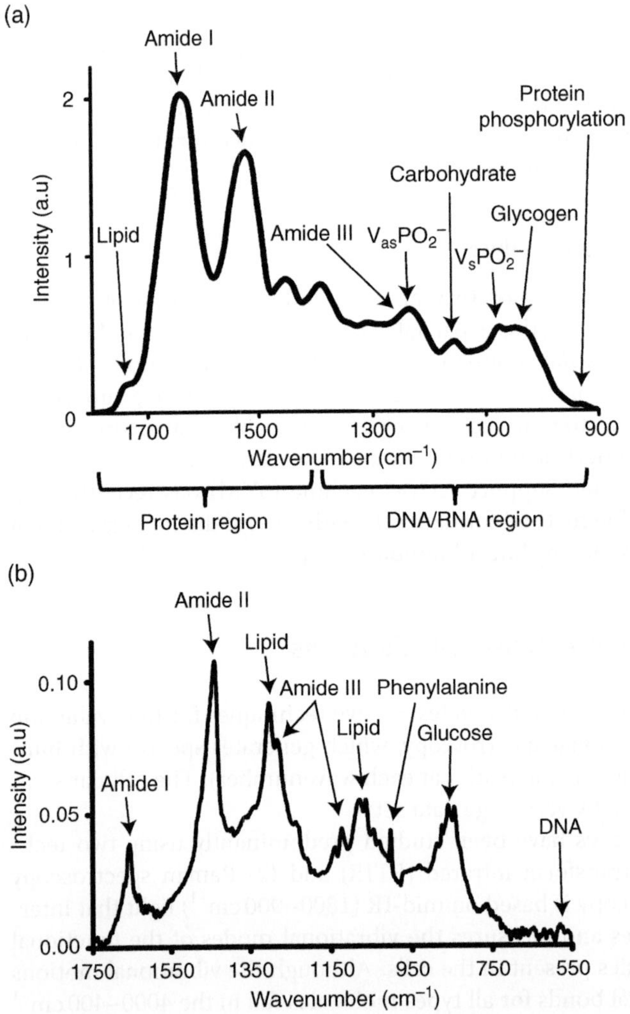

Figure 8.2 Examples of (a) IR and (b) Raman spectra with a tentative peak assignment of biochemically relevant compounds and moieties. (*Source:* [8] Kelly, J.G.; Trevisan, J.; Scott, A.D.; Carmichael, P.L.; Pollock, H.M.; Martin-Hirsch, P.L.; Martin, F.L. *J. Proteome. Res.* **2011**, 10, 1437–1448. Copyright 2017 American Chemical Society.)

vibrational energy level of the molecule, which results in a frequency shift. This shift is known as inelastic or Raman scattering, and, considering that this occurs very rarely [9] Raman spectroscopy generally shows low sensitivity for biological samples. It should be noted however that Raman spectroscopy is well-suited for the analysis of living cells, as it can be conducted using light of longer wavelength (633 or 785 nm), which exhibits low phototoxicity [12].

Moreover, the contribution of water to the Raman spectrum is limited, as the scattering efficiency of water is low [13], which allows analyzing living cells in aqueous media. Given these particular characteristics, the following sections focus on Raman spectroscopy.

8.3.1 Cell Characterization Based on Whole-Cell Analysis by Raman Spectroscopy

Raman spectroscopy has attracted considerable attention over the last decade in the context of tissue engineering (Table 8.1), as this noninvasive and label-free tool finds applications in cellular or subcellular analyses, and can thus be used for the discrimination of simple cell types and stem cells, both for cellular infection as well as for diagnostic purposes.

For instance, Figure 8.3a shows the averaged Raman spectra recorded from human embryonic stem cells (hESCs) and those differentiated into cardiomyocytes (hESC-CMs), as well as human fetal left ventricular cardiomyocytes (FLV-CM) [15]. These spectra do not differ substantially in appearance, but multivariate analyses of the spectra allowed identifying combinations of multiple spectral features that are characteristic for each cell type. Initially, a PCA was carried out. The first six PCs afforded by PCA were subsequently used as input for an LDA. The LDA score plot showed a clear separation between hESCs and FLV-CMs (Figure 8.3b). The hESC-CMs are located between the hESCs and FLV-CMs with partial overlap, but a leave-one-out cross-validation estimated that ~70% of the hESC-CMs can be classified as FLV-CMs. Thus, Raman spectroscopy-based cell profiling offers a noninvasive and label-free approach for the separation of undifferentiated hESCs and their cardiac derivatives.

Similar approaches have also enabled the discrimination between hESCs and their spontaneously differentiated progeny [12], normal and abnormally transformed hMSCs [20], and undifferentiated/differentiated malignant cells [33]. Although analyzed cells were fixed by paraformaldehyde and glutaraldehyde, Ilin *et al.* have recently shown that the differentiation hierarchy of mice hematopoietic stem cells (HSCs) (Figure 8.4a) can be identified by the multivariate analysis of Raman spectra (Figure 8.4b) [30]. Raman spectra from long-term repopulating HSCs (LT-HSCs), short-term repopulating HSCs (ST-HSCs), fully differentiated lymphoid (B cells), and myeloid (granulocytes) cells were used to construct a partial least-squares discriminant analysis (PLS-DA) model. This model was subsequently applied to a test set of spectra acquired from different hematopoietic cells (Figure 8.4c). The cells were classified based on the thresholds for each of the hematopoietic cells (gray dashed lines in Figure 8.4c), affording almost complete classification accuracy. This technique should be valuable, due to the HSCs' ability to self-renew or differentiate into mature blood and immune cell lineages, which potentially offers new clinical approaches to treat blood and bone marrow diseases.

Table 8.1 Representative examples for the characterization of cultured cells using Raman spectroscopy.

Year	Instrumentation	Cell state	Cell types analyzed	Multivariate analyses	Ref.
2006	LTRS	Live	Normal and cancer human hematopoietic cells	PCA	14
2009	RS	Live	Cardiomyocytes-differentiated and undifferentiated human ESC, human cardiomyocytes	PCA, LDA	15
2010	LTRS	Fixed	Normal and cancer human hematopoietic cells, human breast carcinoma cells	SVM	16
2010	RS	Fixed	Normal and cancer human hematopoietic cells, human breast carcinoma cells	SVM	17
2010	RS	Live	Differentiated and undifferentiated human ESC	PCA	12
2011	RS	Live	Normal and immortalized human skin cells	PCA, SVM	18
2011	LTRS	Fixed	Normal and cancer human hematopoietic cells, human breast carcinoma cells	PCA, LDA	19
2012	RS	Live	Normal and transformed human MSCs	PCA, LDA	20
2012	RS	—	Colon tissue samples	PCA, KM, SVM	21
2013	RS	Live	Murine fibroblast cells with/without viral infection	PCA, LDA	22
2013	RS	Live	Human Cancer cell lines, including subcellular analysis	PCA, LDA	23
2013	RS	—	Lymphatic tissue samples	HCA, PCA, VCA	24
2014	RS	Live	Normal and transfected mammalian cells	PLS-DA	25
2014	RS	Live	Proliferating and non-proliferating human liver cancer cells, including subcellular analysis	PCA, HCA, SVM	26
2014	BCARS	Fixed	Differentiated and undifferentiated human MSCs, including subcellular analysis	PCA	27

Table 8.1 (Continued)

Year	Instrumentation	Cell state	Cell types analyzed	Multivariate analyses	Ref.
2014	RS	—	Mouse hair tissue samples	PCA, LDA	28
2015	RS	Fixed	Normal and cancer human lung cells, including subcellular analysis	PCA, LDA	29
2015	RS	Fixed	Differentiated and undifferentiated mice HSCs	PCA, PLS-DA	30
2015	RS	Fixed	Normal and cancer human breast cells	PCA, LDA	31
2015	RS	Live	Human normal cells with/without bacterial infection	PCA, LDA	32
2015	RS	Live	Cancer cells and neural stem cells from rat	PCA	33
2016	RS	Live	Human cancer cell lines	PCA, SVM	34
2016	LTRS	Fixed	Human cancer cell lines	PCA, LDA	35

RS: Raman spectroscopy; LTRS: laser tweezers Raman spectroscopy; BCARS: broadband coherent anti-Stokes Raman scattering; ESC: embryonic stem cell; MSC: mesenchymal stem cell; HSC: hematopoietic stem cell; PCA: principal component analysis; LDA: linear discriminant analysis; SVM: support vector machine; KM: *k*-means-cluster analysis; VCA: vertex component analysis; PLS-DA: partial least-squares discriminant analysis.

8.3.2 Cell Characterization Based on Subcellular Analysis by Raman Spectroscopy

Probing entire cells with Raman spectroscopy allows distinguishing between different cell lines with high specificity and sensitivity. The high spatial resolution from, for example, a narrow and focused laser beam permits investigating the basis of the differences in terms of the subcellular regions within a single cell.

Farhane *et al.* have acquired Raman spectra of three subcellular compartments (cytoplasm, nucleus, and nucleolus), and subsequently used these spectra to compare the spectral profiles between normal (BEAS-2B) and cancerous cell lines (A549 and Calu-1) [29]. A PCA-LDA analysis of the three subcellular compartments revealed that the highest sensitivity and specificity for the discrimination of normal and cancer cell lines was achieved when the spectra from the nucleolar region were used, followed by those of the nucleus region. The authors argue that the effectiveness of Raman spectroscopy on the nucleolus region should probably be attributed to the fact that the size, number, and organization of nucleoli are cell specific, and that nuclear

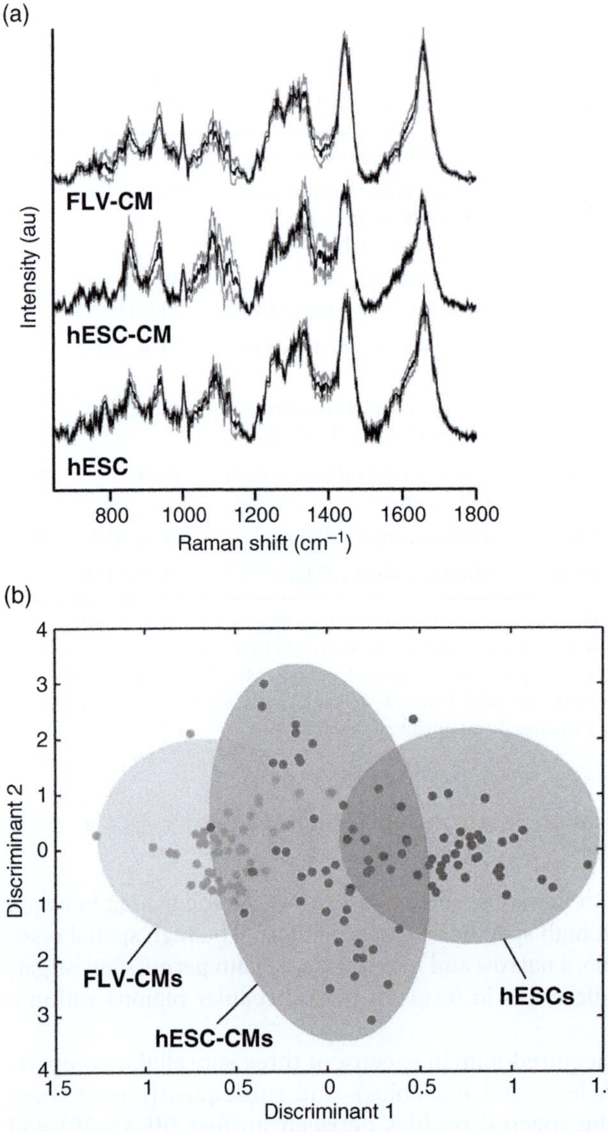

Figure 8.3 Identification of cardiac differentiation of ESCs by Raman spectroscopy. (a) Mean Raman spectra for hESC, hESC-CM, and FLV-CM cells. Black lines indicate the average spectra, whereas gray lines indicate the standard deviation from the average spectra. (b) LDA plot for hESC, hESC-CM, and FLV-CM cells. Shaded ellipsoids superimposed on the data points serve as visual guidelines. (*Source:* [15] Chan, J.W.; Lieu, D.K.; Huser, T.; Li, R.A. *Anal. Chem.* **2009**, 81, 1324–1331. Copyright 2017 American Chemical Society.)

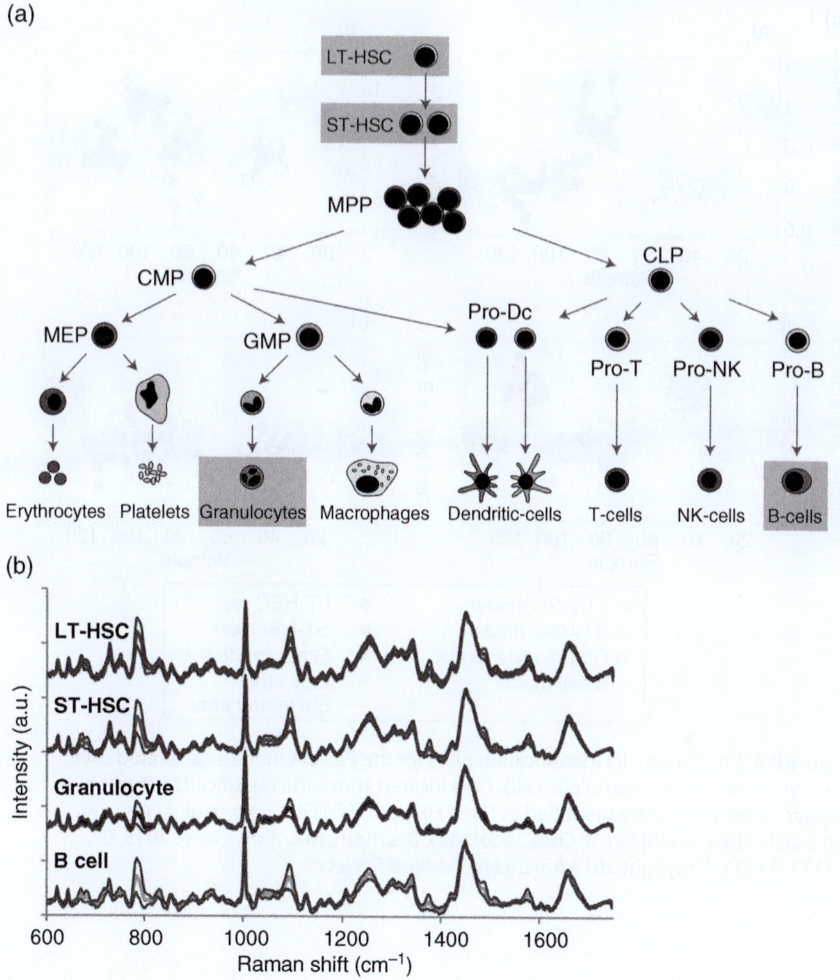

Figure 8.4 Identification of the HSC differentiation hierarchy. (a) Diagram of hematopoiesis [36]. Populations analyzed in this study are highlighted in gray. (b) Average Raman spectra of hematopoietic cells. Gray lines indicate the standard deviation from the average spectra.

proteins play an important role in the nucleoli functions and stability [37]. A different study using cancerous (Calu-1) and normal (epithelial NL20 and fibroblast 35FL) cells from human lung tissue led to a similar conclusion: the discrimination based on the nucleus region should be better than that based on the cytoplasm [23].

However, in contrast to these examples, adipocyte-differentiated, osteoblast-differentiated, and undifferentiated MSCs were not separated well on the PCA

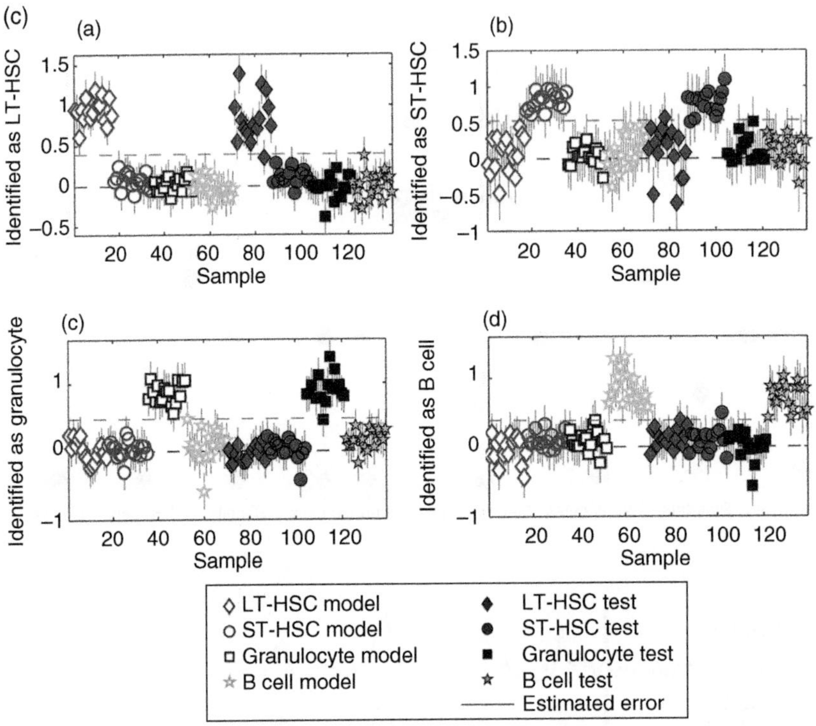

Figure 8.4 (Continued) (c) Identification plots for the PLS-DA models generated using Raman spectra of hematopoietic cells. Cells located above the classification threshold (upper dashed line) were identified as (a) LT-HSCs, (b) ST-HSCs, (c) granulocytes, and (d) B cells. (*Source:* [30] Ilin, Y.; Choi, J.S.; Harley, B.A.; Kraft, M.L. *Anal. Chem.* **2015**, 87, 11317–11324. Copyright 2017 American Chemical Society.)

score plot that was generated from the spectra of the nucleus region compared to those obtained from the cytosol region [27]. In the case of liver cancer cell lines (HepG2) in different cellular growth stages, the SVM classification model based on the Raman spectra of lipid droplet organelles showed the best classification relative to other subcellular regions, including the nucleus and cytoplasm [26]. The authors tentatively assigned the differences in the spectral characteristics of the lipids between these cells to a higher content of fatty acids in the proliferating phase of HepG2 relative to that in the plateau phase [38]. These studies suggest that a judicious selection of the measured region is critical for the accuracy of Raman-based cell profiling. Further investigations are thus required in order to gain an improved understanding of the Raman specificity of targeted subcellular regions.

8.3.3 Raman-Based Cell Characterization Toward Biomedical Applications

In terms of biomedical applications, Raman spectroscopic features have been used to investigate intracellular infections. Understanding the intracellular infection of bacteria and viruses is highly important in order to develop effective preventive and therapeutic strategies. In this context, it is desirable to noninvasively characterize intracellular invaders in their host cell environment or resultant infected cells in order to continuously monitor the infection processes.

Große *et al.* have used Raman spectroscopy to identify the localization of infectious bacteria within human host cells, and to provide information on the metabolic state of bacteria in a noninvasive and label-free manner [32]. After endothelial cells (EA.hy926) had been incubated with *Staphylococcus aureus* for sufficient infection, extracellular bacteria were removed by enzymatic treatment. Subsequently, Raman spectra of the bacteria were extracted from Raman image scans of infected cells at different points in time after the infection (Figure 8.5a). A classification model based on PCA-LDA allowed a classification into intracellular (3h and 9h) or extracellular (0h: before the bacteria enter the host cells, and 24h: after the endothelial cells burst and release the bacteria) bacteria with high identification accuracy (Figure 8.5b). The good separation between intra- and extracellular bacteria in the LDA score plot suggested that the Raman spectra contained the significant biochemical changes between the two different states. The authors tentatively assigned the observed differences to the altering bacterial growth states posterior to the infection. As most available methods for the detection of bacteria depend on destructive procedures that include a separation of the bacteria from the host cell, genetic modifications, or specific labeling, Raman spectroscopic features should be highly useful for the examination of bacterial infection processes. In another example, murine fibroblast cell lines (NIH/3T3) with/without transformations induced by murine sarcoma virus (MuSV), which has been implicated in various types of human and animal tumors, could be discriminated using Raman spectroscopy in combination with a PCA-LDA analysis [22].

The recording of hemograms is recognized as an important routine analysis in hospitals to assist medical diagnosis because the number, size, and morphology of the blood leukocytes in the body often reflect the patient's response to a disease. Raman spectroscopy has been applied to separate the two most abundant leukocytes; that is, neutrophils and $CD4^+$ T-lymphocytes (Figure 8.6) [39]. Based on the spectral differences present in the Raman spectra (Figure 8.6a), two well-separated clusters were observed as shown in the dendrogram (Figure 8.6b) and the PCA score plot (Figure 8.6c). In this article, the authors suggest that this demonstration should pave the way to automated analyses of spectroscopic hemograms.

Raman spectroscopy has also been used to detect different cells in more natural and complex environments. Pudlas *et al.* have employed Raman spectroscopy to

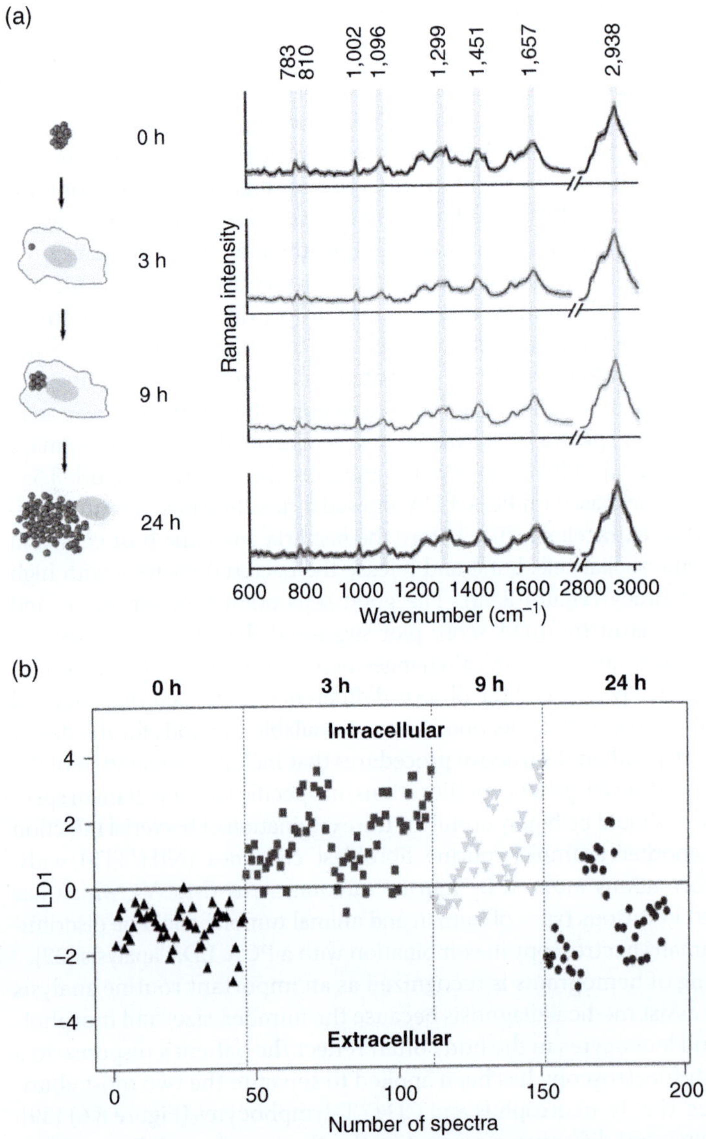

Figure 8.5 Raman spectroscopic analysis of different infection states of *Staphylococcus aureus* in endothelial cells. (a) Schematics of the infection process and corresponding average Raman spectra with standard deviations at different points in time. (b) PCA-LDA score plot. A data set was well separated by the first linear discriminant scores. (*Source:* [32] Grosse, C.; Bergner, N.; Dellith, J.; Heller, R.; Bauer, M.; Mellmann, A.; Popp, J.; Neugebauer, U. *Anal. Chem.* **2015**, 87, 2137–2142. Copyright 2017 American Chemical Society.)

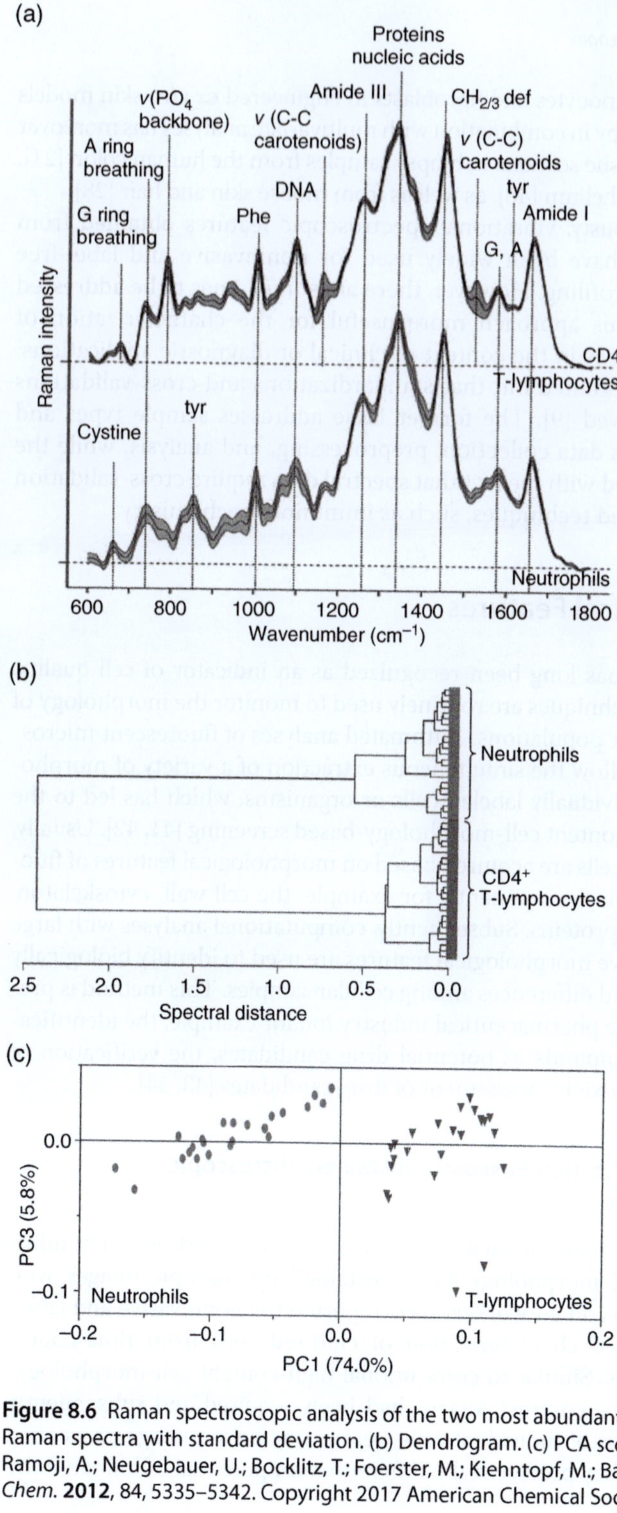

Figure 8.6 Raman spectroscopic analysis of the two most abundant leukocytes. (a) Average Raman spectra with standard deviation. (b) Dendrogram. (c) PCA score plot. (*Source:* [39] Ramoji, A.; Neugebauer, U.; Bocklitz, T.; Foerster, M.; Kiehntopf, M.; Bauer, M.; Popp, J. *Anal. Chem.* **2012**, 84, 5335–5342. Copyright 2017 American Chemical Society.)

discriminate vital keratinocytes and fibroblasts in engineered *ex vivo* skin models [18]. Raman spectroscopy in combination with multivariate analyses has moreover been used to classify tissue sections in biopsy samples from the human colon [21], the human cervical epithelium [40], as well as from mouse skin and hair [28].

As described previously, vibrational spectroscopic features obtained from Raman spectroscopy have been widely used for noninvasive and label-free signature-based cell profiling. However, there are some issues to be addressed in order to render this approach more useful for the characterization of cultured cells and tissues in the context of clinical or diagnostic applications. Sulé-Suso *et al.* have pointed out that standardizations and cross-validations still remain to be solved [9]. The former issue addresses sample types and preparation, as well as data collection, preprocessing, and analysis, while the latter issue is concerned with the fact that spectral data require cross-validation with already established techniques, such as immunohistochemistry.

8.4 Morphological Features

Cellular morphology has long been recognized as an indicator of cell quality. Today, microscopic techniques are routinely used to monitor the morphology of cultured cells and their populations. Automated analyses of fluorescent microscopy images thereby allow the simultaneous extraction of a variety of morphological features of individually labeled cells or organisms, which has led to the development of high-content cell-morphology-based screening [41, 42]. Usually, fluorescent images of cells are acquired based on morphological features of fluorescently stained cellular components, for example, the cell wall, cytoskeleton, organelles, or specific proteins. Subsequently, computational analyses with large amounts of quantitative morphological features are used to identify biologically relevant similarities and differences among cellular samples. This method is predominantly used in the pharmaceutical industry for, for example, the identification of new lead compounds as potential drug candidates, the verification of drug targets, and the toxicity assessment of drug candidates [43, 44].

8.4.1 Cell Characterization Based on Unstained Microscopic Images of Single Cells

Recent advances in computational analysis have enabled the extraction of information regarding cell morphology from unstained microscopic images with unclear outlines. Kato and co-workers have reported the noninvasive and label-free morphology-based characterization of cultured cells from time-course phase-contrast images. Similar to conventional high-content cell-morphology-based screening, phase-contrast images had been acquired and subsequently morphological features of individual cells were extracted via image processing (Table 8.2). Initially, they predicted the differentiation potential of human bone

Table 8.2 Representative definitions of morphology features.

Original parameter	Example	Description
Total area		Area of object including any holes present, regardless of hole filling
Area	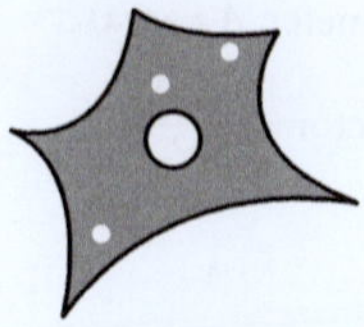	Area of the object without hole area
Perimeter	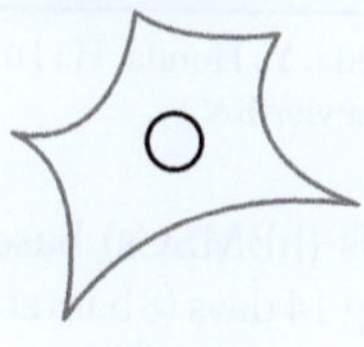	Perimeter of the object measuring from the mid-points of each pixels around the border
Length	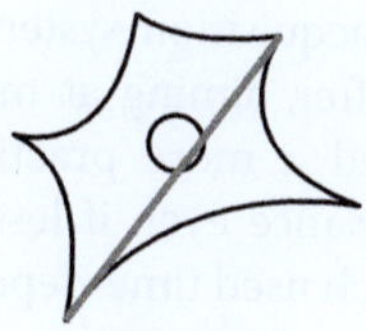	Length of the longest chord through the object
Breadth	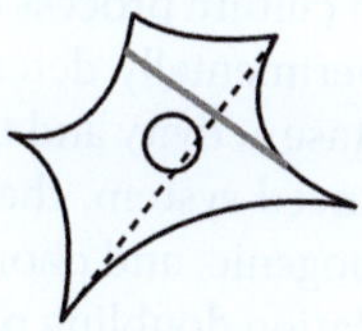	Caliper width of the object perpendicular to the longest chord
Inner radius	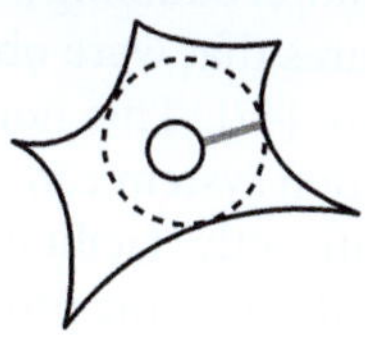	Distance from the centroid to the nearest point along the edge of the object

(Continued)

Table 8.2 (Continued)

Composite parameter	Example	Description
Elliptical form factor	$\text{Elliptical form factor} = \frac{Length}{Breadth}$	The ratio between the length and breadth of an object
Fiber breadth	$\text{Fiber breadth} = \frac{P+\sqrt{P^2-16A}}{4}$ (P = perimeter, A = area)	Width of the object, assuming that it is a fiber
Shape factor	$\text{Shape factor} = \frac{4\neq A}{P^2}$	Value from 0 to 1, which quantifies the resemblance of an object with a circle (circle = 1)

(*Source:* [48] Sasaki, H.; Enomoto, J.; Ikeda, Y.; Honda, H.; Fukuda, J.; Kato, R. *J. Biosci. Bioeng.* **2016**, 121, 117–123. Copyright 2017 Elsevier B.V.)

marrow mesenchymal stem cells (hBMSCs) based on a large collected data set consisting of 9,990 images covering 14 days (8 h intervals), which had been acquired using a fully automated image acquisition system with a multivariate analysis (Ridge regression) [45]. Thereafter, aiming at more user-friendly manual cell characterization, they developed a more practical and efficient model that assures high prediction performance even if less time-course images are collected [46]. In these studies, which used time-dependent morphological features for the osteogenic differentiation culture process of hBMSCs, prediction results were highly correlated with experimentally determined differentiation marker values, such as alkaline phosphatase activity and the calcium deposition rate.

Moreover, using a more advanced system, the (1) multi-lineage differentiation potentials (osteogenic, adipogenic, and chondrogenic) after one month of differentiation and the (2) population doubling potential were predicted on the basis of the morphological features that were obtained during the first 4 days of the hBMSC expansion culture [47]. This noninvasive and label-free morphology-based cell characterization system can also be integrated with microdevices [48], which may significantly facilitate the usually laborious and time-consuming high-throughput screening processes.

8.4.2 Cell Characterization Based on Unstained Microscopic Images of Cell Populations

The examples in the preceding section are based on morphological features of individual cells. However, cells are usually cultured to a state close to confluence on account of the requirement for large amounts of cells in most applications. When cells are in a confluent state, intercellular boundaries are more ambiguous and it is usually impossible to achieve separation of individual cells from unstained phase-contrast images.

Sasaki *et al.* have recently developed an alternative image-processing algorithm for profiling the unlabeled confluent state of cells according to the orientation of cell populations and their heterogeneity [49]. The authors used this image analysis algorithm in combination with SVM to construct three clinically important cell quality control models: (1) remaining lifespan check, (2) manipulation error check, and (3) differentiation potential check.

Colony morphology is the one of the criterions that is used for estimating the health of human pluripotent stem cells (hPSCs), such as hESCs and human induced pluripotent stem cells (hiPSCs). For example, healthy undifferentiated hPSCs typically exhibit a tight, homogeneous texture with a clear circular border [50]. Noninvasive image-based characterization of morphological colony features has recently been proposed to overcome problems associated with the low levels of the quantitative manual assessment of the colony morphology.

Tokunaga *et al.* have constructed living and unlabeled colony image libraries of various hiPSCs for supervised multivariate analysis in order to accurately distinguish between *bona fide* hiPSCs from improperly reprogrammed cells [51]. In this approach, phase-contrast images of live colonies formed by different hiPSC lines (201B7, 253G1, 1H, and 2H), non-hiPSC lines (15B2), and somatic cells (HMECs) had initially been collected (Figure 8.7a). The entire colony images without a prior segmentation of cells were subsequently analyzed by the supervised "wndchrm" (*w*eighted *n*eighbor *d*istances using a *c*ompound *h*ierarchy of algorithms *r*epresenting *m*orphology) algorithm [52]. Figure 8.7b shows the resultant classification accuracy that reflects the degree of morphological dissimilarities relative to completely reprogrammed hiPSC (1H cells). The classifier shows higher levels of accurate cross-validations (maximum classification accuracy = 1.0) when the morphologies of two colony types are substantially different, while two cell types with no feature differences are expected to afford a lower value (0.5). Although the classification accuracy of hiPSC lines (2H, 201B7, 253G1) was ~0.6, the values for non-hiPSCs (15B2 and HMECs) were significantly higher, suggesting a successful discrimination between hiPSCs and non-hiPSCs.

Furthermore, a computational analysis of the colony morphology has been used to distinguish healthy hiPSCs (201B7, 253G1) from their aberrant subclones (201B7–1A, 253-G1-B1), which contain an additional copy of chromosome 12 [53]. Not only was it possible to correctly classify the two hiPSCs and their subclones based on the constructed model, but interestingly, the statistical colony classification based on colony morphology was correlated with that based on global gene-expression profiles. This result indicates that the morphological features of hiPSC colonies are closely associated with phenotypic characteristics of individual cells.

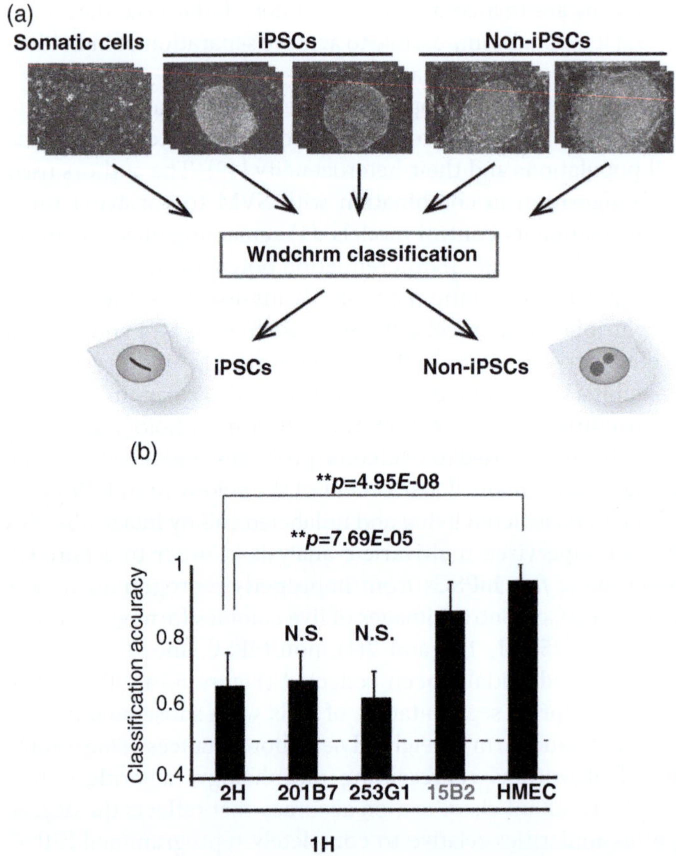

Figure 8.7 Quantitative classification of completely and incompletely reprogrammed hiPSC colonies. (a) Experimental overview. (b) Binary classification of colony images against completely reprogrammed hiPSCs (1H cells). Values represent mean values with standard deviations from 10 independent tests. N.S. = not significant. (*Source:* [51] Naka et al. https://www.ncbi.nlm.nih.gov/pmc/articles/PMC4227026/?report=classic -CC BY 4.0. Tokunaga, K.; Saitoh, N.; Goldberg, I.G.; Sakamoto, C.; Yasuda, Y.; Yoshida, Y.; Yamanaka, S.; Nakao, M. *Sci. Rep.* **2014**, 4, 6996. Copyright 2017 Nature Publishing Group.)

In most of the aforementioned cases, the cell quality could be predicted with high levels of accuracy. By using a large number of morphological features, which were in some cases obtained over the course of several days, this was even possible when using low-quality non-stained cell images. Unlike conventional morphology-based analyses of cell images, label-free methods do not require invasive procedures, such as fluorescent labeling or genetic manipulation, and the evaluated cells can hence be used for subsequent purposes.

8.5 Secreted Molecule Features

The previous sections described noninvasive and label-free signature-based cell profiling using microscopic techniques. However, these approaches still require specialized hardware in order to achieve fast, reproducible, and automated spectrum or image acquisition, as well as software for sophisticated and reliable spectra or image analysis. The analysis of secreted molecule features represents an alternative to achieve a noninvasive and label-free cell characterization. As metabolites are involved in many intracellular functions, they provide important information on the physiological state of the cell. Cell culture metabolomics based on nuclear magnetic resonance spectroscopy or mass spectrometry can be used as a noninvasive and label-free tool for cell characterization, which constitutes a holistic method to analyze the metabolism [54].

Shaham *et al.* have reported the metabolite profiling of cell culture media (the so-called "spent" media) from 60 different cancer cell lines by liquid chromatography-mass spectrometry (LC-MS). This study revealed not only a high correlation between rapid cancer cell proliferation and glycine consumption/synthesis, but also clustering of leukemia cells among other tumor cell lines including, for example, melanoma or colon and renal cancer cells [55].

8.5.1 Cell Characterization Based on Response Signatures

Traditionally, secreted proteins, such as albumin, α-fetoprotein, and transferrin for the hepatocyte function, are used as noninvasive markers for characterizing cultured cells. In recent years, interest has increased in the entire set of secreted proteins, the so-called "secretome," [56] as they reflect cellular characteristics. In particular, the secretome has been recognized as a rich source for the discovery of cancer biomarkers [57] and autocrine/paracrine factors applicable in regenerative medicine [58].

Very recently, a novel approach has been developed for a noninvasive signature-based cell profiling system that uses features of secreted protein responses [59]. This approach employs a cross-reactive sensor array technology, which is based on the pattern recognition of sample signatures for individual analytes obtained from "cross-reactive," rather than specific, interactions between analytes and a set of receptor molecules [2]. For example, Tomita *et al.* have developed sensor arrays that consist of cross-reactive complexes between enzymes and block-copolymers that consist of polyethylene glycol (PEG) and ionic polymers [60, 61] for the identification of cell types and states (Figure 8.8) [59]. This sensing strategy is based on the reversible, electrostatic-driven formation of complexes, which is usually accompanied by a decrease in enzymatic activity (Figure 8.8a). Different complexes thereby exhibit different affinities toward secreted proteins in cell culture media, and accordingly, incubation of the

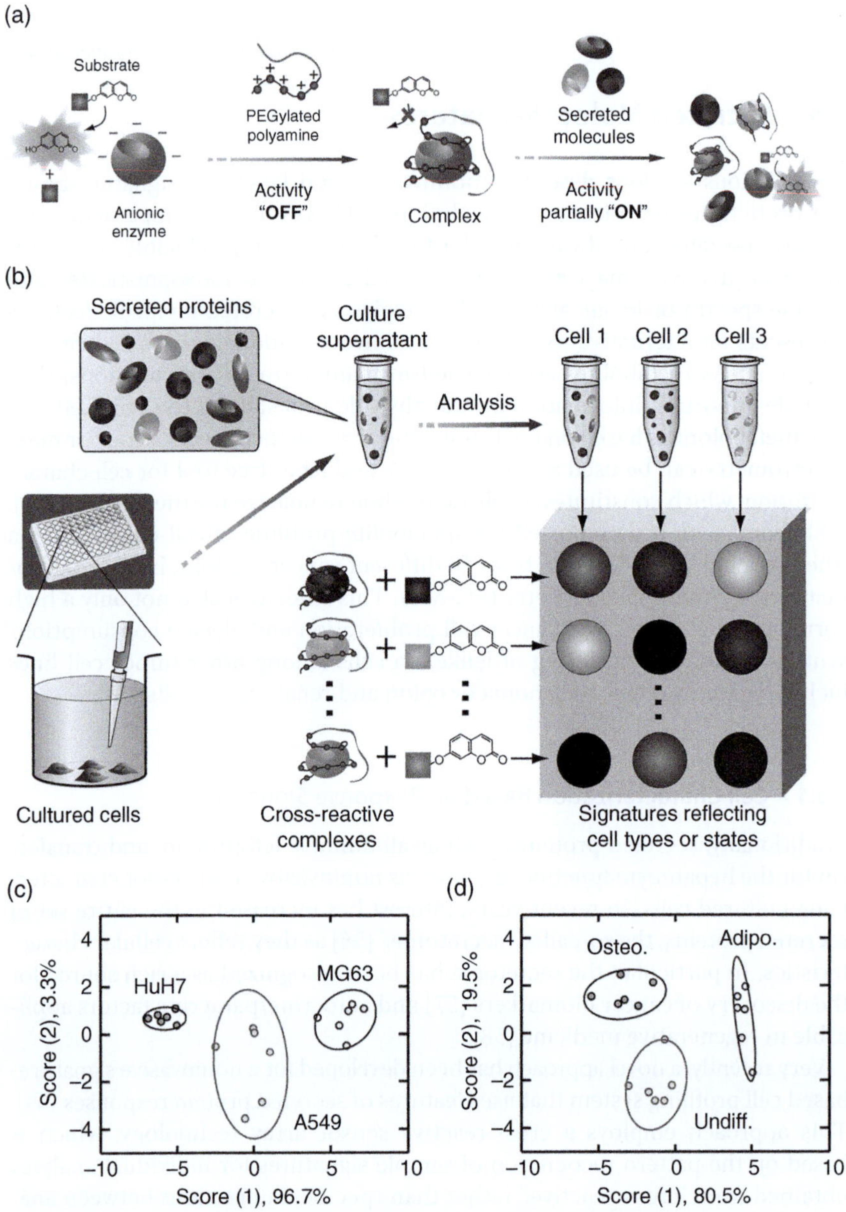

Figure 8.8 Sensor array system for the label-free and noninvasive identification of cell types and lineages. (a) Decrease in the catalytic activity of anionic enzymes through reversible complexation with PEGylated polyamines, and the subsequent partial recovery of activity through competitive interaction with secreted molecules. (b) Culture supernatants collected from cells were mixed with cross-reactive complexes to generate activity patterns reflecting the secretomic signatures for given cell types or lineages. (c) Human cancer cell lines, and (d) differentiated (osteoblast and adipocyte) and undifferentiated hMSCs were analyzed by LDA, and the corresponding LDA plots are shown. Ellipses represent confidence intervals (±1 S.D.) for the individual target cells. (*Source:* [59] Tomita et al. http://pubs.rsc.org/-/content/articlehtml/2015/sc/c5sc01259g-CC BY 3.0. Tomita, S.; Sakao, M.; Kurita, R.; Niwa, O.; Yoshimoto, K. *Chem. Sci.* **2015**, 6, 5831–5836. With permission of The Royal Society of Chemistry.)

culture media with complexes causes enzyme release via competitive interactions (Figure 8.8a). Consequently, it should be possible to generate unique signatures for the recovery of enzyme activity for individual cell types or lineages (Figure 8.8b). A remarkable feature of this approach is that although it is unknown how the complexes interact with secreted proteins, responses reflecting cellular characteristics can be generated. The combined use of six different cross-reactive complexes and LDA allowed not only a discrimination of three different cancer cell lines (A549, MG63, and HuH7) (Figure 8.8c), but also a lineage identification of differentiated hMSCs (Figure 8.8d). This approach thus provides an effective way to characterize cultured cells with common laboratory equipment, which does not require specialized knowledge. Furthermore, unlike specific types of signature-based cell profiling that rely on vibrational spectroscopic and morphological features, this approach does not need to examine the cell itself. Hence, it should be possible to evaluate cells outside the culture environment, which would be compatible with microdevice-based cell culture systems.

8.6 Conclusion and Outlook

As most available cell evaluation methods require staining or extraction of specific molecules, an evaluation of cells is, in many cases, only possible at the end of the cultivation procedure. Moreover, marker molecules that are specific only to the cellular state of interest have so far been only rarely identified, and many methods are based on the detection of one to several marker molecules, which hampers the accurate recognition of cell states that differ slightly with respect to each other. Accordingly, methods for the effective evaluation of cells during culture remain presently scarce. This situation represents a major obstacle in the progress of tissue engineering that aims at cell-based therapy, disease modeling, and drug discovery. As introduced in this chapter, noninvasive and label-free signature-based cell profiling techniques provide a unique opportunity to target an entire cell, as the signature contains information from numerous cellular components.

However, there are still challenges to be overcome before such techniques may find widespread practical use. For example, multivariate analysis is commonly used to extract information associated with a target cell state from the acquired signatures in order to construct a classification model for the identification of cell types or states. The common multivariate analysis techniques that are currently used (e.g., PCA, LDA, and HCA) are relatively basic. In order to unmask the subtle signature differences that denote small biochemical differences between cells, it would be necessary to develop more advanced and sophisticated multivariate analyses tools. Therefore, such computational techniques will become increasingly relevant for research in this context. Furthermore, it is important to validate the reproducibility of the data, since clinical applications usually impose rigorous demands. It is therefore necessary that research is carried out on a sufficiently large scale by different researchers. Nevertheless, it can be expected that ongoing efforts in this research area will provide an approach that may be able to satisfy the requirements of the cell quality assessment method for tissue engineering.

List of Acronyms

BCARS	broadband coherent anti-Stokes Raman scattering
BMSC	bone marrow mesenchymal stem cell
ESC	embryonic stem cell
ESC-CM	embryonic stem cells differentiated into cardiomyocytes
FLV-CM	fetal left ventricular cardiomyocytes
FTIR	Fourier transform infrared
HCA	hierarchical clustering analysis
HSC	hematopoietic stem cell
iPSC	induced pluripotent stem cell
IR	infrared
KM	*k*-means-cluster analysis
LC-MS	liquid chromatography-mass spectrometry
LDA	linear discriminant analysis
LT-HSC	long-term repopulating hematopoietic stem cell
LTRS	laser tweezers Raman spectroscopy
MSC	mesenchymal stem cell
PC	principal component
PCA	principal component analysis
PEG	polyethylene glycol
PLS	partial least-squares
PLS-DA	partial least-squares discriminant analysis
PSC	pluripotent stem cell
RS	Raman spectroscopy
ST-HSC	short-term repopulating hematopoietic stem cell
SVM	support vector machine
VCA	vertex component analysis

References

1 Korenium, T.; Laurikkala, J.; Juhola, M. On principal component analysis, cosine and Euclidean measures in information retrieval. *Inform. Sci.* **2007**, 177, 4893–4905.

2 Askim, J.R.; Mahmoudi, M.; Suslick, K.S. Optical sensor arrays for chemical sensing: The optoelectronic nose. *Chem. Soc. Rev.* **2013**, 42, 8649–8682.

3 Somorjai, R.L.; Dolenko, B.; Baumgartner, R. Class prediction and discovery using gene microarray and proteomics mass spectroscopy data: Curses, caveats, cautions. *Bioinformatics* **2003**, 19, 1484–1491.
4 Kemsley, E.K. Discriminant analysis of high-dimensional data: A comparison of principal components analysis and partial least squares data reduction methods. *Chemom. Intell. Lab. Syst.* **1996**, 33, 47–61.
5 Lichman, M. UCI Machine Learning Repository [http://archive.ics.uci.edu/ml]. Irvine, CA: University of California, School of Information and Computer Science.
6 Jurs, P.C.; Bakken, G.A.; McClelland, H.E. Computational methods for the analysis of chemical sensor array data from volatile analytes. *Chem. Rev.* **2000**, 100, 2649–2678.
7 Anzenbacher, P. Jr; Lubal, P.; Bucek, P.; Palacios, M.A.; Kozelkova, M.E. A practical approach to optical cross-reactive sensor arrays. *Chem. Soc. Rev.* **2010**, 39, 3954–3979.
8 Kelly, J.G.; Trevisan, J.; Scott, A.D.; Carmichael, P.L.; Pollock, H.M.; Martin-Hirsch, P.L.; Martin, F.L. Biospectroscopy to metabolically profile biomolecular structure: A multistage approach linking computational analysis with biomarkers. *J. Proteome. Res.* **2011**, 10, 1437–1448.
9 Sulé-Suso, J.; Forsyth, N.R.; Untereiner, V.; Sockalingum, G.D. Vibrational spectroscopy in stem cell characterisation: Is there a niche? *Trends. Biotechnol.* **2014**, 32, 254–262.
10 Martin, F.L.; Kelly, J.G.; Llabjani, V.; Martin-Hirsch, P.L.; Patel, I.I.; Trevisan, J.; et al. Distinguishing cell types or populations based on the computational analysis of their infrared spectra. *Nat. Protoc.* **2010**, 5, 1748–1760.
11 Levin, I.W.; Bhargava, R. Fourier transform infrared vibrational spectroscopic imaging: Integrating microscopy and molecular recognition. *Annu. Rev. Phys. Chem.* **2005**, 56, 429–474.
12 Schulze, H.G.; Konorov, S.O.; Caron, N.J.; Piret, J.M.; Blades, M.W.; Turner, R.F. Assessing differentiation status of human embryonic stem cells noninvasively using Raman microspectroscopy. *Anal. Chem.* **2010**, 82, 5020–5027.
13 Dorney, J.; Bonnier, F.; Garcia, A.; Casey, A.; Chambers, G.; Byrne, H.J. Identifying and localizing intracellular nanoparticles using Raman spectroscopy. *Analyst* **2012**, 137, 1111–1119.
14 Chan, J.W.; Taylor, D.S.; Zwerdling, T.; Lane, S.M.; Ihara, K.; Huser, T. Micro-Raman spectroscopy detects individual neoplastic and normal hematopoietic cells. *Biophys. J.* **2006**, 90, 648–656.
15 Chan, J.W.; Lieu, D.K.; Huser, T.; Li, R.A. Label-free separation of human embryonic stem cells and their cardiac derivatives using Raman spectroscopy. *Anal. Chem.* **2009**, 81, 1324–1331.
16 Neugebauer, U.; Bocklitz, T.; Clement, J.H.; Krafft, C.; Popp, J. Towards detection and identification of circulating tumour cells using Raman spectroscopy. *Analyst* **2010**, 135, 3178–3182.

17 Neugebauer, U.; Clement, J.H.; Bocklitz, T.; Krafft, C.; Popp, J. Identification and differentiation of single cells from peripheral blood by Raman spectroscopic imaging. *J. Biophotonics* **2010**, 3, 579–587.

18 Pudlas, M.; Koch, S.; Bolwien, C.; Thude, S.; Jenne, N.; Hirth, T.; et al. Raman spectroscopy: A noninvasive analysis tool for the discrimination of human skin cells. *Tissue Eng. Part C Methods* **2011**, 17, 1027–1040.

19 Dochow, S.; Krafft, C.; Neugebauer, U.; Bocklitz, T.; Henkel, T.; Mayer, G.; et al. Tumour cell identification by means of Raman spectroscopy in combination with optical traps and microfluidic environments. *Lab Chip* **2011**, 11, 1484–1490.

20 Harkness, L.; Novikov, S.M.; Beermann, J.; Bozhevolnyi, S.I.; Kassem, M. Identification of abnormal stem cells using Raman spectroscopy. *Stem Cells Dev.* **2012**, 21, 2152–2159.

21 Bielecki, C.; Bocklitz, T.W.; Schmitt, M.; Krafft, C.; Marquardt, C.; Gharbi, A.; et al. Classification of inflammatory bowel diseases by means of Raman spectroscopic imaging of epithelium cells. *J. Biomed. Opt.* **2012**, 17, 076030.

22 Salman, A.; Shufan, E.; Zeiri, L.; Huleihel, M. Detection and identification of cancerous murine fibroblasts, transformed by murine sarcoma virus in culture, using Raman spectroscopy and advanced statistical methods. *Biochim. Biophys. Acta.* 2013, **1830**, 2720–2727.

23 Pijanka, J.K.; Stone, N.; Rutter, A.V.; Forsyth, N.; Sockalingum, G.D.; Yang, Y.; Sule-Suso, J. Identification of different subsets of lung cells using Raman microspectroscopy and whole cell nucleus isolation. *Analyst* **2013**, 138, 5052–5058.

24 Mazur, A.I.; Monahan, J.L.; Miljkovic, M.; Laver, N.; Diem, M.; Bird, B. Vibrational spectroscopic changes of B-lymphocytes upon activation. *J. Biophotonics* **2013**, 6, 101–109.

25 Ilin, Y.; Kraft, M.L. Identifying the lineages of individual cells in cocultures by multivariate analysis of Raman spectra. *Analyst* **2014**, 139, 2177–2185.

26 Tolstik, T.; Marquardt, C.; Matthaus, C.; Bergner, N.; Bielecki, C.; Krafft, C.; et al. Discrimination and classification of liver cancer cells and proliferation states by Raman spectroscopic imaging. *Analyst* **2014**, 139, 6036–6043.

27 Lee, Y.J.; Vega, S.L.; Patel, P.J.; Aamer, K.A.; Moghe, P.V.; Cicerone, M.T. Quantitative, label-free characterization of stem cell differentiation at the single-cell level by broadband coherent anti-Stokes Raman scattering microscopy. *Tissue Eng. Part C Methods.* **2014**, 20, 562–569.

28 Tsai, T.H.; Short, M.A.; McLean, D.I.; Zeng, H.; McElwee, K.; Lui, H. Label-free identification and characterization of murine hair follicle stem cells located in thin tissue sections with Raman micro-spectroscopy. *Analyst* **2014**, 139, 2799–2805.

29 Farhane, Z.; Bonnier, F.; Casey, A.; Maguire, A.; O'Neill, L.; Byrne, H.J. Cellular discrimination using in vitro Raman micro spectroscopy: The role of the nucleolus. *Analyst* **2015**, 140, 5908–5919.

30 Ilin, Y.; Choi, J.S.; Harley, B.A.; Kraft, M.L. Identifying states along the hematopoietic stem cell differentiation hierarchy with single cell specificity via Raman spectroscopy. *Anal. Chem.* **2015**, 87, 11317–11324.

31 Talari, A.C.S.; Evans, C.A.; Holen, I.; Coleman, R.E.; Rehman, I.U. Raman spectroscopic analysis differentiates between breast cancer cell lines. *J. Raman. Spectrosc.* **2015**, 46, 421–427.

32 Grosse, C.; Bergner, N.; Dellith, J.; Heller, R.; Bauer, M.; Mellmann, A.; et al. Label-free imaging and spectroscopic analysis of intracellular bacterial infections. *Anal. Chem.* **2015**, 87, 2137–2142.

33 Barkur, S.; Bankapur, A.; Pradhan, M.; Chidangil, S.; Mathur, D.; Ladiwala, U. Probing differentiation in cancer cell lines by single-cell micro-Raman spectroscopy. *J. Biomed. Opt.* **2015**, 20, 85001.

34 Schie, I.W.; Kiselev, R.; Krafft, C.; Popp, J. Rapid acquisition of mean Raman spectra of eukaryotic cells for a robust single cell classification. *Analyst* **2016**, 141, 6387–6395.

35 Casabella, S.; Scully, P.; Goddard, N.; Gardner, P. Automated analysis of single cells using Laser Tweezers Raman Spectroscopy. *Analyst* **2016**, 141, 689–696.

36 Passegue, E.; Jamieson, C.H.; Ailles, L.E.; Weissman, I.L. Normal and leukemic hematopoiesis: Are leukemias a stem cell disorder or a reacquisition of stem cell characteristics? *Proc. Natl. Acad. Sci. U. S. A. 100. Suppl.* **2003**, 1, 11842–11849.

37 Hinsby, A.M.; Kiemer, L.; Karlberg, E.O.; Lage, K.; Fausboll, A.; Juncker, A.S.; et al. A wiring of the human nucleolus. *Mol. Cell.* **2006**, 22, 285–295.

38 Gao, Y.; Lin, L.P.; Zhu, C.H.; Chen, Y.; Hou, Y.T.; Ding, J. Growth arrest induced by C75, A fatty acid synthase inhibitor, was partially modulated by p38 MAPK but not by p53 in human hepatocellular carcinoma. *Cancer. Biol. Ther.* **2006**, 5, 978–985.

39 Ramoji, A.; Neugebauer, U.; Bocklitz, T.; Foerster, M.; Kiehntopf, M.; Bauer, M.; Popp, J. Toward a spectroscopic hemogram: Raman spectroscopic differentiation of the two most abundant leukocytes from peripheral blood. *Anal. Chem.* **2012**, 84, 5335–5342.

40 Rashid, N.; Nawaz, H.; Poon, K.W.; Bonnier, F.; Bakhiet, S.; Martin, C.; et al. Raman microspectroscopy for the early detection of pre-malignant changes in cervical tissue. *Exp. Mol. Pathol.* **2014**, 97, 554–564.

41 Treiser, M.D.; Yang, E.H.; Gordonov, S.; Cohen, D.M.; Androulakis, I.P.; Kohn, J.; et al. Cytoskeleton-based forecasting of stem cell lineage fates. *Proc. Natl. Acad. Sci. U. S. A.* **2010**, 107, 610–615.

42 Mattiazzi Usaj, M.; Styles, E.B.; Verster, A.J.; Friesen, H.; Boone, C.; Andrews, B.J. High-content screening for quantitative cell biology. *Trends Cell. Biol.* **2016**, 26, 598–611.

43 Misselwitz, B.; Strittmatter, G.; Periaswamy, B.; Schlumberger, M.C.; Rout, S.; Horvath, P.; et al. Enhanced CellClassifier: A multi-class classification tool for microscopy images. *BMC Bioinformatics* **2010**, 11, 30.

44 Caicedo, J.C.; Singh, S.; Carpenter, A.E. Applications in image-based profiling of perturbations. *Curr. Opin. Biotechnol.* **2016**, 39, 134–142.

45 Matsuoka, F.; Takeuchi, I.; Agata, H.; Kagami, H.; Shiono, H.; Kiyota, Y.; et al. Morphology-based prediction of osteogenic differentiation potential of human mesenchymal stem cells. *PLoS One* **2013**, 8, e55082.

46 Matsuoka, F.; Takeuchi, I.; Agata, H.; Kagami, H.; Shiono, H.; Kiyota, Y.; et al. Characterization of time-course morphological features for efficient prediction of osteogenic potential in human mesenchymal stem cells. *Biotechnol. Bioeng.* **2014**, 111, 1430–1439.

47 Sasaki, H.; Takeuchi, I.; Okada, M.; Sawada, R.; Kanie, K.; Kiyota, Y.; et al. Label-free morphology-based prediction of multiple differentiation potentials of human mesenchymal stem cells for early evaluation of intact cells. *PLoS One* **2014**, 9, e93952.

48 Sasaki, H.; Enomoto, J.; Ikeda, Y.; Honda, H.; Fukuda, J.; Kato, R. Comparisons of cell culture medium using distribution of morphological features in microdevice. *J. Biosci. Bioeng.* **2016**, 121, 117–123.

49 Sasaki, K.; Sasaki, H.; Takahashi, A.; Kang, S.; Yuasa, T.; Kato, R. Non-invasive quality evaluation of confluent cells by image-based orientation heterogeneity analysis. *J. Biosci. Bioeng.* **2016**, 121, 227–234.

50 Haney, S.A.; High content analysis of human embryonic stem cell growth and differentiation, in High Content Screening: Science, Techniques and Applications, John Wiley & Sons, Inc., **2008**.

51 Tokunaga, K.; Saitoh, N.; Goldberg, I.G.; Sakamoto, C.; Yasuda, Y.; Yoshida, Y.; Yamanaka, S.; Nakao, M. Computational image analysis of colony and nuclear morphology to evaluate human induced pluripotent stem cells. *Sci. Rep.* **2014**, 4, 6996.

52 Orlov, N.; Shamir, L.; Macura, T.; Johnston, J.; Eckley, D.M.; Goldberg, I.G. WND-CHARM: Multi-purpose image classification using compound image transforms. *Pattern Recognit. Lett.* **2008**, 29, 1684–1693.

53 Kato, R.; Matsumoto, M.; Sasaki, H.; Joto, R.; Okada, M.; Ikeda, Y.; et al.; Furue, M.K. Parametric analysis of colony morphology of non-labelled live human pluripotent stem cells for cell quality control. *Sci. Rep.* **2016**, 6, 34009.

54 Ramm Sander, P.; Hau, P.; Koch, S.; Schutze, K.; Bogdahn, U.; Kalbitzer, H.R.; Aigner, L. Stem cell metabolic and spectroscopic profiling. *Trends Biotechnol.* **2013**, 31, 204–213.

55 Jain, M.; Nilsson, R.; Sharma, S.; Madhusudhan, N.; Kitami, T.; Souza, A.L.; et al. Metabolite profiling identifies a key role for glycine in rapid cancer cell proliferation. *Science* **2012**, 336, 1040–1044.

56 Brown, K.J.; Formolo, C.A.; Seol, H.; Marathi, R.L.; Duguez, S.; An, E.; et al. Advances in the proteomic investigation of the cell secretome. *Expert Rev. Proteomics.* **2012**, 9, 337–345.

57 Schaaij-Visser, T.B.; de Wit, M.; Lam, S.W.; Jimenez, C.R. The cancer secretome, current status and opportunities in the lung, breast and colorectal cancer context. *Biochim. Biophys. Acta.* 2013, **1834**, 2242–2258.
58 Skalnikova, H.K. Proteomic techniques for characterisation of mesenchymal stem cell secretome. *Biochimie* **2013**, 95, 2196–2211.
59 Tomita, S.; Sakao, M.; Kurita, R.; Niwa, O.; Yoshimoto, K. A polyion complex sensor array for markerless and noninvasive identification of differentiated mesenchymal stem cells from human adipose tissue. *Chem. Sci.* **2015**, 6, 5831–5836.
60 Tomita, S.; Yoshimoto, K. Polyion complex libraries possessing naturally occurring differentiation for pattern-based protein discrimination. *Chem. Commun.* **2013**, 49, 10430–10432.
61 Tomita, S.; Niwa, O.; Kurita, R. Artificial modification of an enzyme for construction of cross-reactive polyion complexes to fingerprint signatures of proteins and mammalian cells. *Anal. Chem.* **2016**, 88, 9079–9086.

9

TMS-EEG: Methods and Challenges in the Analysis of Brain Connectivity

Elisa Kallioniemi, Mervi Könönen, and Sara Määttä

Diagnostic Imaging Center, Kuopio University Hospital, Finland

9.1 Introduction

9.1.1 Transcranial Magnetic Stimulation

Transcranial magnetic stimulation (TMS) is a non-invasive brain stimulation method that can be applied to activate neural tissue. TMS is based on electromagnetic induction discovered by Faraday in 1831 and the TMS method was introduced in 1985 by Barker et al. (1985). The basic principle in TMS is that a brief electric current is generated by a large capacitor to a stimulation coil held above the target cortical area (Ilmoniemi, et al., 1999). This short-lasting electric current produces an orthogonal magnetic field around the coil (Amassian, et al., 1992b; Roth, et al., 1991). The magnetic field then penetrates the skull and induces a transient electric field (EF) on the cortex (Ilmoniemi, et al., 1999). The penetrating depth of the magnetic pulse is short and thus, TMS mainly stimulates the upper cortical layers rather than subcortical white matter (Mills, 1991). In connectivity studies, the stimulation coils are commonly figure-of-eight shaped, which produces the maximum induced EF below the center of the coil (Ueno, et al., 1988).

TMS pulses can be applied with different waveforms and the most used ones are monophasic and biphasic. The choice of waveform affects the efficacy, optimal coil orientation, and effects of stimulation (Di Lazzaro, et al., 2008). TMS can be delivered with various pulse sequences. The most common ones are single-pulse, paired-pulse, and repetitive TMS (rTMS). In single-pulse, the pulses are applied with a long interval between them (Ilmoniemi, et al., 1999). This interval should be several seconds, since faster repetition might produce cumulative neural effects (Pellicciari, et al., 2016). Paired-pulse sequence involves repetition of two pulses with a short interval between them followed by a longer pause before repeating the pair (Ilmoniemi, et al., 1999). In rTMS,

Biomedical Engineering Challenges: A Chemical Engineering Insight, First Edition.
Edited by Vincenzo Piemonte, Angelo Basile, Taichi Ito, and Luigi Marrelli.

the pulses are given with a short interval in a train of stimuli (Ilmoniemi, et al., 1999). The choice of pulse sequence affects the obtained neural responses (Fitzgerald, et al., 2006; Reis, et al., 2008).

Targeting in plain TMS is rather inaccurate as there is no knowledge on the underlying brain structure. To improve this, TMS can be combined with an online magnetic resonance imaging (MRI)-guided neuronavigation system (navigated TMS, Figure 9.1) that uses the study subject's own anatomical MRI (Ruohonen and Karhu, 2010). The accuracy in navigated TMS has been validated against direct cortical stimulation (Picht, et al., 2011; Tarapore, et al., 2012). Applying navigated TMS also has other benefits in addition to localization, such as online visualization of the induced EF and storing of the coil location, orientation, and tilting during the stimulation, which means that previous stimuli can be repeated (Ruohonen and Karhu, 2010).

By applying a sufficient stimulation intensity, the TMS-induced EF is able to alter the membrane potential in the neurons. TMS is thought to mainly activate the cortical neurons that have axonal bends or endings within the induced EF (Ilmoniemi, et al., 1999). The orientation of neuronal axons varies in relation to the TMS-induced EF. Thereby, with different coil directions, the induced EF direction varies and different neuronal populations will be activated (Di Lazzaro, et al., 2008).

The sufficient stimulation intensity for each subject is commonly determined by calculating the resting motor threshold (rMT) by motor cortex stimulation. The rMT is calculated at the optimal cortical representation of some target muscle relevant to the specific application. The rMT is the stimulation intensity evoking a motor response (commonly at least 50 μV in peak-to-peak amplitude) in the target muscle in 50% of the stimulations (Rossini, et al., 2015). The rMT is determined by motor responses measured with electromyography (EMG) recorded using surface electrodes. With rMT, the stimulation intensity can be normalized for each subject and depending on the application, either sub- or supra-rMT intensities are applied. It needs to be noted, however, that rMT is dependent on the coil-to-cortex distance (McConnell, et al., 2001) and thus, in the non-motor areas it may not always be applicable. In the non-motor areas, the stimulation intensity could be determined by the amount of induced EF in the target if a navigation system is used.

9.1.2 Electroencephalography

Electroencephalography (EEG) is an electrophysiological method for measuring the electrical activity from the brain. EEG can be recorded both invasively and non-invasively, but in the context of TMS-EEG, we are referring to the non-invasive manner. EEG has an excellent temporal resolution and therefore the principal usage is the evaluation of dynamic neural activity (Britton, et al., 2016). EEG potentials mainly comprise of the summation of excitatory and

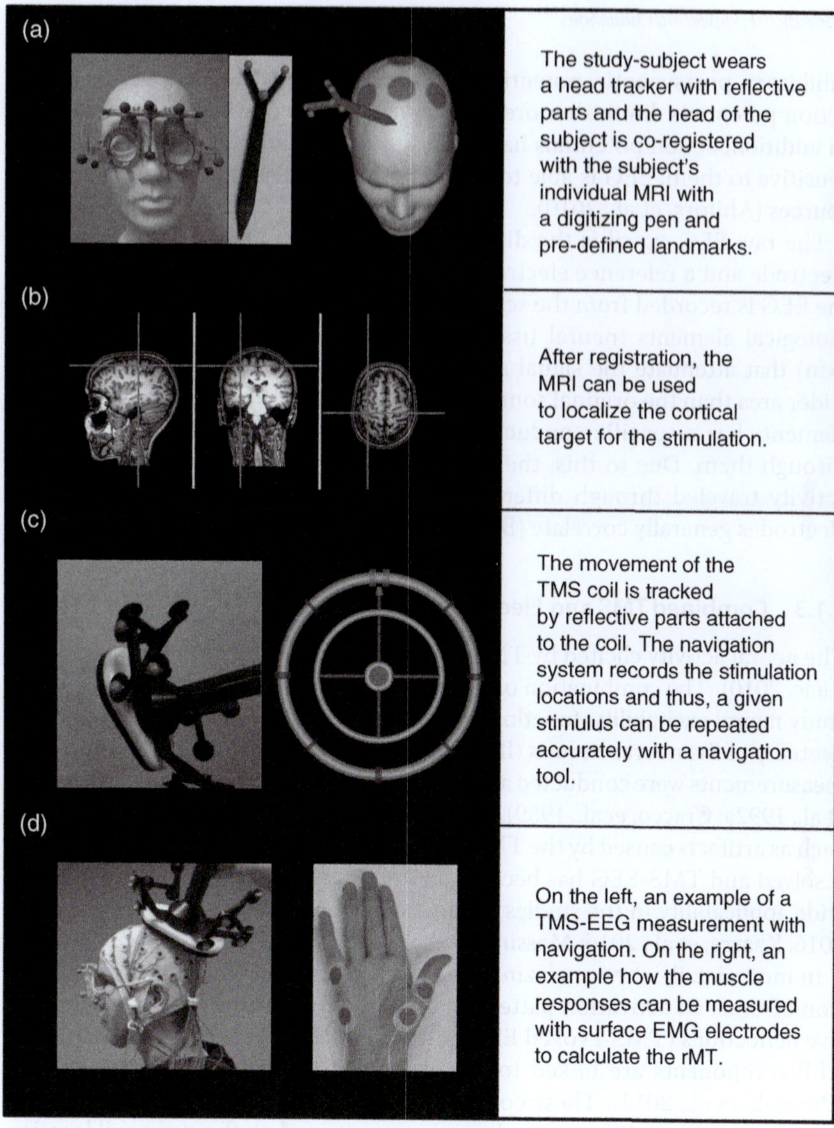

Figure 9.1 The methodology of navigated transcranial magnetic stimulation (an example from one commercial vendor) combined with electroencephalography (TMS-EEG). (a) In order to perform accurate navigation, the subject wears a head tracker with spherical reflective parts. These reflective parts are monitored by an infrared camera. With the help of the head tracker, the subject is co-registered with the subject-specific magnetic resonance image (MRI). **(b)** After co-registration, the MRI can be used to localize the stimulation target. **(c)** The movement of the stimulation coil is also followed by reflective parts. The navigation system records the location, tilt and induced electric field direction of a given TMS pulse and therefore, pulses can be repeated. **(d)** An example of a TMS-EEG measurement setup. If the applied stimulation intensity is normalized according to the resting motor threshold (rMT), TMS pulses are first focused on the motor cortex and evoked motor responses are measured from the target muscle using surface electromyography (EMG) electrodes to calculate the rMT. (*See insert for color representation of the figure.*)

inhibitory postsynaptic potentials (Britton, et al., 2016). EFs generated by action potentials diminish more rapidly than those of postsynaptic potentials. In addition, action potentials have a short duration and thus, EEG is not very sensitive to them. EEG is able to detect both the radial and tangential current sources (Ahlfors, et al., 2010).

The raw EEG signal is the difference between the potential at a recording electrode and a reference electrode as a function of time (Figure 9.2a). Before the EEG is recorded from the scalp, the signals have to travel through multiple biological elements (neural tissue, cerebrospinal fluid, meninges, skull, and skin) that attenuate the signal amplitude and spread the neural activity to a wider area than the original source (Britton, et al., 2016). Each of the biological elements has a specific conductivity, which affects how the signal transmits through them. Due to this, the recorded signal is a linear mixture of neural activity traveled through different elements and the signals in neighboring electrodes generally correlate (Brunner, et al., 2016).

9.1.3 Combined TMS and Electroencephalography

The neural activity elicited by TMS can be measured with EEG (Ilmoniemi and Kicic, 2010). The combination of the methods (TMS-EEG) makes it possible to study neural excitability, functional state of the brain as well as functional connectivity of the neural circuits (Ilmoniemi and Kicic, 2010). The first TMS-EEG measurements were conducted already at the beginning of the 1990s (Amassian, et al., 1992a; Cracco, et al., 1989), but they suffered from several technical issues, such as artifacts caused by the TMS pulse. To date, many of the issues have been resolved and TMS-EEG has become an important neuroimaging method with wide applicability in the studies of functional connectivity (e.g., (Ferreri, et al., 2016; Kaarre, et al., 2016; Massimini, et al., 2005; Määttä, et al., 2017)).

In more detail, when analyzing the averaged TMS-EEG responses as a function of time, an activation pattern that consists of altering positive and negative deflections (TMS-evoked EEG potentials, TEPs) is seen (Figure 9.2b). The TEP components are linked to the activity of the neurotransmitter system (Premoli, et al., 2014). These components are not invariable since in addition to inter-individual differences, the responses depend on the exact coil location (Komssi, et al., 2002) and orientation, on the state of the cortex (Nikulin, et al., 2003) and on the vigilance of the subject (Massimini, et al., 2005). However, provided that the delivery and targeting of TMS is well controlled and stable from pulse to pulse and between experiments, TEPs are highly reproducible (Lioumis, et al., 2009).

9.1.4 Data Acquisition

To acquire TMS-EEG data, specially designed systems need to be used as the standard amplifiers saturate due to the electromagnetic pulse and thus, prevent

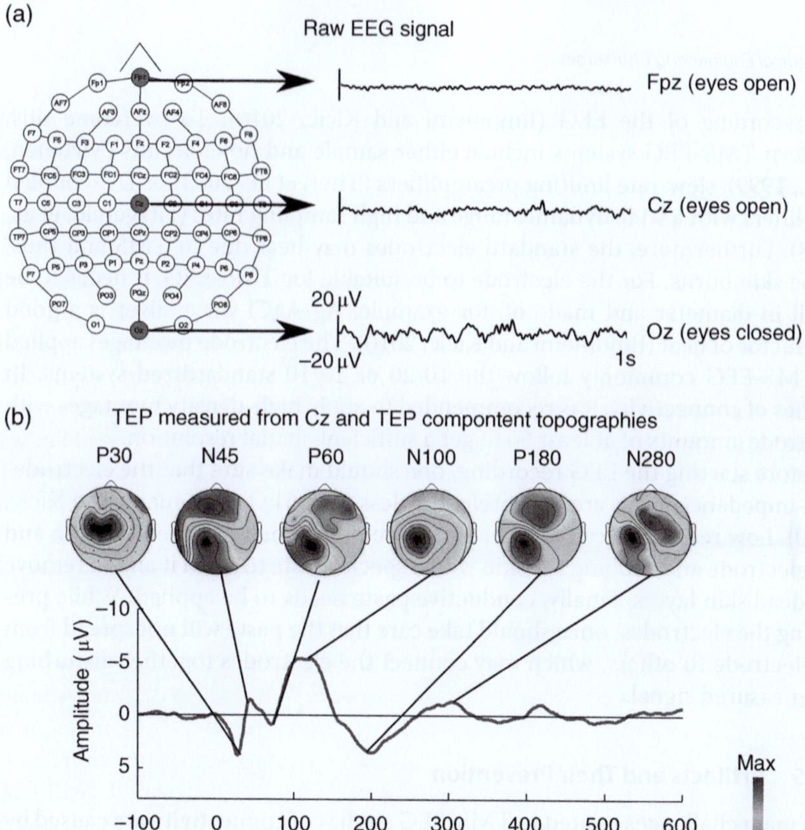

Figure 9.2 EEG signal from a representative subject. (a) An electroencephalography (EEG) electrode montage. Raw EEG signal measured from frontal (Fpz), central (Cz), and posterior (Oz) electrodes is shown. **(b)** A typical TMS evoked potential (TEP) induced by motor cortex stimulation. The TEP is measured from the Cz electrode. The topographical maps present the spread of neural activity at each TEP component (P30 ms, N45 ms, P60 ms, N100 ms, P180 ms, and N280 ms). **(c)** Event related spectral perturbation (ERSP) and inter-trial coherence (ITC) of motor cortex TEPs. In B and C, the TMS pulse was given at a time 0 ms. (*See insert for color representation of the figure.*)

the recording of the EEG (Ilmoniemi and Kicic, 2010). To overcome this, modern TMS-EEG systems include either sample and hold circuits (Virtanen, et al., 1999), slew-rate limiting preamplifiers (Thut, et al., 2003) or DC-coupled amplifiers with a wide dynamic range and high sampling rates (Fitzgerald, et al., 2008). Furthermore, the standard electrodes may heat due to TMS and thus, cause skin burns. For the electrode to be suitable for TMS-EEG, it needs to be small in diameter and made of, for example, Ag/AgCl since silver is a good conductor of heat (Ilmoniemi and Kicic, 2010). The electrode montages applied in TMS-EEG commonly follow the 10/20 or 10/10 standardized systems. In studies of connectivity, it is recommended to apply high-density montages with electrode amounts of at least 60 to get a sufficient spatial resolution.

Before starting the EEG recording, one should make sure that the electrode-skin impedance levels are low, preferably less than 5 kΩ (Ilmoniemi and Kicic, 2010). Low resistance can be achieved by removing hair between the skin and the electrode and rubbing the skin with a special paste to clean it and to remove the dead skin layers. Finally, conductive paste needs to be applied. While preparing the electrodes, once should take care that the paste will not spread from an electrode to others, which may connect the electrodes together disturbing the measured signal.

9.1.5 Artifacts and Their Prevention

The main challenges related to TMS-EEG studies of connectivity are caused by the existence of artifacts in the EEG data. Firstly, the large magnetic field of the TMS pulse causes a substantial spike artifact. This is, however, mainly removed by the modern recording systems. Further, the TMS pulse evokes muscle artifacts, caused by the activation of scalp muscles, and these may last between 10 and 60 ms after the pulse depending on the subject and the stimulated location (Korhonen, et al., 2011; Lioumis, et al., 2009). The muscle artifacts may lead to movement of the electrodes. The movement artifacts can also appear from the pressure of holding the TMS coil against the electrodes. These artifacts can be reduced with thin foam placed between the coil and the electrode (Massimini, et al., 2005). Before another pulse can be given, the capacitors which store the electrical charge in the stimulator need to be recharged and this produces a capacitor recharging artifact (Rogasch, et al., 2017). The effects of the recharging artifact can be avoided by setting the recharging to occur at a time point far from the studied time interval.

The magnetic field of the TMS pulse also interacts with the skin-electrode interface, the EEG electrodes, and the electrode lead wires, which may cause voltage deflections and thus, EEG baseline offset (Rogasch, et al., 2017). These interactions can be minimized by keeping the electrode-skin impedance low, by using TMS compatible electrodes, by skin mini-punctures with the aim of producing an epithelial short-circuit (Julkunen, et al., 2008b), and by organizing the electrode lead wires perpendicular to the induced EF to facilitate the common-mode attenuation of the artifacts (Sekiguchi, et al., 2011).

As the TMS pulse makes a loud click, the pulse will also create an auditory response in the EEG (Nikouline, et al., 1999). To attenuate this, the subject needs to wear headphones playing white noise or noise with the same auditory spectrum as the click to mask it (Massimini, et al., 2005; Nikouline, et al., 1999). Some of the noise, however, is conducted via the skull bones (Nikouline, et al., 1999). TMS pulse may also produce scalp sensations due to the activation of scalp muscles or direct activation of the sensory neurons and these evoke EEG responses (Ilmoniemi and Kicic, 2010). In addition to TMS-EEG specific artifacts, the EEG recordings are also subject to other common EEG artifacts, such as eye blinks and eye movements, muscle activity due to jaw clenching and facial expressions, sweating of the subject, and electrocardiogram and electrical line noise (50 Hz /60 Hz) (Ilmoniemi and Kicic, 2010; Rogasch, et al., 2017).

9.2 Signal Processing Methods

9.2.1 Preprocessing

To receive reliable results, the EEG data needs to be artifact free before the TMS-EEG connectivity analyses. In general, this is rather challenging in all TMS-EEG applications as the TMS-induced artifacts as well as other artifacts are hard to avoid completely, even with proper equipment and other precautions. Therefore, before the actual connectivity analysis the data needs to be properly preprocessed offline to correctly identify and eliminate the artifacts. The preprocessing can be done in several different ways as no standardized pipelines exist yet, but some common main steps can be identified in all of the approaches. These steps include data filtering, artifact removal, segmentation of the continuous signal into trials, baseline correction, and averaging. The order in which the steps are conducted depends mainly on which artifact removal method is chosen. Some freely available toolboxes for TMS-EEG preprocessing exist, such as EEGLAB (Delorme and Makeig, 2004), FieldTrip (Oostenveld, et al., 2011), TMS-EEG (Atluri, et al., 2016) and TESA (Rogasch, et al., 2017).

The main idea in filtering is to remove the non-neural artifactual frequencies from the signal. Commonly, the TMS-EEG data is bandpass filtered which means that both the lowest and highest frequencies defined by some threshold are attenuated from the signal. If line noise is mixed with the data, it needs to be notch-filtered. A notch filter is a band-stop filter with a very narrow stopband. Note that filtering, however, needs to be conducted with caution as the same filters applied in non-TMS-EEG may not always be appropriate for TMS-EEG due to the presence of the sharp TMS pulse artifact. For example, low-pass filters may cause ringing, which widens the duration of the TMS pulse artifact (Rogasch, et al., 2017).

To date, several artifact removal methods have been proposed (e.g., Mutanen, et al., 2015; Mäki and Ilmoniemi, 2011). The simplest approach is to discard the trials containing an artifact but that reduces the amount of data. In addition, the artifactual signals can be attenuated by regression analysis (Croft and Barry, 2000) if the artifact source can be independently recorded, such as those of blinks and eye movements (electro-oculography). If the artifacts occur always at a constant time, such as TMS pulse related artifacts, the contaminated signal can be zero-padded by making the signal amplitude zero during the artifact. Also, baseline correction of the data removes the low-frequency drifts, baseline offsets, and other low-frequency artifacts from the data (Rogasch, et al., 2017).

More advanced artifact removal methods apply blind source separation, such as independent component analysis (Hyvärinen and Oja, 2000) or principal component analysis to decompose the measured signal into different components. Some of the components can be identified to be of artifactual origin and the others of neural origin. After decomposition, the data is reconstructed by using only the neural components. This way, the original number of trials can be kept.

Segmentation of the signal is needed to average the data into time-locked trials containing a pre-stimulus baseline and a post-stimulus time in which the TEP appears. The selection of the baseline time in a trial depends on which connectivity analyses will be conducted. Time domain methods do not need a long baseline whereas some frequency domain methods require a long baseline. Averaging of the signal is needed to increase the signal-to-noise ratio of the data as the amplitudes of the TEPs are relatively small. Furthermore, the ongoing background EEG activity is randomly distributed with respect to the trials and thus, averages out.

9.2.2 Connectivity Analysis Methods in TMS-EEG

The connectivity methods in TMS-EEG comprise of two main approaches: time and frequency domain methods. These can be applied to evaluate the two categories of TMS-EEG connectivity: functional and effective connectivity. In functional connectivity, the aim is to detect brain areas with similar EEG amplitude, frequency, or phase (Miniussi, et al., 2012). Effective connectivity, on the other hand, exploits functional connectivity to determine the direct and indirect effects of a TMS pulse to a neuronal population (Miniussi, et al., 2012). When interpreting the TMS-EEG connectivity results it needs to be taken into account that TEP analyses in both time and frequency domain are reference-dependent and the results are always somewhat influenced by the choice of reference. Two types of references are commonly used, either a common reference electrode (measurement reference) or a calculated average reference.

9.2.3 Time Domain Methods

The time domain analyses answer the question when and where do the neural events occur? The analyses can be conducted locally or globally. In local analyses, region-of-interest (ROI) methods are common. The ROI analyses comprise of averaging the EEG signal over a group of neighboring electrodes. In global analyses, global mean field power (GMFP), topographical, and source localization methods are applied to evaluate the TEP across the whole brain. The TEPs can be evaluated in the time domain by their latency and amplitude.

The GMFP reflects the level of global neuronal activity. The GMFP is the standard deviation of the potentials across all the EEG electrodes as a function of time (Lehmann and Skrandies, 1980). TEPs with large peaks and troughs and steep spatial gradients result in a high value of GMFP, whereas low amplitude TEPs with shallow gradients produce a low GMFP (Brunet, et al., 2011).

Topographical analyses visualize how the TEP activity spreads across the brain at each time point (Nuwer, et al., 1999) (Figure 9.2b). The information from all the electrodes is arranged in two-dimensional topographical maps. These methods commonly use either linear or quadratic interpolation between the neighboring electrodes to create a continuous color-coded EEG potential map (Nuwer, et al., 1999). The maps can also be applied to locate the activity in different frequency bands (Freeman and Quiroga, 2013). Due to volume conduction, the topographical maps, however, cannot be applied for accurate source localization but merely as rough estimates.

More advanced and accurate source localization, such as, dipole modeling methods seek the solution to an inverse problem (Hämäläinen and Ilmoniemi, 1984). These source localization methods can be divided into two main categories depending whether a fixed number of current dipoles estimating the current distribution in the brain is assumed a priori or not. The number of dipoles can range from one to many thousands. Since many different current distributions can explain the measured signals in the EEG electrodes, solving the current distribution is an inverse problem. Finding the best solution involves adding constraints into the model by, for example, taking additional information into account, such as anatomical (brain structure from MRI) and physiological (conductivity of scalp, skull, and brain). The resulting current distribution can then be overlaid on the anatomical MRI. (Grech, et al., 2008; Pascual-Marqui, 1999)

9.2.4 Frequency Domain Methods

The frequency domain methods provide properties of the analyzed signal (such as magnitude and phase) as a function of both frequency and time, or summarize those values over time. Similar to time domain methods, the analyses in the frequency domain can be conducted either locally or globally. When performing the frequency analyses, the sampling rate of the EEG needs to be at least twice the maximum frequency of interest as stated in the Nyquist sampling theorem.

The time-frequency characteristics of TEPs can be assessed with event-related spectral perturbation (ERSP) and inter-trial coherence (ITC) analyses (Figure 9.2c). ERSP describes the TMS-induced changes in the magnitude (in decibels, dB) of EEG frequency spectra as a function of time and it is not sensitive to the phase of the EEG. The ERSP can be thought as event-related synchronization and desynchronization of EEG caused by TMS. ITC, in contrast, measures the phase dynamics related to TMS, irrespective of the EEG amplitude. ITC values range between zero and one, and the value of zero refers to a situation in which there are random phases across the trials and a value of one to a situation in which all the trials have a consistent phase. (Delorme and Makeig, 2004)

In addition to ERSP and ITC, the TEPs can be evaluated by calculating the coherence between different brain regions (French and Beaumont, 1984). This approach identifies whether two regions are consistent in amplitude and phase as a function of frequency (French and Beaumont, 1984). Thereby, coherence is a measure of whether TEPs are in synchrony between different brain regions.

9.3 TMS-EEG Applications in Studies of Connectivity

9.3.1 General Aspects

TMS-evoked EEG responses are a rich source of information. As TMS-EEG bypasses sensory and motor pathways as well as subcortical structures, it provides a powerful technique for the direct investigation of cortical excitability and connectivity (Ilmoniemi, et al., 1997). In this respect, one of the pioneering TMS-EEG experiments on brain connectivity was conducted by Massimini and co-workers (2005) who recorded responses to TMS both in wakefulness and during sleep. Their results showed that, during wakefulness, TMS triggered an initial response at the stimulation site that was followed by a sequence of waves that traveled to connected cortical areas several centimeters away. However, on entering sleep, the brain's response to TMS became a single large wave that did not propagate beyond the stimulation site.

Generally speaking, the very first part of the TEP probably results from the activation of the target area and reflects the excitability of the stimulated cortex. By combining TMS with other neuroimaging methods such as functional MRI, it has been demonstrated that, in addition to affecting brain activity at the site of stimulation, TMS also modulates neural circuits in distant brain regions (Bestmann, et al., 2004; Denslow, et al., 2005). Therefore, it has been suggested that the spatiotemporal distribution of TEPs over the scalp reflects the spread of activation from the stimulated cortical site to ipsilateral and contralateral cortical areas via intra- and interhemispheric corticocortical fibers, and possibly also through subcortical structures, thus revealing the effective connectivity of the brain (Ilmoniemi, et al., 1997; Komssi, et al., 2002; Kähkönen, et al., 2004; Paus, et al., 2001). It is noteworthy that besides the assessment of the general state of the brain (Kähkönen, et al., 2001; Massimini, et al., 2005), concurrent TMS and EEG have the potential to offer insights into how brain areas interact during sensory processing (Bikmullina, et al., 2009); cognition (Bonnard, et al., 2009) or motor control (Ferreri, et al., 2011; Kicic, et al., 2008; Nikulin, et al., 2003).

9.3.2 TMS-Evoked Potentials (TEPs)

Single-pulse TMS on the motor cortex elicits a typical waveform that consists of positive and negative peaks labeled after their polarity and latency, for example P30, N45, P60, N100, P180, and N280 (Bonato, et al., 2006; Ferreri, et al., 2011) (Figure 9.2b). These TEPs represent shifts in the inhibition-excitation balance in cortical circuits following a TMS pulse (Rogasch and Fitzgerald, 2013). The mechanisms underlying the generation of each TEP peak are not completely understood, and contributions from several independent cortical mechanisms are likely. The early P30 component shows a diffuse spatial distribution (Ferreri, et al., 2011). Its origin is not clear but it has been suggested to reflect the activity in the superior wall of the ipsilateral cingulate gyrus, the supplementary motor area (Litvak, et al., 2007), or around ipsilateral sensorimotor/premotor cortex border (Esser, et al., 2006). The generation of P30 is linked to cortical excitatory processes associated with the TMS pulse (Mäki and Ilmoniemi, 2010; Premoli, et al., 2014; Rogasch and Fitzgerald, 2013). The following peak, N45 is maximal in frontal regions in topographical surface maps (Ferreri, et al., 2011; Premoli, et al., 2014). The potential maps and dipole modeling suggests that the generator of N45 is located in the ipsilateral central sulcus (Litvak, et al., 2007; Paus, et al., 2001). By combining high-density TMS-EEG with pharmacology, N45 was recently demonstrated to be linked to gamma-aminobutyric acid $(GABA)_A$ receptor-mediated neurotransmission (Premoli, et al., 2014). The source structure of the next peak, P60 is less studied but based on dipole modeling, the generator of this peak seems to be located in the ipsilateral motor cortex (Määttä, et al., 2017), and this peak may be an indicator of sensorimotor interaction (Ferreri et al., 2012). The N100 is the most dominant peak in the response to motor cortex stimulation (Bender, et al., 2005; Nikulin, et al., 2003). The N100 shows a diffuse spatial distribution and is maximal over the centroparietal site ipsilateral to the stimulation. The N100 has been related to $GABA_B$ergic inhibition (Ferreri, et al., 2011; Premoli, et al., 2014) and to the activity of the cortico–striato–thalamo–cortical loops and long-distance corticocortical connections (Ferrarelli, et al., 2010; Premoli, et al., 2014). The mechanisms contributing to the generation of later TEPs (P180 and N280) have received considerably less attention than those of the earlier potentials. The P180 is maximal in frontal regions in topographical surface maps, and dipole modeling suggests that the generator of the following N280 is located in the ipsilateral centroparietal cortex (Määttä, et al., 2017). The neuropharmacology of these late TEP components is not known, but a link to $GABA_B$-related activity has been suggested (Ferreri and Rossini, 2013). In addition to $GABA_B$, the later TEPs have also been proposed to be related to cholinergic neurotransmission (Ferreri, et al., 2012; Rossini, et al., 2015).

In adults, single-pulse TMS of the frontal cortex elicits a similar sequel of deflections, but the early TEPs (P30, N45, and P60) show a more frontal distribution compared to their motor cortex counter parts, and N100 and P180 have a more central/bilateral distribution (Rogasch, et al., 2014). P30 may also be absent in response to frontal cortex stimulation (Cash, et al., 2016). There are, so far, no pharmacological TMS-EEG studies on the relationships between TEPs to frontal cortex stimulation and GABAergic inhibitory activity. However, in a recent paired-pulse TMS-EEG study, the P60 to frontal cortex stimulation was suppressed by short-interval intracortical inhibition and enhanced by intracortical facilitation, suggesting that this peak is modulated by $GABA_A$ inhibitory and glutaminergic excitatory activity (Cash, et al., 2016). Another paired-pulse TMS-EEG study has suggested that activity at the time interval of 50–100 ms after frontal cortex stimulation may relate to $GABA_B$ inhibitory activity, and that the frontal N100 may represent $GABA_B$-related cortical inhibition (Fitzgerald, et al., 2009; Rogasch, et al., 2015).

The reactivity to TMS differs between motor and prefrontal cortices when the pulse strength is kept the same. The response amplitudes are smaller for prefrontal than primary motor cortex TMS (Lioumis, et al., 2009), and the difference is largest at 30–130 ms after stimulation (Kähkönen, et al., 2004).

9.3.3 TMS-Induced Oscillations

Further insights into the functioning of stimulated brain networks can be obtained through characterizing the oscillatory activity triggered or perturbed by TMS (Fuggetta, et al., 2005; Rosanova, et al., 2009; Thut, et al., 2012). One of the characteristics of rhythmic brain activity is its frequency, and the main frequencies in EEG are delta (0.5–3.5 Hz), theta (3.5–7.5 Hz), alpha (7.5–12.5 Hz), beta (12.5–30 Hz), and gamma (low gamma 25–60 Hz, high gamma >60 Hz). Oscillations in specific frequency bands have been related to distinct states and functions. For example, slow delta wave activity occurs in sleep whereas for instance faster gamma oscillations are known to be relevant for a large number of cognitive processes.

Recent TMS-EEG studies have demonstrated that cortical areas also tend to oscillate at a specific natural frequency when directly perturbed by TMS. More specifically, TMS applied to motor cortex induces transient synchronization of spontaneous activity of cortical neurons within the beta (Paus, et al., 2001) and alpha (Veniero, et al., 2011) frequency ranges. In the occipital cortex, TMS also evokes dominant alpha-band oscillations whereas in the parietal cortex beta-band oscillations and in the frontal cortex fast gamma/beta-band oscillations are evoked (Rosanova, et al., 2009). These electrical rhythms triggered by TMS probably reflect overall circuit properties at the level of whole cortical areas and connected thalamic and subcortical nuclei (Rosanova, et al., 2009). Thus, various factors, including GABAergic inhibitory modulation of the evoked oscillations (Cho, et al., 2015; Ferrarelli, et al., 2008; Fukui, et al., 2010; Julkunen, et al., 2013), and neural network connectivity (Haider, et al., 2010; Lippe, et al., 2009; Miyauchi, et al., 2016; Moldakarimov, et al., 2015; Vakorin, et al., 2011) may contribute to the generation of TMS-triggered brain rhythms.

9.3.4 Clinical Perspectives

Several recent studies have presented TMS-EEG measures that provide new insights into the neurobiology of major neuropsychiatric disorders, including Alzheimer's disease, schizophrenia, and altered states of consciousness. Future work is needed to clarify the role of these findings in characterizing the etiology and clinical course of these brain disorders, and whether these findings may ultimately serve to identify endophenotypes or biomarkers for neurological and psychiatric disorders.

9.3.4.1 Alzheimer's Disease

The combination of TMS with EEG has given rise to a new line of investigations in Alzheimer's disease (AD). The method offers a means to characterize significant deviations from the physiological aging and a promise for detecting markers of early stage of disease or its progression. Important neuropathological features associated with AD include cortical and subcortical brain atrophy due to neuronal loss, and frontal lobes are known to be markedly involved in the pathophysiology of this disease. A study utilizing TEPs after frontal cortex stimulation in healthy young and elderly individuals as well as in patients with AD demonstrated that the EEG response to TMS of the frontal cortex was not affected by physiological aging but is markedly altered by cognitive impairment (Casarotto, et al., 2011). In that study, analysis was focused on the early and local response to TMS, an indicator of cortical excitability. In patients with AD, excitability as indexed by significant current density (SCD) over 10–45 ms post-stimulus was reduced compared with healthy young and healthy elderly controls.

In spite of the fact that motor symptoms manifest late in the course of AD, neuropathological changes in the motor cortex parallel those in other brain areas affected by this disease. A recent TMS-EEG study of motor cortex demonstrated a strong cortical hyperexcitability in early AD, despite the lack of clinically evident motor manifestations (Ferreri, et al., 2016). In this study, the GMFP analysis revealed an increase in TEP amplitude between 20 and 150 ms post-stimulus in AD patients, and topographically the differences were evident in the stimulated sensorimotor cortices. More specifically, the analysis showed a significant group difference in the P30 wave and in the P60 wave amplitude. Furthermore, the results revealed also an adjunctive peak in AD around 80 ms that was absent in healthy age-matched controls. The authors speculated that this additional peak may represent a reverberant local circuit in the sensorimotor system, possibly supported by neural degeneration leading to disconnection and/or aberrant connectivity, while the concomitant higher local sensorimotor activity at around P30–P60 time window could represent a compensatory mechanism for the decreased functional connectivity (Ferreri, et al., 2016). Contrary to these recent results, however, an earlier study found significantly reduced TMS-evoked P30 in the AD subject while subjects with mild cognitive impairment showed slightly increased activity in the P30 time window (Julkunen, et al., 2008a). One possible explanation for the discrepancy between the results from these studies could stem from the patient selection: in a recent study (Ferreri, et al., 2016), the AD patients were newly diagnosed and not on acetylcholinesterase inhibitors or other medication known to influence corticospinal excitability whereas in an earlier study (Julkunen, et al., 2008a) all the AD patients were receiving acetylcholinesterase inhibitor medication.

9.3.4.2 Schizophrenia

Several differences have been noticed when TMS-EEG responses were compared between patients with schizophrenia and healthy subjects. After frontal cortex TMS, schizophrenia patients showed a decrease within 100 ms post-stimulus in GMFP and when analyzed in frequency domain, EEG-evoked responses in the gamma oscillations were reduced (Ferrarelli, et al., 2008). Similarly, in a different study when measures of cortical activity (SCD) and connectivity (significant current scattering, SCS) were computed, these parameters were reduced in premotor/prefrontal areas in schizophrenia patients whereas there was no difference in SCD and SCS after TMS of parietal/motor cortices. In prefrontal areas, where these measures were most defective, SCD was negatively correlated with performance in the Penn Word memory delayed test whereas higher SCS values were associated with more errors in the Penn Conditional Exclusion Test (Ferrarelli, et al., 2015). Of note is that in both of the aforementioned studies, a time window of 0–300 ms post-stimulus was used and no later responses were analyzed.

When TMS was applied to motor cortex and a longer (1 s) time window was used, however, no difference in the initial response was seen but instead, an excessive spread of neuronal excitation induced by TMS was found (Frantseva, et al., 2014). The response propagation was enhanced both in time and in space. This was associated with increased oscillatory activity primarily in gamma, beta, and delta ranges, and these oscillatory abnormalities were interpreted to reflect increased interhemispheric functional connectivity. Furthermore, increased cortical conductivity in gamma frequency band between 400 and

700 ms was positively correlated with positive schizophrenia symptoms, while negative symptoms were positively correlated with functional cortical conductivity in theta and delta bands at 200 ms post-TMS (Frantseva, et al., 2014).

9.3.4.3 Disorders of Consciousness

While the results from the TMS-EEG studies presented here show group-level differences between patients and healthy controls, they are less reliable when it comes to detecting reproducible and graded changes in single patients due to large inter-individual variance. In overcoming this shortage, one of the most promising fields of applications of TMS-EEG is in disorders of consciousness (Ferrarelli, et al., 2008; Massimini, et al., 2005; Napolitani, et al., 2014; Rosanova, et al., 2012). A profound weakness in clinical practice is the lack of a reliable method to objectively assess the level of patients' consciousness under conditions such as during intraoperative anesthesia or after severe brain injury. To defeat this problem, a theory-driven index of the level of consciousness called the perturbational complexity index (PCI) was recently introduced (Casali, et al., 2013). The authors calculated PCI by perturbing the cortex with TMS to engage distributed interactions in the brain (integration) and then compressing the spatiotemporal pattern of these EEG responses to measure their algorithmic complexity. According to the authors, PCI provides a data-driven metric that can discriminate the level of consciousness in single subjects under different levels of consciousness, for example wakefulness, dreaming, midazolam-, xenon-, and propofol-induced loss of consciousness and the vegetative /unresponsive wakefulness state (Casali, et al., 2013).

9.4 Conclusions and Future Trends

TMS-EEG allows to safely and noninvasively probe virtually any cortical area making it an excellent technique for connectivity studies. So far, TMS-EEG has contributed substantially to basic and cognitive neuroscience by providing information with high temporal precision that is not accessible with any other method. Also, an increasing number of studies have proven the potential of TMS-EEG in differentiating the healthy and diseased neural states, thus enabling significant advances in the understanding of neurobiology of many major neurologic and psychiatric disorders. In the future, these lines of research should also be extended to pediatric populations. Despite some methodological challenges related to the acquisition and data analyses still exist, such as the TMS-induced artifact, several improvements have been made and many of the challenges can be overcome with special care.

To utilize the full potential of TMS-EEG, future studies would benefit from advancements in the acquisition devices and signal processing methods combined with multi-level data analysis, as well as from standardizing the signal processing pipelines. Furthermore, factors influencing the TMS-EEG data such as inter-individual differences in the EEG properties immediately preceding the moment of stimulation should be addressed and controlled in more detail. In this context, worth mentioning is the brain-state dependent closed-loop methodology in which the TMS delivery is controlled by the ongoing EEG activity, and inter-individual differences in the excitability and connectivity of brain networks are taken into account. Altogether, the advancements and suggestions presented here may be important for the adoption of the TMS-EEG in the future even for clinical diagnostics purposes.

List of Acronyms

AD	Alzheimer's disease
EEG	electroencephalography
EF	electric field
EMG	electromyography
ERSP	event-related spectral perturbation
GABA	gamma-aminobutyric acid
GMFP	global mean field power
ITC	inter-trial coherence
MRI	magnetic resonance imaging
PCI	perturbational complexity index
rMT	resting motor threshold
ROI	region-of-interest
rTMS	repetitive transcranial magnetic stimulation
SCD	significant current density
SCS	significant current scattering
TEP	transcranial magnetic stimulation evoked potential
TMS	transcranial magnetic stimulation
TMS-EEG	combined transcranial magnetic stimulation and electroencephalography

References

Ahlfors, S.P., Han, J., Belliveau, J.W., Hämäläinen, M.S. (2010) Sensitivity of MEG and EEG to source orientation. Brain Topogr, 23:227–232.

Amassian, V.E., Cracco, R.Q., Maccabee, P.J., Cracco, J.B. (1992a) Cerebello-frontal cortical projections in humans studied with the magnetic coil. Electroencephalography and Clinical Neurophysiology, 85:265–272.

Amassian, V.E., Eberle, L., Maccabee, P.J., Cracco, R.Q. (1992b) Modelling magnetic coil excitation of human cerebral cortex with a peripheral nerve immersed in a brain-shaped volume conductor: the significance of fiber bending in excitation. Electroencephalography and Clinical Neurophysiology, 85:291–301.

Atluri, S., Frehlich, M., Mei, Y., Garcia Dominguez, L., Rogasch, N.C., Wong, W., et al. (2016) TMSEEG: A MATLAB-based graphical user interface for processing electrophysiological signals during transcranial magnetic stimulation. Front Neural Circuits, 10:78.

Barker, A.T., Jalinous, R., Freeston, I.L. (1985) Non-invasive magnetic stimulation of human motor cortex. Lancet, 1:1106–1107.

Bender, S., Basseler, K., Sebastian, I., Resch, F., Kammer, T., Oelkers-Ax, R., Weisbrod, M. (2005) Electroencephalographic response to transcranial magnetic stimulation in children: Evidence for giant inhibitory potentials. Ann Neurol, 58:58–67.

Bestmann, S., Baudewig, J., Siebner, H.R., Rothwell, J.C., Frahm, J. (2004) Functional MRI of the immediate impact of transcranial magnetic stimulation on cortical and subcortical motor circuits. The European Journal of Neuroscience, 19:1950–1962.

Bikmullina, R., Kicic, D., Carlson, S., Nikulin, V.V. (2009) Electrophysiological correlates of short-latency afferent inhibition: a combined EEG and TMS study. Exp Brain Res, 194:517–526.

Bonato, C., Miniussi, C., Rossini, P.M. (2006) Transcranial magnetic stimulation and cortical evoked potentials: a TMS/EEG co-registration study. Clin Neurophysiol, 117:1699–1707.

Bonnard, M., Spieser, L., Meziane, H.B., de Graaf, J.B., Pailhous, J. (2009) Prior intention can locally tune inhibitory processes in the primary motor cortex: direct evidence from combined TMS-EEG. Eur J Neurosci, 30:913–923.

Britton, J.W., Frey, L.C., Hopp, J.L.E.A. (2016) An introductory text and atlas of normal and abnormal findings in adults, children and infants. In: St. Louis, E.K., Frey, L.C., eds. *Electroencephalography (EEG)*: Chicago, IL: American Epilepsy Society.

Brunet, D., Murray, M.M., Michel, C.M. (2011) Spatiotemporal analysis of multichannel EEG: CARTOOL. Computational Intelligence and Neuroscience, ID: 813870.

Brunner, C., Billinger, M., Seeber, M., Mullen, T.R., Makeig, S. (2016) Volume conduction influences scalp-based connectivity estimates. Front Comput Neurosci, 10:121.

Casali, A.G., Gosseries, O., Rosanova, M., Boly, M., Sarasso, S., Casali, K.R., et al. (2013) A theoretically based index of consciousness independent of sensory processing and behavior. Science Translational Medicine, 5:198ra105.

Casarotto, S., Maatta, S., Herukka, S.K., Pigorini, A., Napolitani, M., Gosseries, O., et al. (2011) Transcranial magnetic stimulation-evoked EEG/cortical potentials in physiological and pathological aging. Neuroreport, 22:592–597.

Cash, R.F., Noda, Y., Zomorrodi, R., Radhu, N., Farzan, F., Rajji, T.K., et al. (2016) Characterization of glutamatergic and GABAA-mediated neurotransmission in motor and dorsolateral prefrontal cortex using paired-pulse TMS-EEG. Neuropsychopharmacology: Official Publication of the American College of Neuropsychopharmacology, DOI: 10.1038/npp.2016.133.

Cho, R.Y., Walker, C.P., Polizzotto, N.R., Wozny, T.A., Fissell, C., Chen, C.M., Lewis, D.A. (2015) Development of sensory gamma oscillations and cross-frequency coupling from childhood to early adulthood. Cereb Cortex, 25:1509–1518.

Cracco, R.Q., Amassian, V.E., Maccabee, P.J., Cracco, J.B. (1989) Comparison of human transcallosal responses evoked by magnetic coil and electrical stimulation. Electroencephalography and Clinical Neurophysiology, 74:417–424.

Croft, R.J., Barry, R.J. (2000) Removal of ocular artifact from the EEG: a review. Neurophysiol Clin, 30:5–19.

Delorme, A., Makeig, S. (2004) EEGLAB: an open source toolbox for analysis of single-trial EEG dynamics including independent component analysis. J Neurosci Methods, 134:9–21.

Denslow, S., Lomarev, M., George, M.S., Bohning, D.E. (2005) Cortical and subcortical brain effects of transcranial magnetic stimulation (TMS)-induced movement: an interleaved TMS/functional magnetic resonance imaging study. Biol Psychiatry, 57:752–760.

Di Lazzaro, V., Ziemann, U., Lemon, R.N. (2008) State of the art: Physiology of transcranial motor cortex stimulation. Brain Stimulation, 1:345–362.

Esser, S.K., Huber, R., Massimini, M., Peterson, M.J., Ferrarelli, F., Tononi, G. (2006) A direct demonstration of cortical LTP in humans: a combined TMS/EEG study. Brain Research Bulletin, 69:86–94.

Ferrarelli, F., Massimini, M., Peterson, M.J., Riedner, B.A., Lazar, M., Murphy, M.J., et al. (2008) Reduced evoked gamma oscillations in the frontal cortex in schizophrenia patients: a TMS/EEG study. Am J Psychiatry, 165:996–1005.

Ferrarelli, F., Massimini, M., Sarasso, S., Casali, A., Riedner, B.A., Angelini, G., et al. (2010) Breakdown in cortical effective connectivity during midazolam-induced loss of consciousness. Proceedings of the National Academy of Sciences of the United States of America, 107:2681–2686.

Ferrarelli, F., Riedner, B.A., Peterson, M.J., Tononi, G. (2015) Altered prefrontal activity and connectivity predict different cognitive deficits in schizophrenia. Hum Brain Mapp, 36:4539–4552.

Ferreri, F., Pasqualetti, P., Määttä, S., Ponzo, D., Ferrarelli, F., Tononi, G., et al. (2011) Human brain connectivity during single and paired pulse transcranial magnetic stimulation. Neuroimage, 54:90–102.

Ferreri, F., Ponzo, D., Hukkanen, T., Mervaala, E., Kononen, M., Pasqualetti, P., et al. (2012) Human brain cortical correlates of short-latency afferent inhibition: a combined EEG-TMS study. Journal of Neurophysiology, 108:314–323.

Ferreri, F., Rossini, P.M. (2013) TMS and TMS-EEG techniques in the study of the excitability, connectivity, and plasticity of the human motor cortex. Reviews in the Neurosciences, 24:431–442.

Ferreri, F., Vecchio, F., Vollero, L., Guerra, A., Petrichella, S., Ponzo, D., et al. (2016) Sensorimotor cortex excitability and connectivity in Alzheimer's disease: A TMS-EEG Co-registration study. Hum Brain Mapp, 37:2083–2096.

Fitzgerald, P.B., Daskalakis, Z.J., Hoy, K., Farzan, F., Upton, D.J., Cooper, N.R., Maller, J.J. (2008) Cortical inhibition in motor and non-motor regions: a combined TMS-EEG study. Clinical EEG and Neuroscience, 39:112–117.

Fitzgerald, P.B., Fountain, S., Daskalakis, Z.J. (2006) A comprehensive review of the effects of rTMS on motor cortical excitability and inhibition. Clin Neurophysiol, 117:2584–2596.

Fitzgerald, P.B., Maller, J.J., Hoy, K., Farzan, F., Daskalakis, Z.J. (2009) GABA and cortical inhibition in motor and non-motor regions using combined TMS-EEG: a time analysis. Clin Neurophysiol, 120:1706–1710.

Frantseva, M., Cui, J., Farzan, F., Chinta, L.V., Perez Velazquez, J.L., Daskalakis, Z.J. (2014) Disrupted cortical conductivity in schizophrenia: TMS-EEG study. Cereb Cortex, 24:211–221.

Freeman, W.J., Quiroga, Q. (2013) *Imaging Brain Function with EEG: Advanced Temporal and Spatial Analysis of Electroencephalography Signals*. Springer. pp. 21–35.

French, C.C., Beaumont, J.G. (1984) A critical review of EEG coherence studies of hemisphere function. Int J Psychophysiol, 1:241–254.

Fuggetta, G., Fiaschi, A., Manganotti, P. (2005) Modulation of cortical oscillatory activities induced by varying single-pulse transcranial magnetic stimulation intensity over the left primary motor area: a combined EEG and TMS study. Neuroimage, 27:896–908.

Fukui, I., Burger, R.M., Ohmori, H., Rubel, E.W. (2010) GABAergic inhibition sharpens the frequency tuning and enhances phase locking in chicken nucleus magnocellularis neurons. J Neurosci, 30:12075–12083.

Grech, R., Cassar, T., Muscat, J., Camilleri, K.P., Fabri, S.G., Zervakis, M., et al. (2008) Review on solving the inverse problem in EEG source analysis. J Neuroeng Rehabil, 5:25.

Haider, B., Krause, M.R., Duque, A., Yu, Y., Touryan, J., Mazer, J.A., McCormick, D.A. (2010) Synaptic and network mechanisms of sparse and reliable visual cortical activity during nonclassical receptive field stimulation. Neuron, 65:107–121.

Hyvärinen, A., Oja, E. (2000) Independent component analysis: algorithms and applications. Neural Netw, 13:411–430.

Hämäläinen, M.S., Ilmoniemi, R.J. (1984) Interpreting measured magnetic fields of the brain: estimates of current distributions. Technical Report TKK-F-A559, Helsinki University of Technology.

Ilmoniemi, R.J., Kicic, D. (2010) Methodology for combined TMS and EEG. Brain Topogr, 22:233–248.

Ilmoniemi, R.J., Ruohonen, J., Karhu, J. (1999) Transcranial magnetic stimulation – a new tool for functional imaging of the brain. Crit Rev Biomed Eng, 27:241–284.
Ilmoniemi, R.J., Virtanen, J., Ruohonen, J., Karhu, J., Aronen, H.J., Näätänen, R., Katila, T. (1997) Neuronal responses to magnetic stimulation reveal cortical reactivity and connectivity. Neuroreport, 8:3537–3540.
Julkunen, P., Jauhiainen, A.M., Westeren-Punnonen, S., Pirinen, E., Soininen, H., Könönen, M., et al. (2008a) Navigated TMS combined with EEG in mild cognitive impairment and Alzheimer's disease: a pilot study. J Neurosci Methods, 172:270–276.
Julkunen, P., Pääkkönen, A., Hukkanen, T., Könönen, M., Tiihonen, P., Vanhatalo, S., Karhu, J. (2008b) Efficient reduction of stimulus artefact in TMS-EEG by epithelial short-circuiting by mini-punctures. Clin Neurophysiol, 119:475–481.
Julkunen, P., Säisänen, L., Könönen, M., Vanninen, R., Kälviäinen, R., Mervaala, E. (2013) TMS-EEG reveals impaired intracortical interactions and coherence in Unverricht–Lundborg type progressive myoclonus epilepsy (EPM1). Epilepsy Res, 106:103–112.
Kaarre, O., Kallioniemi, E., Könönen, M., Tolmunen, T., Kekkonen, V., Kivimöki, P., et al. (2016) Heavy alcohol use in adolescence is associated with altered cortical activity: a combined TMS-EEG study. Addiction Biology, DOI: 10.1111/adb.12486.
Kicic, D., Lioumis, P., Ilmoniemi, R.J., Nikulin, V.V. (2008) Bilateral changes in excitability of sensorimotor cortices during unilateral movement: combined electroencephalographic and transcranial magnetic stimulation study. Neuroscience, 152:1119–1129.
Komssi, S., Aronen, H.J., Huttunen, J., Kesäniemi, M., Soinne, L., Nikouline, V.V., et al. (2002) Ipsi- and contralateral EEG reactions to transcranial magnetic stimulation. Clin Neurophysiol, 113:175–184.
Korhonen, R.J., Hernandez-Pavon, J.C., Metsomaa, J., Mäki, H., Ilmoniemi, R.J., Sarvas, J. (2011) Removal of large muscle artifacts from transcranial magnetic stimulation-evoked EEG by independent component analysis. Med Biol Eng Comput, 49:397–407.
Kähkönen, S., Kesaniemi, M., Nikouline, V.V., Karhu, J., Ollikainen, M., Holi, M., Ilmoniemi, R.J. (2001) Ethanol modulates cortical activity: direct evidence with combined TMS and EEG. Neuroimage, 14:322–238.
Kähkönen, S., Wilenius, J., Komssi, S., Ilmoniemi, R.J. (2004) Distinct differences in cortical reactivity of motor and prefrontal cortices to magnetic stimulation. Clin Neurophysiol, 115:583–588.
Lehmann, D., Skrandies, W. (1980) Reference-free identification of components of checkerboard-evoked multichannel potential fields. Electroencephalography and clinical neurophysiology, 48:609–621.
Lioumis, P., Kicic, D., Savolainen, P., Mäkelä, J.P., Kähkönen, S. (2009) Reproducibility of TMS-evoked EEG responses. Hum Brain Mapp, 30:1387–1396.

Lippe, S., Kovacevic, N., McIntosh, A.R. (2009) Differential maturation of brain signal complexity in the human auditory and visual system. Frontiers in Human Neuroscience, 3:48.

Litvak, V., Komssi, S., Scherg, M., Hoechstetter, K., Classen, J., Zaaroor, M., et al. (2007) Artifact correction and source analysis of early electroencephalographic responses evoked by transcranial magnetic stimulation over primary motor cortex. Neuroimage, 37:56–70.

Massimini, M., Ferrarelli, F., Huber, R., Esser, S.K., Singh, H., Tononi, G. (2005) Breakdown of cortical effective connectivity during sleep. Science, 309:2228–2232.

McConnell, K.A., Nahas, Z., Shastri, A., Lorberbaum, J.P., Kozel, F.A., Bohning, D.E., George, M.S. (2001) The transcranial magnetic stimulation motor threshold depends on the distance from coil to underlying cortex: a replication in healthy adults comparing two methods of assessing the distance to cortex. Biol Psychiatry, 49:454–459.

Mills, K.R. (1991) Magnetic brain stimulation: a tool to explore the action of the motor cortex on single human spinal motoneurones. Trends Neurosci, 14:401–405.

Miniussi, C., Bortoletto, M., Thut, G., Veniero, D. (2012) Chapter 5, Accessing cortical connectivity using TMS: EEG co-registration TMS and connectivity. In: Chen, R., Rothwell, J.C., eds. *Cortical Connectivity*. Berlin Heidelberg Springer-Verlag.

Miyauchi, E., Kitajo, K., Kawasaki, M. (2016) TMS-induced theta phase synchrony reveals a bottom-up network in working memory. Neurosci Lett, 622:10–14.

Moldakarimov, S., Bazhenov, M., Sejnowski, T.J. (2015) Feedback stabilizes propagation of synchronous spiking in cortical neural networks. Proceedings of the National Academy of Sciences of the United States of America, 112:2545–2550.

Mutanen, T.P., Kukkonen, M., Nieminen, J.O., Stenroos, M., Sarvas, J., Ilmoniemi, R.J. (2015) Recovering TMS-evoked EEG responses masked by muscle artifacts. Neuroimage, 139:157–166.

Mäki, H., Ilmoniemi, R.J. (2010) EEG oscillations and magnetically evoked motor potentials reflect motor system excitability in overlapping neuronal populations. Clin Neurophysiol, 121:492–501.

Mäki, H., Ilmoniemi, R.J. (2011) Projecting out muscle artifacts from TMS-evoked EEG. Neuroimage, 54:2706–2710.

Määttä, S., Könönen, M., Kallioniemi, E., Lakka, T., Lintu, N., Lindi, V., et al. (2017) Development of cortical motor circuits between childhood and adulthood: A navigated TMS-HdEEG study. Hum Brain Mapp, 38:2599–2615.

Napolitani, M., Bodart, O., Canali, P., Seregni, F., Casali, A., Laureys, S., et al. (2014) Transcranial magnetic stimulation combined with high-density EEG in altered states of consciousness. Brain Injury, 28:1180–1189.

Nikouline, V., Ruohonen, J., Ilmoniemi, R.J. (1999) The role of the coil click in TMS assessed with simultaneous EEG. Clin Neurophysiol, 110:1325–1328.

Nikulin, V.V., Kicic, D., Kähkönen, S., Ilmoniemi, R.J. (2003) Modulation of electroencephalographic responses to transcranial magnetic stimulation: evidence for changes in cortical excitability related to movement. The European Journal of Neuroscience, 18:1206–1212.

Nuwer, M.R., Lehmann, D., da Silva, F.L., Matsuoka, S., Sutherling, W., Vibert, J.F. (1999) IFCN guidelines for topographic and frequency analysis of EEGs and EPs. The International Federation of Clinical Neurophysiology. Electroencephalogr Clin Neurophysiol Suppl, 52:15–20.

Oostenveld, R., Fries, P., Maris, E., Schoffelen, J.M. (2011) FieldTrip: Open source software for advanced analysis of MEG, EEG, and invasive electrophysiological data. Comput Intell Neurosci, 2011:156869.

Pascual-Marqui, R.D. (1999) Review of methods for solving the EEG inverse problem. International Journal of Bioelectromagnetism, 1:75–86.

Paus, T., Sipila, P.K., Strafella, A.P. (2001) Synchronization of neuronal activity in the human primary motor cortex by transcranial magnetic stimulation: an EEG study. Journal of Neurophysiology, 86:1983–1990.

Pellicciari, M.C., Miniussi, C., Ferrari, C., Koch, G., Bortoletto, M. (2016) Ongoing cumulative effects of single TMS pulses on corticospinal excitability: An intra- and inter-block investigation. Clin Neurophysiol, 127:621–8.

Picht, T., Schmidt, S., Brandt, S., Frey, D., Hannula, H., Neuvonen, T., et al. (2011) Preoperative functional mapping for rolandic brain tumor surgery: comparison of navigated transcranial magnetic stimulation to direct cortical stimulation. Neurosurgery, 69:581–8; discussion 588.

Premoli, I., Castellanos, N., Rivolta, D., Belardinelli, P., Bajo, R., Zipser, C., et al. (2014) TMS-EEG signatures of GABAergic neurotransmission in the human cortex. J Neurosci, 34:5603–5612.

Reis, J., Swayne, O.B., Vandermeeren, Y., Camus, M., Dimyan, M.A., Harris-Love, M., et al. (2008) Contribution of transcranial magnetic stimulation to the understanding of cortical mechanisms involved in motor control. The Journal of Physiology, 586:325–351.

Rogasch, N.C., Daskalakis, Z.J., Fitzgerald, P.B. (2015) Cortical inhibition of distinct mechanisms in the dorsolateral prefrontal cortex is related to working memory performance: a TMS-EEG study. Cortex; A Journal Devoted to the Study of the Nervous System and Behavior, 64:68–77.

Rogasch, N.C., Fitzgerald, P.B. (2013) Assessing cortical network properties using TMS-EEG. Hum Brain Mapp, 34:1652–1669.

Rogasch, N.C., Sullivan, C., Thomson, R.H., Rose, N.S., Bailey, N.W., Fitzgerald, P.B., et al. (2017) Analysing concurrent transcranial magnetic stimulation and electroencephalographic data: A review and introduction to the open-source TESA software. Neuroimage, 147:934–951.

Rogasch, N.C., Thomson, R.H., Farzan, F., Fitzgibbon, B.M., Bailey, N.W., Hernandez-Pavon, J.C., et al. (2014) Removing artefacts from TMS-EEG recordings using independent component analysis: importance for assessing prefrontal and motor cortex network properties. Neuroimage, 101:425–439.

Rosanova, M., Casali, A., Bellina, V., Resta, F., Mariotti, M., Massimini, M. (2009) Natural frequencies of human corticothalamic circuits. J Neurosci, 29:7679–7685.

Rosanova, M., Gosseries, O., Casarotto, S., Boly, M., Casali, A.G., Bruno, M.A., et al. (2012) Recovery of cortical effective connectivity and recovery of consciousness in vegetative patients. Brain, 135:1308–1320.

Rossini, P.M., Burke, D., Chen, R., Cohen, L.G., Daskalakis, Z., Di Iorio, R., et al. (2015) Non-invasive electrical and magnetic stimulation of the brain, spinal cord, roots and peripheral nerves: Basic principles and procedures for routine clinical and research application. An updated report from an I.F.C.N. Committee. Clin Neurophysiol, 126:1071–1107.

Roth, B.J., Saypol, J.M., Hallett, M., Cohen, L.G. (1991) A theoretical calculation of the electric field induced in the cortex during magnetic stimulation. Electroencephalography and Clinical Neurophysiology, 81:47–56.

Ruohonen, J., Karhu, J. (2010) Navigated transcranial magnetic stimulation. Neurophysiol Clin, 40:7–17.

Sekiguchi, H., Takeuchi, S., Kadota, H., Kohno, Y., Nakajima, Y. (2011) TMS-induced artifacts on EEG can be reduced by rearrangement of the electrode's lead wire before recording. Clin Neurophysiol, 122:984–990.

Tarapore, P.E., Tate, M.C., Findlay, A.M., Honma, S.M., Mizuiri, D., Berger, M.S., Nagarajan, S.S. (2012) Preoperative multimodal motor mapping: a comparison of magnetoencephalography imaging, navigated transcranial magnetic stimulation, and direct cortical stimulation. J Neurosurg, 117:354–362.

Thut, G., Miniussi, C., Gross, J. (2012) The functional importance of rhythmic activity in the brain. Current Biology: CB, 22:R658–663.

Thut, G., Northoff, G., Ives, J.R., Kamitani, Y., Pfennig, A., Kampmann, F., et al. (2003) Effects of single-pulse transcranial magnetic stimulation (TMS) on functional brain activity: a combined event-related TMS and evoked potential study. Clin Neurophysiol, 114:2071–2080.

Ueno, S., Tashiro, T., Harada, K. (1988) Localized stimulation of neural tissues in the brain by means of a paired configuration of time-varying magnetic fields. J Appl Phys: 5862–5864.

Vakorin, V.A., Lippe, S., McIntosh, A.R. (2011) Variability of brain signals processed locally transforms into higher connectivity with brain development. J Neurosci, 31:6405–6413.

Veniero, D., Brignani, D., Thut, G., Miniussi, C. (2011) Alpha-generation as basic response-signature to transcranial magnetic stimulation (TMS) targeting the human resting motor cortex: a TMS/EEG co-registration study. Psychophysiology, 48:1381–1389.

Virtanen, J., Ruohonen, J., Näätänen, R., Ilmoniemi, R.J. (1999) Instrumentation for the measurement of electric brain responses to transcranial magnetic stimulation. Med Biol Eng Comput, 37:322–326.

10

Thermal Treatments of Tumors: Principles and Methods

P. Saccomandi[1], E. Schena[1], M. Diana[2], J. Marescaux,[2] and G. Costamagna[2]

[1] *Unit of Measurements and Biomedical Instrumentation, Department of Engineering, Università Campus Bio-Medico di Roma, Italy*
[2] *Institute of Image-Guided Surgery, IHU-Strasbourg, France*

10.1 Introduction

This chapter describes the use of thermal treatments for the minimally invasive removal of localized tumors. Minimally invasive techniques have gained a widespread recognition for tumor treatment as an alternative to surgical resection and to treat patients who are not candidates for surgery [1,2]. Thermal ablations are particular classes of minimally invasive techniques, which rely on the change of tissue temperature to induce a controlled coagulative necrosis of the tumor. They cover a multidisciplinary field, which takes advantages from the collaboration between physicians, engineers, and physicists for the development of a patient-safe and effective technique, as well as the tuning of all the methods used for its monitoring and improvement.

A brief introduction about the processes and physical principles of all thermal techniques is given, along with the description of the main tools and devices involved, and the main clinical applications. Finally, the main approaches to monitoring of thermal effects are discussed.

10.2 Effects of Temperature on Living Tissue

The long-term career of heat as a means of therapeutic treatment of diseases dates back to the ancient Greeks, Romans, Egyptians, and to ancient civilizations in India, China, and Scandinavia thousands of years ago. The first known

Biomedical Engineering Challenges: A Chemical Engineering Insight, First Edition.
Edited by Vincenzo Piemonte, Angelo Basile, Taichi Ito, and Luigi Marrelli.

use of heat for healing was carried out by the Egyptian aruspice named Imhotep (2655–2600 BC). The Edwin Smith Papyrus from around 1700 BC witnesses that ancient Egyptians used hot blades and sticks to burn breast cancer [3]. The first official use of heat in the medical practice can be traced back to Hippocrates (460–370 BC), who considered local hyperthermia to be very effective, as expressed in his famous citation [4]: "Those who cannot be cured by medicine, can be cured by surgery. Those cannot be cured by surgery, can be cured by heat. Those cannot be cured by heat are to be considered incurable." Since these first steps, heat has been extensively used for centuries as both local and regional hyperthermia [5].

A similar story can be told for the use of low temperature for medical treatments. As a matter of fact, the benefit of cold for the decrease of pain and inflammation has been well-known since ancient Egyptian times. The first documented use of cold for the ablation of tumor masses dates back to the 1865, when James Arnott used salt/ice mixtures to produce tissue necrosis for the palliation of breast, uterine, and skin cancers [6]. A few decades later, liquid air and liquid carbon dioxide showed their usefulness as cryogens, until the 1960s, when liquid nitrogen turned out to be the suitable material to induce the desired hypothermia.

Since ancient times, it has been clear that change in tissue temperature can have a strong and benefic impact on health, and effort has been made to investigate the biological mechanisms underlying the phenomena of tissue destruction caused by cold and heat.

10.2.1 Hyperthermal Tissue Destruction

During hyperthermal treatments, tissues undergo metabolic and molecular changes. The two main alterations that tissues experience are as follows: (1) a conformational change of macromolecules (caused by a heat-induced rupture of bonds between molecules and their surrounding groups and environment), and (2) a modified rate of metabolic reactions (e.g., inhibition of DNA, RNA, and protein synthesis) [4]. The application of heat to cells causes damage at membrane and subcellular levels, hence inducing necrosis of cells themselves, as well as injury to surrounding cells. As shown in Figure 10.1, the thermally damaged lesion can be identified in three main zones: the internal and central one, close to the tip of the applicator, which undergoes ablation-induced coagulative necrosis (see Figure 10.1a); a peripheral region of sublethal hyperthermia, which results from the thermal conduction of heat, yielded by the high temperature gradient between the center ($T > 100\,^{\circ}C$) and the periphery (body temperature, ~37 °C) (see Figure 10.1.b); an external and surrounding zone, which is unaffected by ablation (see Figure 10.1.c).

The direct thermal damage on tissues is related to several factors (e.g., the energy deposited inside tissues, the rate and modality of ablation, thermal sensitivity represented by conductivity, specific heat, and absorption of the tissue itself [7]).

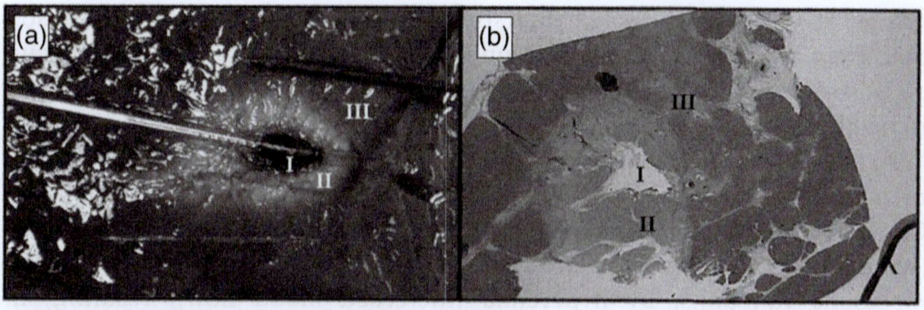

Figure 10.1 Effects of thermal ablation on tissue: (a) Macroscopic effect of diode laser-induced thermal ablation of healthy ex vivo animal liver; (b) Specimen from a histopathological examination (zoom × 10) 24 h after Nd:YAG laser ablation on in vivo animal pancreas. Although the laser settings differed between the two treatments, it is possible to distinguish similar effects: (I) the central zone of irreversible damage (in both cases, cells vaporized as a result of the raised temperature), (II) a zone of coagulated necrosis, and (III) the surrounding uninjured parenchyma. Also, a thin (1–2 mm) and watershed zone of early inflammatory response surrounding coagulated tissue is detectable [11]. (*Source:* Di Matteo, F. et al. EUS-guided Nd: YAG laser ablation of normal pancreatic tissue: a pilot study in a pig model. *Gastrointest Endosc* 2010, 72, 358–363.) (*See insert for color representation of the figure.*)

The effects of temperature and application time on tissues are the following: at temperatures around 43–45 °C, irreversible cell damage occurs after a prolonged time exposure between 30 and 60 minutes. This exposure, which can potentially induce irreversible thermal damage, decreases exponentially until 60 °C. Temperatures above 60 °C are extremely cytotoxic; they trigger an inactivation of vital enzymes up to a rapid protein denaturation, which leads to coagulative necrosis, and the time required to achieve irreversible damage decreases exponentially. This relationship between tissue damage originating from temperature function and exposure time is usually obtained by using the Arrhenius integral, a model at the origin of the theory of chemical reaction kinetics, and used to estimate the rate process of tissue thermal damage [8].

The increasing temperature has a direct effect on membrane integrity, permeability, and fluidity, inducing the dysfunction of actin filaments and microtubules, metabolite accumulation, and intracellular fluid movements causes cytolysis [9]. In particular, mitochondria dysfunction has been correlated with hyperthermia, since the temperature raised can promote the leakage of protons through the inner mitochondrial membrane [10]. Even DNA is affected by hyperthermia. The phenomenon is due to the denaturation of crucial replication enzymes, such as DNA polymerase α, DNA polymerase β, and to the denaturation of the polymerase substrate chromatin. Some studies have demonstrated that tumor tissue is more sensitive to raised temperatures than normal tissue. This could be due to increased tumor metabolic stresses above normal levels, and to its acidic interstitial environment [7]. It stands to reason that, even after the direct application of hyperthermia, delayed effects occur, because of the thermal conduction of heat and vascular damage which causes ischemia.

To sum up, the phenomena occurring in the central zone are the following: membrane collapse, protein denaturation, and mitochondria dysfunction. On the other hand, the surrounding zone affected by sublethal damage (temperatures ranging between 43 and 45 °C) is characterized by compromised metabolic functions, by hyperemia (increased blood flow), resulting in increased oxygenation which sensitizes tumor tissue to radiation and may increase the formation of reactive oxygen species, such as free radicals. Additionally, hyperemia furthers the accumulation of liposomally delivered chemotherapeutic agents. The exposed hyaluronic acid and markers of endothelial injury stimulate the expression of vascular adhesion molecules and chemokines which attract immune cells. This zone is also characterized by the presence of inflammatory infiltrates, for instance neutrophils, macrophages, natural killer cells, dendritic cells, and some lymphocytes.

10.2.2 Cold Temperature for Tissue Destruction

Cold temperature can induce cell damage and necrosis by means of a direct cell injury, and indirect mechanisms, identified as vascular injury and ischemia, apoptosis, and immunomodulation [12]. The direct mechanism is responsible for tissue dehydration. This phenomenon can be explained as follows: the application of the cryoprobe inside the tissue causes a quick drop in temperature until −40 °C, and the absorption of heat by means of the probe results in the formation of ice crystals inside cells and in tonicity increase in the extracellular space. The consequent osmotic pressure entails the dehydration of surrounding cells. Simultaneously, the accumulation of solutes in the extracellular space impairs cytoplasmic enzymes and destabilizes the cellular membrane [13,14]. Cell dehydration and the high extracellular solute concentration is called the solution effect injury. During thaw, the intracellular compartment becomes hypertonic, and fluid shift causes the cell to burst [15]. The vascular injury is related to vasoconstriction and impairment to the microvasculature, which is the cause of micro-thrombosis and stasis. Such phenomena lead to ischemic death to the targeted area, and this brings about coagulative necrosis.

While coagulative necrosis directly induced by cold occurs at the center of cryoablative lesions, apoptosis occurs in a peripheral zone of sublethal cold temperatures, probably induced by reversible damage [16]. The simultaneous presence of both necrosis and apoptosis impacts immunomodulation itself induced by cryoablation. As far as immunomodulation is concerned, blood vessels supply immune cell infiltrates. Both increased and reduced anti-tumor immunity can be induced by cryoablation. It was experimentally proven that organ-specific and tumor-specific serum antibodies are present after cryoablation [17,18], and nowadays there is active research into the immunomodulation potentialities of cryoablation, even over other hyperthermal techniques [19–21].

10.3 Physical Principles of Thermal Treatments

Thermal treatments based on principles described previously and spread in the clinical practice are classified into hyperthermal ablation (Laser Ablation: LA, Microwave Ablation: MWA, high-intensity focused ultrasound: HIFU, and Radio Frequency Ablation: RFA) and cryoablation. The common feature of all these techniques is the removal of tumors using the induction of tissue temperature change, although specific working principles are different.

10.3.1 Hyperthermal Treatments

Hyperthermal ablation is obtained by using Laser, Microwave, Radiofrequency, and HIFU sources. Figure 10.2 shows that each treatment is based on the interaction of different forms of energy (mechanical, electrical, magnetic, light) and/or using different frequencies to induce the desired therapeutic effect on tissues.

We present a brief description of the working principles of each technique, from the ones characterized by lower frequencies (RFA and HIFU) to the ones characterized by higher frequencies (MWA and LA).

10.3.1.1 High-Intensity Focused Ultrasound Ablation

HIFU ablation is based on the focusing of an ultrasound wave on the target tumor region. The wave is produced by means of an oscillating piezoelectric crystal (frequencies ranging from 0.2 to 3.5 MHz) from a generator outside the body. This principle translates into the capacity to deliver energy into the human body transcutaneously and without any physical contact between the tissue and the delivery probe [22]. HIFU transducers deliver ultrasound with power densities in the range of 100–10,000 $W{\cdot}cm^{-2}$ to the focal region, with peak compression pressures up to 70 MPa, and peak rarefaction pressures up to 20 MPa (these values depend on the target tissue). Thermal and mechanical mechanisms are predominant simultaneous mechanisms involved in tissue damage. Concerning the mechanical phenomena, stable cavitation, inertial cavitation, and micro-streaming are responsible for the oscillation of bubble size when exposed to a low-pressure acoustic field (stable cavitation) and violent oscillations in the bubble and its consecutive rapid growth during the rarefaction phase when they reach their size of

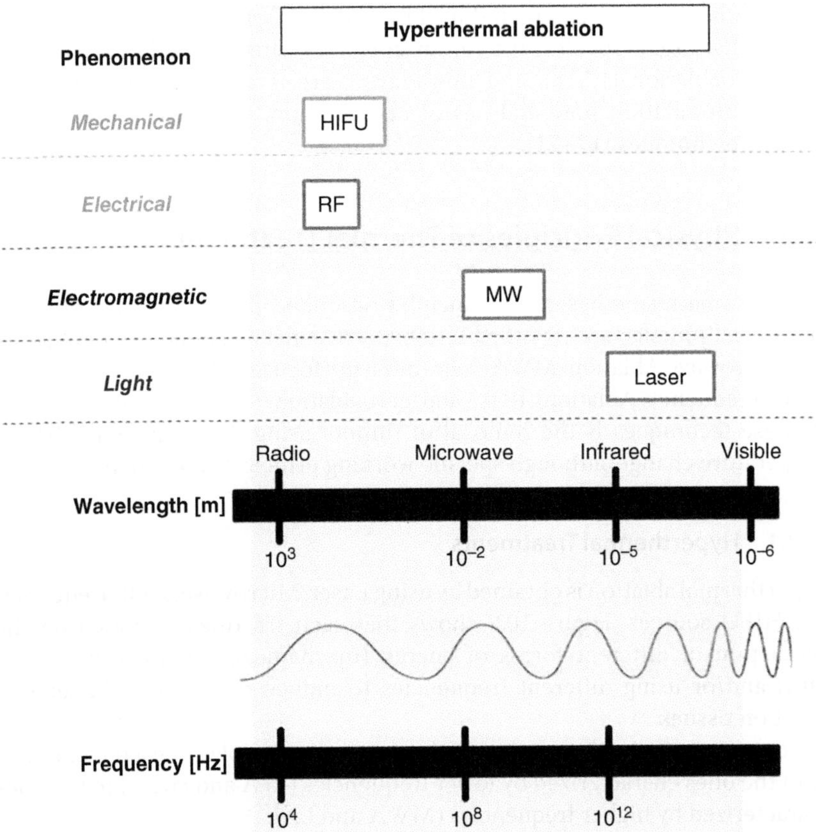

Figure 10.2 Different kinds of hyperthermal treatment whose energy source is characterized by different frequencies (or wavelengths). (*See insert for color representation of the figure.*)

resonance (inertial cavitation). The oscillating motion of stable cavitation causes the rapid movement of fluid near the bubble (micro-streaming), and induces cell apoptosis. Currently, HIFU is used for the ablation of tumors in liver, prostate, breast, and kidney [23,24], and other FDA-approved studies are under investigation for application of the method to the palliative treatment of bone metastasis [25].

10.3.1.2 Radiofrequency Ablation (RFA)

RFA relies on electrical conduction through the tissue. The RF current passes through the tissue because of the abundant presence of ionic fluid. The typical RF values used for this thermal treatment is in the order of magnitude of 1 MHz. It allows for the oscillation of the electric field, hence the oscillatory movements of ions, with a velocity proportional to the intensity of the electric field. As a consequence, a biological tissue results into a conductive medium, therefore characterized by a proper electrical conductivity σ (ω), where ω is the frequency of the electrical current flowing through the tissue. The value of σ (ω) is tissue-specific (for instance, the typical value of σ (ω) is 0. 5, 1, and 3.3 $S \cdot m^{-1}$ for soft tissue, lung and kidney, respectively [26]). The mechanism of tissue heating with RFA is based on frictional (or resistive) energy loss associated with the ionic current, and coagulative necrosis and sublethal damage described in Section 10.1 are the consequences on the tissue. Usually, if the temperature is higher than 100 °C, the heating effect is limited, because of increased tissue impedance. A general (monopolar) treatment of RFA requires the following: a RF generator, that is, a system which can generate a RF current; an interstitial electrode, that is, the tool inserted, usually percutaneously, inside the tumor to be resected or a group of electrodes; a return electrode fixed on the patient's skin, which is used to close the electrical circuit and to collect the current dispersed inside the body during RF treatment. The RF current can also be applied in bipolar mode, with the main difference of using multiple interstitial electrodes, with the pros of producing a more effective heating in the area between the electrodes and a reduced dependence on background conductivity. It also allows to prevent the use of ground pads. However, the bipolar mode requires additional electrode insertions, the zone of heating is limited and often requires saline infusion to improve results.

In addition to other thermal techniques, a relevant strength of the RFA technique is the feasibility to adapt the geometry of the electrode to the shape of the tumor to be treated, as well as the use of multi-probes to increase the ablation area [27,28]. The main constraint is related to the heat-sink effect which manifests itself when the tumor is close to a blood vessel responsible for heat dissipation.

At present, RFA is widespread in the clinical practice for tumor treatment in several organs (e.g., liver, lung, kidney, pancreas [29–31]) and also for other and emerging therapies, to interrupt pain signals in spinal nerves in chronic low back pain [32].

10.3.1.3 Microwave Ablation (MWA)

Among the microwave spectrum, the most commonly used frequencies for microwave ablation procedures are 915 MHz and 2.45 GHz. When MW radiation, with a power up to 100 W, is applied to the tissue, the so-called dielectric heating occurs. The interaction between polar water molecules of the biological medium and the electromagnetic field applied through an antenna forces dipoles to continuously realign with such electromagnetic field, thereby producing frictional energy that is then converted into heat. The transmission of electromagnetic energy is determined by the dielectric permittivity and magnetic permeability of the media in which the waves propagate [33–35]. Commercial MWA systems are relatively insensitive to tissue features (e.g., impedance, perfusion, etc.), making this technology more attractive than LA and RFA [36,37]. In addition, microwaves propagate through all types of tissues, including water vapor and dehydrated, charred, and desiccated tissues created during the ablative process [33].

A typical system for MWA is constituted by a MW generator and a cooling water system connected to a pump, intended to make cool water circulate within the MW antenna. The antenna is inserted, usually through percutaneous access, inside the tumor, and the cool water is useful to decrease the temperature in the tissue just around the emitting part of the antenna. The frequencies of 915 MHz and 2.45 GHz ensure a penetration of approximately 2–4 cm inside the tissue, suitable for the size of tumor masses which can usually undergo MWA. Additionally, new devices allow to achieve a more reproducible and controlled ablation area, by controlling the change of dielectric properties during treatment, thanks to the integration of some levels of control (e.g., thermal control, field control, and wavelength control) [38]. MWA was recently introduced for the treatment of prostate, kidney, lung, and liver tumor [39,40], and some data are also available about pancreatic cancer [41,42].

10.3.1.4 Laser Ablation (LA)

A biological tissue is a dielectric medium whose average refractive index is usually higher than the one of air [43]. As a result, when light interacts with the tissue, part of it undertakes a partial reflection at the tissue/air interface, whereas part of it penetrates the tissue. The penetrating light is subjected to two main phenomena, namely absorption and scattering, which are responsible for the increase in tissue temperature [44]. Indeed, LA is based on the conversion of laser light into heat. A low-energy photon (e.g., near infrared) incident on biological molecules entails an absorption of energy and a consequent vibrational energy transition, resulting in temperature increase of the absorbing medium. The heating of the tissue by means of light depends on the absorption capability of the tissue itself, which is physically described by optical properties (in particular, absorption, scattering, and anisotropy coefficients) [43–45]. Tissue optical properties depend on the wavelength of the incident light, and are specific for each tissue, tumors, and biological components. Inside the spectral band between 650 and 1300 nm (the so-called "therapeutic window"), wavelengths (in the 940–1100 nm range) are favored for LA purposes, since almost all tissues exhibit a good balance between absorption and penetration, where this last one is in the order of some millimeters. The main tools required to perform LA are the following: a laser source, emitting in the near-infrared region of the spectrum; an energy delivering device, made up of an optical fiber, conveniently designed to modulate light energy deposition into the tissue; a system for water cooling and the annexed pump in case of water-cooled fiber optic applicator. Treatment efficacy is related to several parameters such as laser settings (power, energy, time of exposure), laser wavelength, emission modalities of the applicator, as well as absorption characteristics of the target tissue.

In 1983, Bown was the first physician who described the use of LA for brain tumor ablation [46]. LA was used for the treatment of liver, thyroid, and prostate tumors [47]. Afterward, a couple of decades of spread in the clinical practice, the use of LA was restrained by the prominence of RFA and MWA. Notwithstanding, to date LA is living a renaissance, thanks to the pioneering application for the treatment of pancreatic cancer [48], and thanks to nanomedicine and nanotechnology. As a matter of fact, LA can be used in combination with biocompatible nanoparticles that play a crucial role in enhancing the selectivity of thermal treatment. Injected into the target, nanoparticles absorb laser light more efficiently than the tissue without them [49,50], and locally induce tumor temperature increase; on the other hand, the tissue without nanoparticles should not experience any significant temperature increase due to direct light absorption, since near-infrared laser wavelengths can be chosen to allow for a maximum light penetration through tissue. Even though the use of nanoparticles for medical purposes has been investigated since the end of the 1990s [51], research into the development and optimization of nanoparticles functions and biocompatibility is intense and challenging [52], and active investigation is still mandatory to tailor their use in clinical LA.

A schematic representing all thermal therapies for tumor ablation is reported in Figure 10.3.

10.3.2 Cryoablation

As already debated, cryoablation cannot be classified among hyperthermal therapies, since it uses cold to induce coagulative necrosis and tumor apoptosis. This technique is performed percutaneously under imaging guidance and provides the therapeutic destruction of tumor by means of freezing. Ice formation within the extracellular space causes an osmotic gradient, responsible for tissue dehydration. At the iceball boundary, the temperature is 0 °C, whereas lethal values between −50 and −20 °C are achieved within 5 mm inside the iceball edge. The procedure is performed by means of a cryoprobe and a cryogenic freezing unit: the unit allows for a high-pressure gas (e.g., Argon, at a temperature of −196 °C) to circulate within the lumen of the cryoprobe. The low pressure within the lumen causes a rapid expansion of gas, which results in a temperature decrease and in the formation of an iceball around the probe tip [54]. Cryoablation is mainly known for the treatment of liver, prostate, and lung cancer.

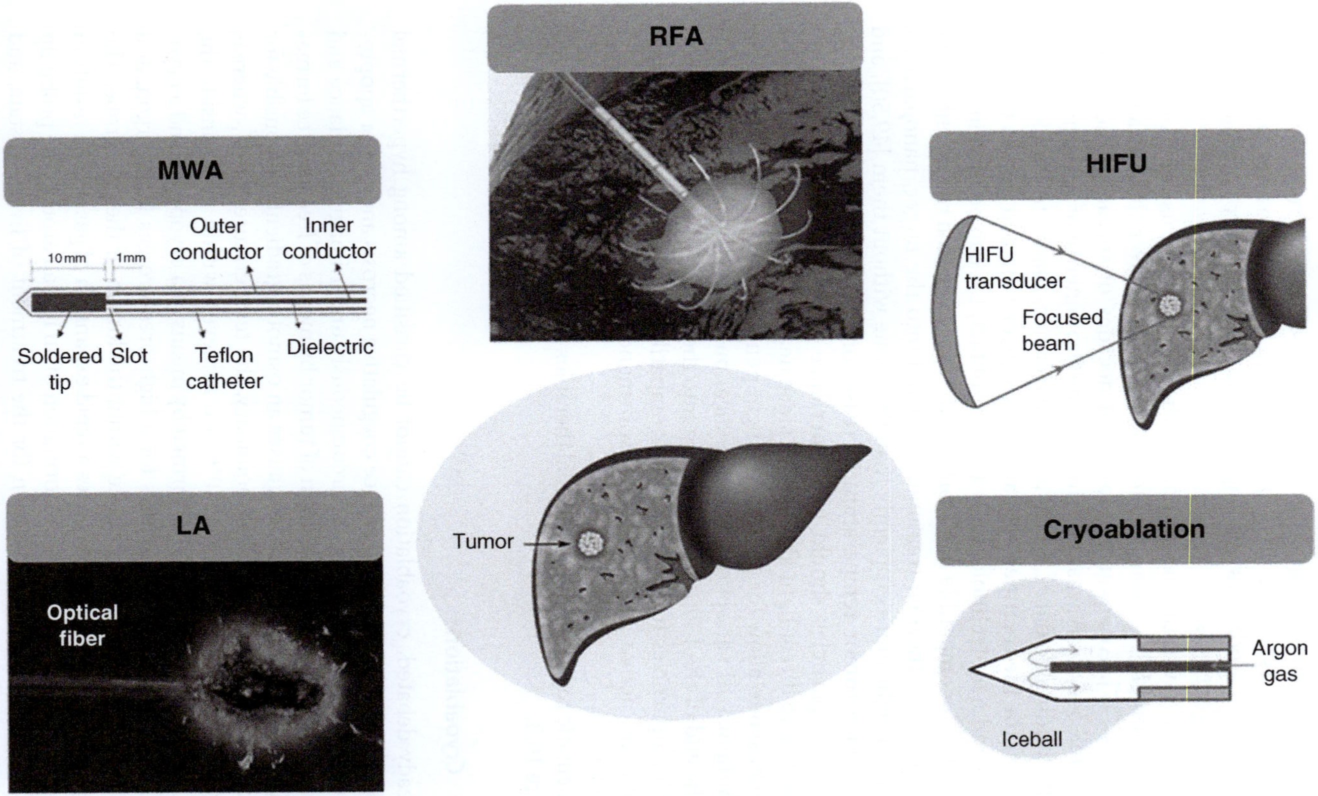

Figure 10.3 Minimally invasive thermal treatment for tumor removal: laser ablation (LA), microwave ablation (MWA), radiofrequency ablation (RFA), high-intensity focused ultrasound (HIFU), and cryoablation [53]. (*Source:* Schena, E. et al. Fiber Optic Sensors for Temperature Monitoring during Thermal Treatments: An Overview. Sensors 2016, 16, 1144.) (*See insert for color representation of the figure.*)

10.4 Mathematical Modeling of Thermal Therapies

Mathematical models are very useful in the field of thermal treatment, to simulate and predict the effects of the specific tumor therapy; they can support the clinician in the choice of optimal source settings (usually, power, treatment time, and type of energy delivery tool) to be used to achieve the desired goal, that is, the removal of the whole tumor mass and the absence of thermal injury to the surrounding healthy tissue.

Basically, temperature distribution within a biological medium can be described by the so-called "bioheat equation," put forward by the physician H.H. Pennes [55] who proposed a model of heat production rate by human tissue in 1948.

The bioheat equation is a heat-transfer model, expressed as:

$$\rho \cdot c \frac{\partial T(x,y,z,t)}{\partial t} = \nabla\left(k \nabla T(x,y,z,t)\right) + Q_b + Q_m + Q_s \tag{10.1}$$

where ρ is tissue density [$\mathrm{kg \cdot m^{-3}}$], c is the tissue-specific heat [$\mathrm{J \cdot kg^{-1} \cdot K^{-1}}$], and k is tissue heat conductivity [$\mathrm{W \cdot m^{-1}\, K^{-1}}$]. T (x, y, z, t) is the tissue temperature, expressed as a function of spatial coordinates, x, y, z, and of time, t. Usually, as a simplified hypothesis, the tissue is assumed to be homogeneous and isotropic.

Other terms in (10.1) are the following:

Q_b [$\mathrm{W \cdot m^{-3}}$], the heat contribution due to blood perfusion per volume unit, which can be expressed by:

$$Q_b = \rho_b \cdot c_b \cdot w_b \left(T(x,y,z,t) - T_b\right) \tag{10.2}$$

where ρ_b is blood density [$\mathrm{kg \cdot m^{-3}}$], c_b the blood-specific heat [$\mathrm{J \cdot kg^{-1} \cdot K^{-1}}$], w_b the blood perfusion rate per volume unit [$\mathrm{s^{-1}}$], and T_b blood temperature outside the treatment site;

Q_m [$\mathrm{W \cdot m^{-3}}$], the metabolic heat generation due to the oxidative process of lipids, proteins, and carbohydrates;

Q_s [$\mathrm{W \cdot m^{-3}}$], the heat source term due to the energy source used to perform ablation.

In case of HIFU treatment simulation, the acoustic propagation model describing the mechanical effects of an ultrasound on the tissue, can be simplified into the term in Eq. (10.4), after taking into account all combined effects of diffraction, absorption, and non-linearity:

$$Q_s = 2\alpha\left(f_0\right) \cdot I \tag{10.3}$$

where I is the mean wave intensity, for a monochromatic wave with frequency f_0, and α is the absorption coefficient of the tissue at f_0 [56].

To simulate RFA,

$$Q_s = J \cdot E \tag{10.4}$$

where J is the current density [$A \cdot m^{-2}$] and E is the electric field intensity [$V \cdot m^{-1}$] delivered by the RF antenna [57].

For MWA, the external heat source (Q_s) is equal to the resistive heat generated by the electromagnetic field (E):

$$Q_s = 0.5\mathrm{Re}\left[(\sigma - j\omega\varepsilon)E \cdot E^*\right] \tag{10.5}$$

where σ [$S \cdot m^{-1}$] and ε [$H \cdot m^{-1}$] are tissue conductivity and permeability, respectively, and ω is the frequency of the source (Re is the real part of the number).

In case of LA, the heat source term is due to photon absorption caused by laser-tissue interaction, expressed as:

$$Q_s = \mu_{eff} \cdot I(x,y) \cdot e^{-\mu_{eff} \cdot z} \tag{10.6}$$

where μ_{eff} is the effective absorption coefficient of the tissue, and $I(x, y)$ the laser irradiance [$W \cdot m^{-2}$], usually represented by a two-dimensional Gaussian distribution:

$$I(x,y) = I_0 \cdot \exp\left(-\frac{x^2 + y^2}{2\sigma_r^2}\right) \tag{10.7}$$

where $I_0 = \dfrac{P}{2 \cdot \pi \cdot r^2}$, P is the output laser power [W] and σ_r is the geometrical parameter of the applicator, related to the fiber radius r [58].

Finally, in case of simulation of cryoablation effects, Eq. (10.1) is modified as follows: the term Q_s is absent, and the terms Q_b and Q_m are neglected during the freezing phase, since the rate and metabolic heat generation quickly drop to zero [59].

Often, these equations cannot be solved directly, hence a number of software, both commercial and open source, for instance implementing Finite Element Modeling (FEM) simulations, are used for the solution [60,61]. Examples of temperature distribution obtained with simulation of thermal ablation in tissue are shown in Figure 10.4.

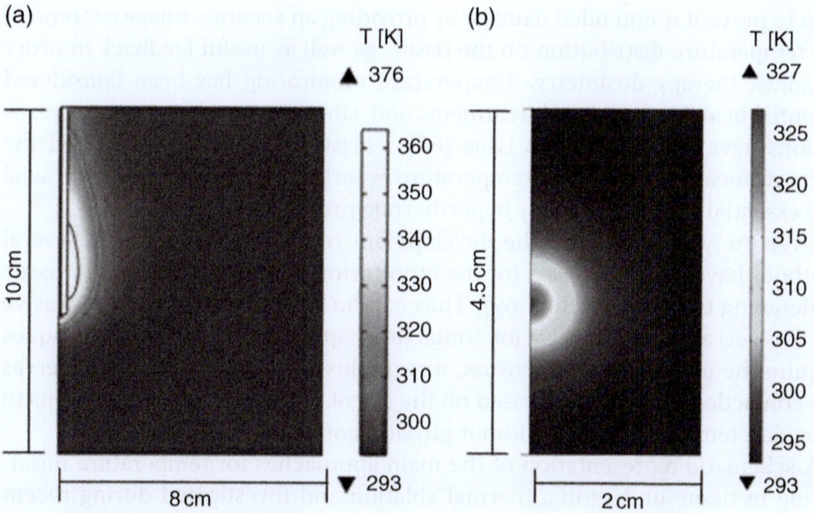

Figure 10.4 Finite Element Modeling (FEM) simulations of temperature distribution induced by (a) MWA and (b) LA within liver tissue. (*See insert for color representation of the figure.*)

10.5 Temperature Monitoring During Thermal Treatments

The goal of all thermal therapies is the destruction of the tumor mass involving the ablation of a safety margin (usually 5 mm around the lesion), while sparing healthy tissue and structures that surround the target. Even if it is very simple in theory, this goal is still very challenging, since the ablative technique has to be highly controllable, selective, and patient-specific in order to obtain an optimal therapeutic outcome.

One of the main challenges is the fine control of the evolving tissue temperature during the treatment, in order to have a real-time parameter that can guide the clinician in the treatment performance, answering the following questions: Are the settings (power, application time...) effective for the tumor to be removed? Can the heat-sink phenomenon impair ablation? Should the ablation be continued? Should the settings be changed during the ongoing treatment in order to make the therapy more effective?

Since tumor destruction is related to the temperature change and its history [62], the real-time knowledge of tissue temperature may be particularly beneficial for the adjustment of laser settings applied during thermal ablation [63]. If a reliable model is necessary to steer the clinician toward the optimal treatment, additional information about tumor temperature evolution is essential to improve the prediction of the heat-transfer model in real time, since unpredictable physiological events can occur [64]. Indeed, a good thermometry can help to prevent unintended damage by providing an accurate image or trend of the temperature distribution on the tissue, as well as useful feedback in order to adjust therapy dosimetry. Temperature monitoring has been considered essential in all hyperthermal treatments and, since the seventies, some investigations have focused on this issue [65]: Christensen and Cetas defined the accurate measurement of the temperature experienced by the organ as a crucial and essential ingredient of any hyperthermia procedure [66–68].

Over the years and with the development of ablation techniques, several methods have been proposed for the monitoring of temperature within tissue undergoing thermal ablation [69]. They can be basically divided into invasive (or contact) and non-invasive (or contactless) approaches. Invasive techniques require the use of physical sensors, directly inserted into the tissue, whereas the contactless methods are based on the use of diagnostic imaging systems to provide a temperature map without physical contact [70].

A schematic representation of the main approaches for temperature monitoring in tissue undergoing thermal ablation and investigated during recent decades is shown in Figure 10.5.

10.5.1 Invasive (Contact) Thermometric Techniques

The need to introduce the sensing element within the organ is the main drawback of these approaches, which require a direct contact with the measurement site (Figure 10.5a). In addition, since the information concerning the outcome of thermal therapy is related to the distribution of tissue temperature, a certain number of sensors is required. The previously mentioned features contribute to the classification of these techniques as invasive, but the research is evolving toward the minimization of the number of probes required for the temperature map estimation, as well as the minimization of invasiveness, by developing multi-point measurement systems or by embedding the sensing elements inside the applicator shaft.

The main families of sensors that are used to perform temperature measurements during thermal ablations are thermocouples and fiber optic sensors. A *thermocouple* consists of two different metal wires joined in two junctions. The electromotive force of the open circuit is a non-linear function of the temperature difference between the two junctions. Thermocouples are widely used due to some valuable characteristics: they are mostly cost-effective, quite accurate, and have a wide measurement range. In addition, the small size and the rapid response make them suitable for localized monitoring of temperature with fast changes. Some commercial ablation devices for RFA and MWA incorporate many thermocouples for the monitoring of temperature during the treatment, as well as to provide an end-point for the treatment: in some advanced systems, the clinician can program the value of a temperature threshold, and the energy delivery is adjusted in order to maintain that value for the

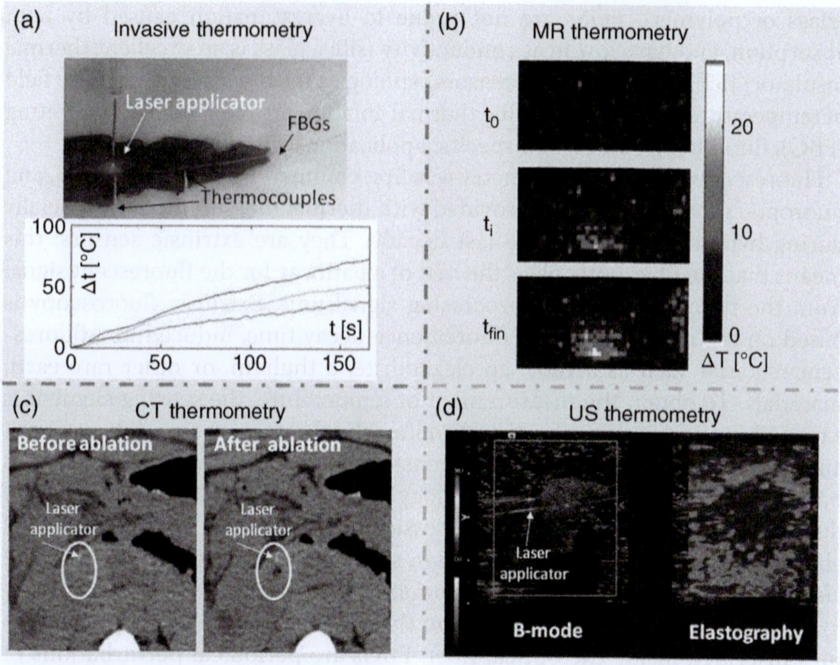

Figure 10.5 Temperature monitoring during thermal treatments. (a) Invasive thermometry, based on the use of sensors (Fiber Bragg Gratings and thermocouples), which require contact with the target intended for thermal treatment; trends of temperature increase measured by FBG sensors placed inside the ex vivo organ undergoing LAs. Non-invasive thermometry based on: (b) MR imaging: MR scans are acquired at different times during the ablation procedure, and the temperature increase in each pixel of the image is obtained by comparing the successive scans and the change of signal intensity in each pixel; (c) CT-scans, where the change of tissue temperature is related to the change of CT-number of the image (when the temperature increases, the CT-number decreases, hence the pixel values turn from gray to dark gray), as clearly visible from the images obtained before and at the end of the ablation; (d) US imaging, where the increase of tissue temperature can be monitored by means of the classical B-mode, resulting into an increase of echogenicity, or, more efficiently, by means of elastography, where the variation of tissue stiffness can be used as an indirect parameter to estimate tissue temperature. (*See insert for color representation of the figure.*)

desired time [71]. Thermocouples can present some concerns in their use during LA. As a matter of fact, their metallic conductors highly absorb the radiation emitted by laser, and this phenomenon causes a local temperature increase, causing a significant overestimation of actual tissue temperature [72–74].

Fiber optic sensors are characterized by attractive features, such as biocompatibility, immunity from electromagnetic interferences, flexibility, and the possibility of having long fiber cables to transport the light signal from the measurement site to the processing site. In addition, thanks to their constitution (glass or polymer), FOSs are not prone to overestimation caused by light absorption, and have low heat conductivity (silica glass is an excellent thermal insulator) [53]. The fiber optic sensors typologies most widely used in the field of temperature monitoring during thermal ablation are the Fiber Bragg Grating (FBG), fluoroptic sensors, and specific applications with interferometers.

Fluorescence-based thermometry was first commercialized in 1978 [75], and fluoroptic sensors have been provided with thermal measurements, especially during hyperthermia, over the last decade. They are extrinsic sensors; this means that the fiber optic plays the role of a pathway for the fluorescent signal from the fluoroscope to the processing signal unit. Extrinsic fluoroscopy is based on the measurement of fluorescence decay time, induced in a fluorescent material such as a ruby, an alexandrite, a thulium, or other rare earth materials. To obtain the measurement of temperature, the sensing element is excited by a light pulse; after this stimulus, the fluorescent signal decays with an exponential pattern, and the time constant of the exponential trend depends on temperature [76,77]. This technology was used during all thermal procedures, mostly during experimental sessions (e.g., *ex vivo* or *in vivo* animal ablation, for the assessment of ablation systems performances) [78–81], and also with some applications during human treatments [82,83].

FBGs are intrinsic sensors, based on the thermal sensitivity of optical gratings inscribed inside the optical fiber. FBGs are periodical perturbations of refraction index of an optical fiber [1,2] photo-inscribed within the core of the fiber. A polychromatic light emitted by a laser source, usually in the IR spectrum (1500–1600 nm), screens the grating, which acts like an optical filter: it transmits all the wavelengths from the source, with the exception of a narrowband centered around the Bragg wavelength (λ_B). The change of temperature experienced by the grating (hence, the change of the temperature of the surrounding medium) induces modifications of its refractive index, resulting in the shift of λ_B. Since each grating is characterized by its own λ_B, this type of optical sensor has the advantage of providing a multi-point measuring system inside the same fiber. Consequently, several FBGs can be embedded within the core of the same fiber, reducing the invasiveness of the measure to a minimum.

Another specific strength of the FBGs is the built-in self-referencing capability: the capability to encode an absolute parameter directly in wavelength terms, provides an output which does not depend directly on total light intensity, on the loss in connecting fibers and couplers, or the power source used [86].

Gratings can have different lengths, from a few centimeters (2–5 cm) to a few millimeters (up to 1 mm), and are inscribed into fibers with an external diameter in the order of hundreds of micrometers (250–500 μm). The new developments in FBG sensors, and particularly the consolidation fabrication methods based on drawing tower [87], have reduced costs and improved the spatial resolution of FBG sensors from 0.5 to 4 sensors/cm, within the same fiber. There can be two types of FBG: uniform and chirped. The main difference is the step of the grating: uniform ones are characterized by a step that is uniform (the step size is constant), while the chirped grating is not uniform, since the step size changes depending on the fiber axis space, and usually this relationship is linear. The main advantage of these chirped FBGs seems to be that they provide both the indirect measurement of temperature change (as the uniform FBGs described previously), and information concerning the thermal spatial gradient [88]. FBGs have been largely used in *ex vivo* and *in vivo* experimentation, on and for the monitoring of temperature distribution in several organs undergoing LA [89–95], MWA [96], RFA [97], and cryoablation [98]. No literature has been found concerning their use during HIFU, probably because the use of invasive temperature probes reduces the advantage of non-invasive thermal therapies such as HIFU. Since FBGs are sensitive to both temperature and strain, the main concern of their use for temperature monitoring during *in vivo* trials regards potential measurement errors caused by breathing and all potential physiological movements. Studies showed errors of up to 2 °C. This issue can be minimized by inserting the fiber inside a needle, or a protective sleeve, which can limit the strain experienced by the sensor [99,100].

10.5.2 Non-Invasive (Contactless) Thermometric Techniques

The other face of thermometry during thermal ablation consists of techniques based on images provided by clinical diagnostic systems, such as magnetic resonance (MR), computed tomography (CT), ultrasound (US), and, recently, the shear wave elastography (SWE). They are defined as contactless, therefore non-invasive, as they do not require the insertion of any mechanical parts into the patient, but rely on other means to look inside the body. The biggest advantage is related to the possibility to provide a 2D or 3D mapping of temperature distribution, which can be a strong assistance for the clinician, who can see the treatment effects on time-evolving images, and to superimpose this mapping to the anatomical structures containing and surrounding the target.

Magnetic resonance imaging (MRI) is quite complex, and basically relies on the principle that water molecules, which spin naturally inside the cells, change their magnetization according to an external magnetic field. After the transmission, and the successive interruption, of a radio frequency signal to the medium, the spin proton reacquires its initial position and direction. Two temporal parameters, that is, spin-lattice relaxation time T1 and transversal relaxation time T2, quantify the mechanisms of relaxation of the spin proton to recover the equilibrium status, and provide information about the structure of the tissue, hence they are used to form the patient's image. This is only one among several other principles that are used to form MR images, like proton shift resonance (PRF), diffusion, and others (readers are welcome to refer to other sources to have further details about the principles of MRI [101,102]). Here, it is essential to highlight how the use of MRI for thermometry purposes is increasing, and still highly attractive. The dependence of MR parameters to temperature has been firstly assessed by Bloemberg *et al.* [103] in 1948, and since 1983 MRI-based thermometry has been considered as a non-invasive approach for internal temperature estimation, when Parker *et al.* [104] observed variations in local temperature within water and blood samples. The imaging method is safe, and different parameters, such as T1 relaxation time and PRF, present a good thermal sensitivity. T1 change due to temperature depends on tissue type (e.g., 1–2% $°C^{-1}$ in liver, 1.4% $°C^{-1}$ in bovine muscle, and 0.97% $°C^{-1}$ in fat), hence a previous calibration for the specific tissue is required. In addition, this thermal sensitivity is constant only within a certain temperature range (30–70 °C). PRF has a thermal sensitivity of about −0.01 $ppm \cdot °C^{-1}$ within a temperature range of −15–100 °C. Of course, MR thermometry is feasible when thermal ablation is performed with MR-compatible tools, hence it is suitable for LA and HIFU [95]. More particularly, to guide and monitor HIFU ablation, MRI offers clear advantages over other imaging modalities. MRI has unparalleled soft tissue contrast and provides high-resolution imaging in any orientation to plan treatment and evaluating treatment effects (Figure 10.5b). In addition, MRI is the only currently available technique with proven capabilities to create quantitative temperature maps [105].

CT thermometry is also emerging, although there is an issue related to ionizing radiation and the related dose. Basically, the change of CT-number is related to the change of tissue temperature because of the decrease of electronic tissue density with the temperature increase (Figure 10.5c). The thermal sensitivity is approximately −0.5–0.6 $HU \cdot °C^{-1}$, and it is tissue-specific (it ranges from −0.54 $HU \cdot °C^{-1}$ for *ex vivo* swine liver to −0.5 $HU \cdot °C^{-1}$ for the *ex vivo* pancreas) [91].

The main advantage of CT thermometry is the optimal contrast and spatial resolution useful for both the guidance and the monitoring of the ablation treatment. It is even superior to the image quality, which can be obtained with MRI. The main concern is related to the presence of artifacts due to metallic objects (e.g., the shaft of the RF electrode and MWA antenna), which totally confuses the image and impairs CT-based thermometry. In fact, this thermometric approach is based on the dependence between the change of CT-number (a property of the image) and the temperature. The presence of artifacts due to metallic objects significantly modifies the image (in fact, the artifact is clearly visible), hence it is hard to assess the correlation between image intensity changes and temperature. Research concerning the assessment of CT thermometry on different organs, and with other techniques (e.g., dual source

CT, cone beam) are under investigation to establish the best setting of the system, in terms of kV and current to apply to the X-ray tube [106–108].

Ultrasound imaging relies on the interaction of a high-frequency sound waves (commonly ranging from 2 to 15 MHz) transmitted into the body by means of an ultrasound probe. When the wave interacts with the tissue, some wave energy is reflected back to the probe: the time delay between the transmitted ultrasound and the received ultrasound is an indirect measurement of the characteristics of the medium. The image is formed by echoes generated by the reflection of ultrasound waves at boundaries of tissue. US thermometry dates back to 1979, when a group of the University of Arizona [109] carried out the first *in vitro* and *in vivo* experiments to assess the feasibility of this technique in canine organs. Since this first study, aggressive efforts have been put in to develop and improve this technique to monitor the temperature evolution in soft tissue undergoing thermal ablation.

Non-invasive ultrasound thermometry is based on the temperature dependence of acoustic properties of tissue [110]. Changes in sound speed and thermal expansion with temperature cause echo shifts in the backscattered signal. Changes in backscattered energy with temperature are primarily caused by thermal effects on the backscatter coefficient [111,112]. Other ultrasound temperature estimation methods are based on texture features of B-mode ultrasound images, ultrasound contrast agents [113] tissue shear modulus, and others (Figure 10.5d). Most of the existing ultrasound thermometry methods are applied only within the temperature range of conventional hyperthermia (43–45 °C), the temperature dependence of echo-strain varies strongly between tissues. The present status of the research defines US thermometry suitable for the monitoring of hyperthermia, but not thermal ablation. At higher temperatures, especially when the tissue is coagulated, temperature monitoring can become problematic [114], even if the boundary, shape, and size of ablation zones may be monitored by US also when the tissue temperature exceeds 45 °C. Even if it is very attractive because US thermometry has the potential to be the safest image-based technique, and it is based on a cost-effective technology widespread in most clinical places, it did not stand out when compared to MR.

However, in recent years, a novel approach, the SWE technique, emerged as a potential tool for real time and non-invasive thermometry [115]. Quantitative elastography is used to estimate the mechanical properties of tissue, and can be defined as "virtual palpation." This method is useful in the diagnosis of some pathologies, where mechanical properties (in particular, elasticity) strongly vary depending on the state of the tissue. Elastography proved to be an effective tool for the monitoring of thermal effects due to ablation, because of the change of tissue stiffness. In particular, the thermal lesion is characterized by higher stiffness with respect to normal tissue.

An elastography can be performed with different imaging systems, for example, US and MRI. In case of US elastography, a quasi-static or dynamic stress is applied to the tissue, resulting in a displacement; this displacement is assessed from the RF data sent back by the tissue; this displacement is used to estimate elasticity parameters of the tissue, like Young's modulus, or the shear wave speed and modulus [116]. The increased stiffness of an ablated lesion entails a decrease of the strain. As a result, the elastogram displays different tissue elasticity (see Figure 10.5c), which corresponds to the increased echogenicity at the site of the ablation exhibited by the B-mode ultrasound image [117], but is characterized by better delineation.

The technique known as supersonic shear wave imaging (SSI) provides a combination of radiation force induced by an ultrasonic beam and an ultrafast imaging sequence (5000 frames/s and more) capable of catching the propagation of the resulting shear waves in real time. The method allows to retrieve the shear velocity (Vg [$m{\cdot}s^{-1}$]) for each pixel of the image. The elasticity map is calculated using a simple and robust time-of-flight technique, which computes the value of the shear wave group velocity on each point on the image. This process results in the computation of a complete 2D shear modulus map (μ) according to the relationship:

$$\mu = \rho V_g^{\,2} \tag{10.8}$$

where ρ is the medium density and μ [Pa] is the shear modulus at the given level of static compression.

Since both ρ and Vg are thermal dependent, μ can be used as an indirect measurement of temperature (e.g., for an *ex vivo* animal muscle undergoing HIFU, the thermal sensitivity of the SW speed is about -1 $kPa{\cdot}°C^{-1}$ in the range 30–45 °C) [118]. The big advantage of this technique is the ability to provide information about thermal damage since it has been shown that when thermal damage occurs, tissue stiffness was found to increase up to four-fold.

This emerging technique is very promising, but is feasible mainly during contactless thermal treatments (as HIFU or potentially contactless LA, for instance, mediated with nanoparticles), because the presence of an ablation tool could interfere with the propagation of sheared wave through the tissue under examination.

10.6 Conclusions

Minimally invasive techniques have been well recognized as a valid alternative to conventional surgery in the treatment of focal malignancies. Several promising solutions aiming at improving the clinical success of these techniques have been investigated over the last decades.

This chapter provides an overview of the essential physics of the thermal treatments used to ablate solid tumor, along with new approaches that have been proposed to further improve clinical success of these therapies. Nevertheless, the scientific and industrial research in this field is still highly active: it fosters the exploration of novel solutions to achieve more selective treatments by improving the energy delivery and absorption, the predictive abilities of mathematical models, as well as the real-time monitoring of the treatment's effects. Many efforts are made to merge these three aspects, with the aim to design technological tools able to finely set and control in real time the therapy's outcome, as well as to provide insights about the long-term outcome of the patient undergone the thermal treatment.

List of Acronyms

BC	Before Christ
CT	Computed Tomography
DNA	Deoxyribonucleic Acid
FBG	Fiber Bragg Grating
FDA	Food and Drug Administration
FEM	Finite Element Modeling
FOS	Fiber Optic Sensor
HIFU	High-Intensity-Focused Ultrasound
IR	Infrared
LA	Laser Ablation
MRI	Magnetic Resonance Imaging
MWA	Microwave Ablation
RF	Radiofrequency
RFA	Radiofrequency Ablation
RNA	Ribonucleic Acid
SSI	Shear Wave Imaging
SW	Shear Wave
SWE	Shear Wave Elastography
US	Ultrasound

List of Symbols

T	temperature [°C]
t	time [s]
x, y, z	spatial coordinates [m]
ρ	tissue density [$\text{kg}\cdot\text{m}^{-3}$]
c	tissue-specific heat [$\text{J}\cdot\text{kg}^{-1}\cdot\text{K}^{-1}$]

k	tissue heat conductivity $[\mathrm{W \cdot m^{-1} K^{-1}}]$
ρ_b	blood density $[\mathrm{kg \cdot m^{-3}}]$
c_b	blood-specific heat $[\mathrm{J \cdot kg^{-1} \cdot K^{-1}}]$
w_b	blood perfusion rate per volume unit $[\mathrm{s^{-1}}]$
T_b	blood temperature [°C]
Q	heat per volume unit $[\mathrm{W \cdot m^{-3}}]$
μ_{eff}	effective absorption coefficient $[\mathrm{m^{-1}}]$
α	absorption coefficient $[\mathrm{m^{-1}}]$
J	current density $[\mathrm{A \cdot m^{-2}}]$
E	electric field intensity $[\mathrm{V \cdot m^{-1}}]$
σ	tissue conductivity $[\mathrm{S \cdot m^{-1}}]$
ε	tissue permeability $[\mathrm{H \cdot m^{-1}}]$
ω	angular frequency $[\mathrm{rad \cdot s^{-1}}]$
$I(x,y)$	the laser irradiance $[\mathrm{W \cdot m^{-2}}]$
λ	wavelength [nm]
f	frequency [Hz]
σ_r	standard deviation of Gaussian function [m]
r	radius [m]
V_g	shear velocity $[\mathrm{m \cdot s^{-1}}]$
μ	shear modulus [Pa]

References

1 Ahmed, M.; Brace, C.L.; Lee, F.T.; Goldberg, S.N. Principles of and advances in percutaneous ablation. *Radiology* **2011**, 258, 351–369.

2 Goldberg, S.N.; Grassi, C.J.; Cardella, J.F.; Charboneau, J.W.; Dodd, G.D.; Dupuy, D.E.; *et al.* Image-guided tumor ablation: standardization of terminology and reporting criteria 1. *Radiology* **2005**, 235, 728–739.

3 Gas, P. Essential facts on the history of hyperthermia and their connections with electromedicine. *Przegląd Elektrotechniczny* **2011**, 87, 37–40.

4 Seegenschmiedt, M.H.; Fessenden, P.; Vernon. C.C. *Thermoradiotherapy and Thermochemotherapy: Biology, Physiology, Physics.* Springer Science & Business Media, **2012**.

5 Pang, C.L.K. *Hyperthermia in Oncology*, CRC Press, **2015**.

6 Chaussy, C. *Therapeutic Energy Application in Urology: Standards and Recent Developments*, Georg Thieme Verlag, **2005**.

7 Nikfarjam, M.; Muralidharan, V.; Christophi, C. Mechanisms of focal heat destruction of liver tumors. *J. Surg. Res.* **2005**, 127, 208–223.

8 Thomsen, S.; Pearce, J.A. Thermal damage and rate processes in biologic tissues. in *Optical Thermal Response of Laser Irradiated Tissue*, 2nd Edn, Springer, pp. 487–549, **2011**.

9 Fajardo, L.F.; Egbert, B.; Marmor, J.; Hahn, G.M. Effects of hyperthermia in a malignant tumor. *Cancer* **1980**, 45, 613–623.
10 Willis, W.T.; Jackman, M.R.; Bizeau, M.E.; Pagliassotti, M.J.; Hazel, J.R. Hyperthermia impairs liver mitochondrial function in vitro. *Am J Physiol* **2000**, 278, R1240–R1246.
11 Di Matteo, F.; Martino, M.; Rea, R. et al. EUS-guided Nd: YAG laser ablation of normal pancreatic tissue: a pilot study in a pig model. *Gastrointest Endosc* **2010**, 72, 358–363.
12 Chu, K.F.; Dupuy, D.E. Thermal ablation of tumours: biological mechanisms and advances in therapy. *Nat Rev Cancer* **2014**, 14, 199–208.
13 Erinjeri, J.P.; Clark, T.W. Cryoablation: mechanism of action and devices. *J Vasc Interv Radiol* **2010**, 21, S187–S191.
14 Lovelock, J.E. The haemolysis of human red blood-cells by freezing and thawing. *Biochimica et biophysica acta* **1953**, 10, 414–426.
15 Sabel, M.S. Cryo-immunology: a review of the literature and proposed mechanisms for stimulatory versus suppressive immune responses. *Cryobiology* **2009**, 58, 1–11.
16 Baust, J.G.; Gage, A.A. The molecular basis of cryosurgery. *BJU Int.* **2005**, 95, 1187–1191.
17 Ablin, R.J. Cryosurgery of the rabbit prostate. Comparison of the immune response of immature and mature bucks. *Cryobiology* **1974**, 11, 416–422.
18 Ablin, R.J.; Reddy, K.P. Cryosurgery of the monkey (macaque) prostate. II. Apparent immunopathologic alterations following cryostimulation. *Cryobiology* **1977**, 14, 205–214.
19 Ahmad, F.; Gravante, G.; Bhardwaj, N. et al. Changes in interleukin-1β and 6 after hepatic microwave tissue ablation compared with radiofrequency, cryotherapy and surgical resections. *Am J Surg* **2010**, 200, 500–506.
20 Jansen, M.C.; van Hillegersberg, R.; Schoots, I.G. *et al.* Cryoablation induces greater inflammatory and coagulative responses than radiofrequency ablation or laser induced thermotherapy in a rat liver model. *Surgery* **2010**, 147, 686–695.
21 Chapman, W.C.; Debelak, J.P.; Pinson, C.W. *et al.* Hepatic cryoablation, but not radiofrequency ablation, results in lung inflammation. *Ann. Surg.* **2000**, 231, 752–761.
22 Diana, M.; Schiraldi, L.; Liu, Y.Y. et al. High intensity focused ultrasound (HIFU) applied to hepato-biliopancreatic and the digestive system—current state of the art and future perspectives. *Hepatob Surg Nutrit* **2016**, 5, 239.
23 Kennedy, J.E. High-intensity focused ultrasound in the treatment of solid tumours. *Nat Rev Cancer* **2005**, 5, 321–327.
24 Dubinsky, T.J.; Cuevas, C.; Dighe, M.K.; Kolokythas, O.; Hwang, J.H. High-intensity focused ultrasound: current potential and oncologic applications. *Am J Roentgenol* **2008**, 190, 191–199.

25 Huisman, M.; Lam, M.K.; Bartels, L.W. et al. Feasibility of volumetric MRI-guided high intensity focused ultrasound (MR-HIFU) for painful bone metastases. *Journal of Therapeutic Ultrasound* **2014**, 2, 1.

26 Ahmed, M.; Liu, Z.; Humphries, S. et al. Computer modeling of the combined effects of perfusion, electrical conductivity, and thermal conductivity on tissue heating patterns in radiofrequency tumor ablation. *Int J Hyperther* **2008**, 24, 577–588.

27 Goldberg, S.N. Radiofrequency tumor ablation: principles and techniques. *Eur J Ultrasound* **2001**, 13, 129–147.

28 Pereira, P.L.; Trübenbach, J.; Schenk, M. et al. Radiofrequency ablation: in vivo comparison of four commercially available devices in pig livers. *Radiology* **2004**, 232, 482–490.

29 Livraghi, T.; Meloni, F.; Di Stasi, M. et al. Sustained complete response and complications rates after radiofrequency ablation of very early hepatocellular carcinoma in cirrhosis: is resection still the treatment of choice? *Hepatology* **2008**, 47, 82–89.

30 Girelli, R.; Frigerio, I.; Salvia, R.; et al. Feasibility and safety of radiofrequency ablation for locally advanced pancreatic cancer. *Brit J Surg* **2010**, 97, 220–225.

31 Dupuy, D.E. ; Zagoria, R.J. ; Akerley, W. et al. Percutaneous radiofrequency ablation of malignancies in the lung. *Am J Roentgenol* **2000**, 174, 57–59.

32 Leggett, L.E.; Soril, L.J.; Lorenzetti, D.L. et al. Radiofrequency ablation for chronic low back pain: a systematic review of randomized controlled trials. *Pain Res Manag* **2014**, 19, e146–e153.

33 Brace, C. L. Microwave tissue ablation: biophysics, technology, and applications. *Crit Rev Biomed Eng* **2010**, 38, 65–68.

34 Rossmann, C.; Haemmerich, D. Review of temperature dependence of thermal properties, dielectric properties, and perfusion of biological tissues at hyperthermic and ablation temperatures. *Crit Rev Biomed Eng* **2014**, 42, 467–492.

35 Lazebnik, M.; Popovic, D.; McCartney, L. et al. A large-scale study of the ultrawideband microwave dielectric properties of normal, benign and malignant breast tissues obtained from cancer surgeries. *Phys Med Biol* **2007**, 52, 6093.

36 Wright, A.S.; Sampson, L.A.; Warner, T.F. et al. Radiofrequency versus microwave ablation in a hepatic porcine model 1. *Radiology* **2005**, 236, 132–139.

37 Dodd, G.D.; Dodd, N.A.; Lanctot, A.C. et al. Effect of variation of portal venous blood flow on radiofrequency and microwave ablations in a blood-perfused bovine liver model. *Radiology* **2013**, 267, 129–136.

38 Ierardi, A.M.; Mangano, A.; Floridi, C. et al. A new system of microwave ablation at 2450 MHz: preliminary experience. *Updates Surg* **2015**, 67, 39–45.

39 Illing, R.O.; Kennedy, J.E. ; Wu, F. et al. The safety and feasibility of extracorporeal high-intensity focused ultrasound (HIFU) for the treatment of liver and kidney tumours in a Western population. *Brit J Cancer* **2005**, 93, 890–895.

40 Crouzet, S.; Murat, F.J.; Pasticier, G. et al. High intensity focused ultrasound (HIFU) for prostate cancer: current clinical status, outcomes and future perspectives. *Int J Hyperther* **2010**, 26, 796–803.

41 Simon, C.; Dupuy, D.; Mayo-Smith, W. Microwave ablation: principles and applications 1. *Radiographics* **2005**, 25, S69–S83.

42 Carrafiello, G.; Laganà, D.; Mangini, M. et al. Microwave tumors ablation: principles, clinical applications and review of preliminary experiences. *Int J Surg* **2008**, 6, S65–9, 2008.

43 Tuchin, V.V.; Tuchin, V. *Tissue Optics: Light Scattering Methods and Instruments for Medical Diagnosis*. Vol. 642. Bellingham: SPIE Press, **2007**.

44 Welch, A.J.; van Gemert M.J.C. *Optical-Thermal Response of Laser-Irradiated Tissue*, Springer, **2011**.

45 Jacques, S.L. Optical properties of biological tissues: a review. *Phys Med Biol* **2013**, 58, R37.

46 Bown, S.G. Phototherapy in tumors. *World J Surg* **1983**, 7, 700–709.

47 Stafford, R J.; Fuentes, D.; Elliott, A.; Weinberg, J.S.; Ahrar, K. Laser-induced thermal therapy for tumor ablation. *Crit Rev Biomed Eng* **2010**, 38, 79–100.

48 Di Matteo, F.; Picconi, F.; Martino, M. et al. Endoscopic ultrasound-guided Nd: YAG laser ablation of recurrent pancreatic neuroendocrine tumor: a promising revolution? *Endoscopy* **2013**, 46, E380–E381.

49 Mooney, R.; Schena, E.; Saccomandi, P. et al. Gold nanorod-mediated near-infrared laser ablation: *in vivo* experiments on mice and theoretical analysis at different settings. *Int J Hyperther* **2017**, 33, 150–159.

50 Mooney, R.; Roma, L.; Zhao, D. et al. Neural stem cell-mediated intratumoral delivery of gold nanorods improves photothermal therapy. *ACS Nano* **2014**, 8, 12450–12460.

51 Bruchez, M.; Moronne, M.; Gin, P. et al. Semiconductor nanocrystals as fluorescent biological labels. *Science* **2008**, 281, 2013–2016.

52 Li, Y.; Lin, T.Y.; Luo, Y. et al. A smart and versatile theranostic nanomedicine platform based on nanoporphyrin. *Nat Communications* **2014**, 5.

53 Schena, E.; Tosi, D.; Saccomandi, P. et al. Fiber optic sensors for temperature monitoring during thermal treatments: an overview. *Sensors* **2016**, 16, 1144.

54 Tatli, S.; Acar, M.; Tuncali, K.; Morrison, P.R.; Silverman, S. Percutaneous cryoablation techniques and clinical applications. *Diagnostic Interv Radiol* **2010**, 16, 90–95.

55 Pennes, H.H. Analysis of tissue and arterial blood temperatures in the resting human forearm. *J Appl Physiol* **1948**, 1, 93–122.

56 Curra, F.P.; Mourad, P.D.; Khokhlova, V.A. et al. Numerical simulations of heating patterns and tissue temperature response due to high-intensity focused ultrasound. *IEEE T Ultras Ferr* **2000**, 47, 1077–1089.

57 Tungjitkusolmun, S.; Staelin, S.T.; Haemmerich, D. et al. Three-dimensional finite-element analyses for radio-frequency hepatic tumor ablation. *IEEE T Bio-Med Eng* **2002**, 49, 3–9.

58 Saccomandi, P.; Schena, E.; Caponero, M.A. et al. Theoretical analysis and experimental evaluation of laser-induced interstitial thermotherapy in ex vivo porcine pancreas. *IEEE T Bio-Med Eng* **2012**, 59, 2958–2964.

59 Dodi R.; Ferraguti, F.; Ristolainen, A. et al. Planning and simulation of percutaneous cryoablation. *AASRI Procedia* **2014**, 6, 118–122.

60 SOFA. Website, available at: https://www.sofa-framework.org/about/features/ (accessed November 2017), 2015–2017.

61 COMSOL. Website, available at: https://www.comsol.com/ (accessed November 2017), 2017.

62 Germer, C.T.; Roggan, A.; Ritz, J.P. et al. Optical properties of native and coagulated human liver tissue and liver metastases in the near infrared range. *Laser Surg Med* **1998**, 23, 194–203.

63 Lepetit-Coiffé, M.; Laumonier, H.; Seror, O. et al. Real-time monitoring of radiofrequency ablation of liver tumors using thermal-dose calculation by MR temperature imaging: initial results in nine patients, including follow-up. *Eur Radiol* **2010**, 20, 193–201.

64 Paulides, M. M.; Stauffer, P.R.; Neufeld, E. et al. Simulation techniques in hyperthermia treatment planning. *Int J Hyperther* **2013**, 29, 346–357.

65 Bowman, R.R. A probe for measuring temperature in radio-frequency-heated material. *IEEE T Microw Theory* **1976**, 24, 43–45.

66 Christensen, D.A. Thermal dosimetry and temperature measurements. *Cancer Res* **1979**, 39, 2325–2327.

67 Cetas T.C.; Connor, W.G. Thermometry considerations in localized hyperthermia. *Med Phys* **1978**, 5, 79–91.

68 Cetas, T.C.; Connor, W.G.; Manning, M.R. Monitoring of tissue temperature during hyperthermia therapy. *Ann NY Acad Sci* **1980**, 335, 281–297.

69 Saccomandi, P.; Schena, E.; Silvestri, S. Techniques for temperature monitoring during laser-induced thermotherapy: An overview. *Int J Hyperther* **2013**, 29, 609–619.

70 Schena, E.; Saccomandi, P. and Fong, Y. Laser ablation for cancer: past, present and future. *J Funct Biomaterials* **2017**, 8, 19.

71 Saccomandi P.; Schena, E.; Diana, M. et al. Multipoint temperature monitoring in liver undergoing Computed Tomography-guided Radiofrequency Ablation with fiber Bragg grating probes. *Engineering in Medicine and Biology Society, EMBC, 2016 IEEE Annual International Conference, 38th ed.*

72 Anvari, B.; Motamedi, M.; Torres, J. H.; Rastegar, S.; Orihuela, E. Effects of surface irrigation on the thermal response of tissue during laser irradiation. *Laser Surg Med* **1994**, 14, 386–395.

73 Reid, A.D.; Gertner, M.R.; Sherar, M.D. Temperature measurement artefacts of thermocouples and fluoroptic probes during laser irradiation at 810 nm. *Phys Med Biol* **2001**, 46, N149–N157.

74 Schena, E.; Majocchi, L. Assessment of temperature measurement error and its correction during Nd: YAG laser ablation in porcine pancreas. *Int J Hyperther* **2014**, 30, 328–334.

75 LumaSense Technologies. Available Online: http://www.lumasenseinc.com/ (accessed on 20 November 2015).

76 Grattan, K.T.V.; Zhang, Z.Y. *Fiber Optic Fluorescence Thermometry*. Springer Science & Business Media: Berlin, Germany, **1995**.

77 Zhang, Z.; Grattan, K.; Palmer, A. Fiber-optic high-temperature sensor based on the fluorescence lifetime of alexandrite. *Rev Sci Instrum* **1992**, 63, 3869.

78 Viallon, M.; Terraz, S.; Roland, J. Observation and correction of transient cavitation-induced PRFS thermometry artifacts during radiofrequency ablation, using simultaneous ultrasound/MR. *Med Phys* **2010**, 37, 1491–1506.

79 Davidson, S.R.H.; Vitkin, I.A.; Sherar, M.D.; Whelan, W.M. Characterization of measurement artefacts in fluoroptic temperature sensors: Implications for laser thermal therapy at 810 nm. *Lasers Surg Med* **2005**, 36, 297–306.

80 Rubio, M.; Hernández, A. Coaxial slot antenna design for microwave hyperthermia using finite-difference time-domain and finite element method. *Open Nanomed J* **2011**, 3, 2–9.

81 Favazza, C.P.; Gorny, K.R.; King, D.M.; Rossman, P.J.; Felmlee, J.P.; Woodrum, D.A.; Mynderse, L.A. An investigation of the effects from a urethral warming system on temperature distributions during cryoablation treatment of the prostate: A phantom study. *Cryobiology* **2014**, 69, 128–133.

82 Van den Bosch, M.; Daniel, B. MRI-guided radiofrequency ablation of breast cancer: Preliminary clinical experience. *J Magn Reson Imaging* **2008**, 27, 204–208.

83 Chen, J.C.; Moriarty, J.A.; Derbyshire, J.A. et al. Prostate cancer: MR imaging and thermometry during microwave thermal ablation-initial experience. *Radiology* **2000**, 214, 290–297.

84 Rao, Y.J. In-fiber Bragg grating sensors. *Meas Sci Technol* **1997**, 8, 355–375.

85 Kersey, A.D.; Davis, M.A.; Patrick, H.J. et al. Fiber grating sensors. *J Lightwave Technol* **1997**, 15, 1442–1463.

86 Udd, E. *Fiber Optic Sensors*, Wiley, Hoboken, NJ, **1992**.

87 Askins, C.G.; Putman, M.A.; Williams, G.M.; Friebele, E.J. Stepped-wavelength optical-fiber Bragg grating arrays fabricated in line on a draw tower. *Opt Lett* **1994**, 19, 147.

88 Napoleoni, F.; Caponera, M.; Polimadei, A. et al. Monitoring of thermal treatment by linearly chirped fiber Bragg grating sensors: Feasibility

assessment during laser ablation on ex vivo liver. *Engineering in Medicine and Biology Society (EMBC), 2016 IEEE 38th Annual International Conference of the. IEEE,* **2016**.

89 Saccomandi, P.; Schena, E.; Giurazza, F. et al. Temperature monitoring and lesion volume estimation during double-applicator laser-induced thermotherapy in ex vivo swine pancreas: A preliminary study. *Lasers Med Sci* **2014**, 29, 607–614.

90 Saccomandi, P.; Lupi, G.; Schena, E. et al. Influence of FBG sensors length on temperature measures in laser-irradiated pancreas: Theoretical and experimental evaluation. *Conf. Proc. Annu. Int. Conf. IEEE Eng. Med. Biol. Soc.* **2013**, 2013, 3737–3740.

91 Schena, E.; Saccomandi, P.; Giurazza, F. et al. Experimental assessment of CT-based thermometry during laser ablation of porcine pancreas. *Phys Med Biol* **2013**, 58, 5705–5716.

92 Polito, D.; Caponero, M.A.; Polimadei, A. et al. A needle-like probe for temperature monitoring during laser ablation based on FBG: Manufacturing and characterization. *J Med Device* **2015**, 9, doi:10.1115/1.4030624.

93 Cappelli, S.; Saccomandi, P.; Massaroni, C. et al. Magnetic Resonance-compatible needle-like probe based on Bragg grating technology for measuring temperature during Laser Ablation. In *Proceedings of the 37th Annual International Conference of the IEEE Engineering in Medicine and Biology Society (EMBC)*, Milan, Italy, 25–29 August 2015; pp. 1287–1290.

94 Liu, Y.; Chen, W.; Yu, H.; Gassino, R.; Braglia, A.; Olivero, M.; Perrone, G.; Vallan, A. *All-fiber probe for laser-induced thermotherapy with integrated temperature measurement capabilities*. In *Proceedings of the Optical Fibers and Sensors for Medical Diagnostics and Treatment Applications XV CA*, USA, 7–8 February **2015**.

95 Allegretti, G.; Saccomandi, P.; Giurazza, F. et al. Magnetic resonance-based thermometry during laser ablation on ex vivo swine pancreas and liver. *Med Eng Phys* **2015**, 37, 631–641.

96 Demura, K. An easy-to-use microwave hyperthermia system combined with spatially resolved mr temperature maps: phantom and animal studies. *J Surg Res* **2006**, 135, 179–186.

97 Samset, E.; Mala, T.; Ellingsen, R.; Gladhaug, I.; Søreide, O.; Fosse, E. Temperature measurement in soft tissue using a distributed fibre Bragg-grating sensor system. *Minim Invasive Ther Allied Technol* **2001**, 10, 89–93.

98 Saccomandi, P.; Frauenfelder, G.; Massaroni, C. et al. Temperature monitoring during radiofrequency ablation of liver: In vivo trials. *Engineering in Medicine and Biology Society (EMBC),* **2016** *IEEE 38th Annual International Conference of the. IEEE,* 2016.

99 Cavaiola, C.; Saccomandi, P.; Massaroni, C. et al. Error of a temperature probe for cancer ablation monitoring caused by respiratory movements: ex vivo and in vivo analysis. *IEEE Sens J* **2016**, 16, 5934–5941

100 Di Santo, N.; Cavaiola, C.; Saccomandi, P. et al. Feasibility assessment of an FBG-based probe for distributed temperature measurements during laser ablation. *Medical Measurements and Applications (MeMeA), 2016 IEEE International Symposium on. IEEE,* **2016**.

101 Buxton, R.B. *Introduction to Functional Magnetic Resonance Imaging: Principles and Techniques*. Cambridge University Press, **2009**.

102 Plewes, D.B.; Kucharczyk, W. Physics of MRI: a primer. *J Magn Reson Imaging* **2012**, 35, 1038–54.

103 Bloembergen, N.; Purcell, E. M.; Pound, R.V. Relaxation effects in nuclear magnetic resonance absorption. *Phys Rev* **1948**, 73, 679–712.

104 Parker, D.L.; Smith, V.; Sheldon, P.; Crooks, L.E.; Fussell, L. Temperature distribution measurements in two-dimensional NMR imaging. *Med Phys* **1983**, 10, 321–325.

105 Jolesz, F. A. MRI-guided focused ultrasound surgery. *Annu Rev Med* **2009**, 60, 417.

106 Schena, E.; Fani, F.; Saccomandi, P. et al. Feasibility assessment of CT-based thermometry for temperature monitoring during thermal procedure: Influence of ROI size and scan setting on metrological properties. *Annual International Conference of the IEEE Engineering in Medicine and Biology Society. IEEE Engineering in Medicine and Biology Society. Annual Conference*, **2015**.

107 Li, M.; Abi-Jaoudeh, N.; Kapoor, A. et al. Towards cone-beam CT thermometry. *SPIE Med Im* **2013**. International Society for Optics and Photonics.

108 DeStefano, Z.; Abi-Jaoudeh, N.; Li, M. et al. CT thermometry for cone-beam CT guided ablation. *SPIE Med Im* **2016**. International Society for Optics and Photonics.

109 Bowen, T.; Connor, W.G.; Nasoni, R.L.; Pifer, A.E.; Sholes, R.R. Measurement of the temperature dependence of the velocity of ultrasound in soft tissue. In: Linzer, M., ed. *Ultrasonic Tissue Characterization II*. National Bureau of Standards, **1979**, pp. 57–61.

110 Martin, A.R. Temperature dependence of ultrasonic backscattered energy in motion compensated images. *IEEE Trans Ultrason Ferr* 2005, 52, **1644–1652**.

111 Xia, J.; Li, Q.; Liu, H.L. et al. An approach for the visualization of temperature distribution in tissues according to changes in ultrasonic backscattered energy. *Comput Math Methods Med* **2013**, 682827.

112 Tsui, P.H.; Chien, Y.T. ; Liu, H.L. et al. Using ultrasound CBE imaging without echo shift compensation for temperature estimation. *Ultrasonics* **2012**, 52, 925–935.

113 Mulvana, H.; Stride, E.; Hajnal, J.V. et al. Temperature dependent behavior of ultrasound contrast agents. *Ultrasound Med Biol* **2010**, 36, 925–934.

114 Geng, X.; Zhou, Z.; Li, Q. et al. Comparison of ultrasound temperature imaging with infrared thermometry during radio frequency ablation. *Japanese J Appl Phys* **2014**, 53, 047001.

115 Zhou, Z.; Wu, W.; Wu, S. et al. A survey of ultrasound elastography approaches to percutaneous ablation monitoring. *Proceedings of the Institution of Mechanical Engineers, Part H: Journal of Engineering in Medicine* **2014**, 228(10), 1069–1082.

116 DeWall, R.J. Ultrasound elastography: principles, techniques, and clinical applications. *Crit Rev Biomed Eng* **2013**, 41, 1–19.

117 Varghese, T.; Zagzebski, J.A., Lee, F.T. Elastographic imaging of thermal lesions in the liver in vivo following radiofrequency ablation: preliminary results. *Ultrasound Med Biol* **2002**, 28, 1467–1473.

118 Arnal, B.; Pernot, M.; Tanter, M. Monitoring of thermal therapy based on shear modulus changes: I. shear wave thermometry. *IEEE Trans Ultrason Ferr* **2011**, 58, 369–378.

Index

a

acute respiratory distress syndrome (ARDS) intensive care 68
Alzheimer's disease 187–188
artificial kidney
  bio-printing 21
  chronic kidney disease (CKD) 9
  depression 10
  dialysis (*see* dialysis)
  end-stage renal disease 10
  epidemiologic data 9
  implantable artificial kidney 16–17
  incident and prevalence rate 9
  kidney tissue engineering 17–20
  mortality rates 10
  National Health costs 10–11
  non-invasive technique 21
  nuclear transfer techniques 21
  renal replacement devices 13, 15–16
artificial liver
  hemodialysis 28
  liver regeneration and functional recovery 27
  mortality rates 27
  non-bio artificial liver (*see* non-bio artificial liver)
  tissue engineering, for liver construction 40–41
  whole organ engineering 42–43
artificial neural networks (ANN) 78
artificial pancreas (AP)
  black-block approach 85–90
  closed-loop control 76
  model-predictive-control (MPC) 77
  PB approach 90
  phenomenological models
    background and two-compartmental models 78–79
    three-compartment models 79–85
  translational research and industrial development 76

b

bacteriophage-based therapy 5
bioartificial kidney (BAK) 18
bioartificial livers (BAL). *see also* artificial liver
  bioartificial liver support system
    AMC-BAL 36–37
    controlled and randomized clinical trials 37
    extracorporeal liver assist device (ELAD) 35
    hybrid liver support system (LSS) 36
    modular extracorporeal liver support system (MELS) 36

Biomedical Engineering Challenges: A Chemical Engineering Insight, First Edition.
Edited by Vincenzo Piemonte, Angelo Basile, Taichi Ito, and Luigi Marrelli.

bioartificial livers (BAL). *see also* artificial liver (*cont'd*)
cell source for
embryonic development 39
functional hepatocytes 39
hepatocytes 37
MSC 38
porcine endogenous retrovirus (PERV) 37
stem cells 38
bioartificial liver support system 35–37
bioheat equation 209
bio-printing 21
biotechnology 109, 111
biotin ligase 128
black-block approach 85–90
and ANN approach
glucose concentration, prediction of 88
glucose prediction 86
linear and nonlinear dynamics 90
neural network inverse model 88
predictive control neural models 85
simple feedforward neural network 87
blood oxygenators
bubble oxygenator 58
bubble-type oxygenators 57
carbon dioxide removal 56
clinical extra-corporeal blood treatments 58
coil-type membrane oxygenator 59
computational fluid dynamics (CFD) approach 63
ECMO technology development 55
extra-corporeal life support 57
film-type oxygenators 57
gas-liquid contactors 56
hollow-fiber membrane oxygenator 59
mass and momentum transferr 62
mass transfer analysis 64
membrane contactor performance 64
membrane-type oxygenators 57
Monod–Wyman–Changeux model 60
oxygen dissociation curves, carbon dioxide partial pressures 62
rotating film oxygenator 58
transport phenomena 62
blood purification, with albumin dialysis
fractionated plasma separation and adsorption 32
hepa wash 32
molecular adsorbent recirculating system (MARS) 31–32
single-pass albumin dialysis 30–31
Bolton–Hunter reagent 125
bubble oxygenator 58
bubble-type oxygenators 57

C

^{11}C-acetate 124
cell and tissue organization 108
cell therapy devices 15
cellular morphology
cell populations, unstained microscopic images 162–164
computational analyses 160
single cells, unstained microscopic images 160–162
chemical engineering (CE) 2
chemical exchange saturation transfer (CEST) 122
chemical-field effect transistors (ChemFETs) 21
coil-type membrane oxygenator 59
computational fluid dynamics (CFD) approach 63
computerized tomography (CT) 21

conformational switching 105
connectivity analysis methods 182
continuous glucose monitoring (CGM) 75
continuous subcutaneous insulin infusion (CSII) methods 75, 83
conventional techniques 145
^{11}C-phosphatidylcholine 124
cross-fertilization process 2
cryoablation 207
CT thermometry 216–217
^{64}Cu-diacetyl-bis (N^4-methylthiosemicarbazone) (ATSM) 124

d

depression 10
diabetes 75
diagnostic technique 6
dialysis
  in America 10
  in Europe 10
  first experimental dialysis 15
  in Germany 10
  National Health costs 11
  psychological consequences 10
  transport costs 11

e

ECMO systems
  bubble column bioreactors 67
  gas-liquid contactors 66
  gas-liquid equipment 65
  Oxygen Transfer Rate (OTR) 67
elastography 218
electroencephalography 176–178
end-stage renal disease (ESRD) treatment 15
extended-hours therapies 16
extracorporeal bioartificial kidney 18
extracorporeal circuit 18
Extra-Corporeal Life Support 55
extracorporeal liver assist device (ELAD) 35

f

^{18}F-fluoro-2-deoxyglucose (FDG) 123
^{18}F-fluoromisonidazole (FMISO) 124
fiber bragg grating (FBG) 214–215
film-type oxygenators 57
finite element modeling (FEM) simulations 211
fluorescence imaging
  fluorescent proteins
    Aequorea Victoria 126
    bimolecular fluorescence complementation (BiFC) 127
    brightness and fluorescence emission properties 128
    cell cycle imaging 128
    FRET analysis 127
    genetic fusion 127
    protein engineering approaches 127
    protein fragment complementation assay (PCA) 127
    protein–protein interactions 127
  small organic fluorophores
    BAPTA ligand 130
    chemistry-based labeling methods 129
    covalent modification methods 128
    FlAsH and ReAsH 128
    fluorescence turn-on feature 130
    His- or flag-tagged membrane protein 128
    hydroxymethyldiethylrhodol (HMDER) 131
    hydroxymethyl rhodamine green (HMRG) 131
    intramolecular spirocyclization 130
    lipoic acid ligase and biotin ligase 128
    non-covalent coiled-coil interaction 128
    photoaffinity labeling method 129

fluorescence imaging (*cont'd*)
SNAP tag substrate 129
tag-metal chelate interaction 128
xanthene and benzene moieties 130
fractal geometry approach 4
frequency domain methods 183–184

g
genetic fusion 127
glucose-sensitive hydrogels 5

h
haemodialysis 13, 15
hemodiafiltration (HDF) 28
hemodialysis 28
hepatocyte transplantation 41
hierarchical clustering analysis (HCA) 148–149
high-intensity focused ultrasound ablation 203–204
hollow-fiber gas-liquid contactors 67
hollow-fiber membrane oxygenator 59
hybrid organs 3
hydroxymethyldiethylrhodol (HMDER) 131
hydroxymethyl rhodamine green (HMRG) 131
hyperpolarization 122
hyperthermal tissue destruction
Arrhenius integral 201
direct thermal damage 200
heat and vascular damage 202
metabolic and molecular changes 200
mitochondria dysfunction 201
oxygenation 202
protein denaturation 201
hyperthermal treatments
cryoablation 207
HIFU ablation 203–204
laser ablation (LA) 206–208
microwave ablation (MWA) 205–206
radiofrequency ablation (RFA) 204–205

i
immunocytochemical staining 19
implantable artificial kidney (IBAK) 16–17
interdisciplinary approach 19
International Monetary Fund 22
invasive (contact) thermometric techniques
extrinsic fluoroscopy 214
fiber bragg grating (FBG) 214–215
fiber optic sensors 213–214
fluorescence-based thermometry 214
temperature map estimation 212
thermocouples 212–213

k
kidney tissue engineering 17–20
augmentation renal tissue 19
bioartificial kidney (BAK) 18
cell therapy 19
extracorporeal bioartificial kidney 18
extracorporeal circuit 18
immunocytochemical staining 19
interdisciplinary approach 19
nuclear transfer techniques 20
kidney transplant
cost 11
post-transplant costs 12–14
statistics 11–12
transplantation costs 12

l
laser ablation (LA) 206–208
linear discriminant analysis (LDA) 148
lipoic acid ligase 128
liver construction 40–41
liver regeneration 27
L-type amino acid transporter 1 (LAT1) 124

m
magnetic resonance imaging (MRI) 21, 215–216

^{19}F-MRI 121–122
^{1}H-MRI 120–121
magnetization transfer 122–123
mass transfer analysis 64
membrane-type oxygenators 57
microelectromechanical systems (MEMS) fabrication techniques 17
microfluidic devices 3
microwave ablation (MWA) 205–206
model-predictive-control (MPC) 77
molecular adsorbent recirculating system (MARS) 31–32
Monod–Wyman–Changeux model 60
multivariate analyses
hierarchical clustering analysis (HCA) 148–149
linear discriminant analysis (LDA) 148
partial least-squares (PLS) 149
principal component analysis (PCA) 147–148
support vector machine (SVM) 149
muscular system 2

n

nano-machines 98
non-bio artificial liver
activated charcoal 28
blood purification, with albumin dialysis
fractionated plasma separation and adsorption 32
hepa wash 32
molecular adsorbent recirculating system (MARS) 31–32
single-pass albumin dialysis 30–31
clinical observations 33–35
hemodiafiltration (HDF) 28
mechanism 28
molecular weight (MW) substances 28
PE and HDF
high-flow dialysate continuous HDF (CHDF) 29–30
high-volume therapeutic PE 29
online HDF 30
plasma exchange (PE) 28
selective plasma filtration therapy
biologic-detoxifilter/plasma filter 32
plasma filtration with dialysis 33
selective plasma-exchange therapy 32–33
non-invasive (contactless) thermometric techniques
CT thermometry 216–217
elastography 218
magnetic resonance imaging (MRI) 215–216
supersonic shear wave imaging (SSI) 218
ultrasound imaging 217
nonlinear autoregressive model with exogenous input (NARX) 88
non-Newtonian behavior 4
nuclear transfer techniques 20

o

organic synthesis 97–98
oxygenation 202
oxygen transfer rate (OTR) 67

p

partial least-squares (PLS) 149
peritoneal albumin dialysis 35
peritoneal dialysis 13
perturbational complexity index (PCI) 189
PET. *see* positron emission tomography (PET)
phenomenological models
background and two-compartmental models 78–79
three-compartment models 79–85
photoaffinity labeling method 129
plasma exchange (PE) 28

porcine endogenous retrovirus (PERV) 37
porcine hepatocytes 37
positron emission tomography (PET) 123–124
post-transplant costs 12–14
primary human hepatocytes 37–38
principal component analysis (PCA) 147–148
protein denaturation 201
protein synthetic biology
  complex networks formalism 106
  conformational switching 105
  dyneins 100–101
  kinesin motor domain 100
  kinesin superfamily (KIF) 100
  microenvironment conditions 102
  myosins 101–102
  protein contact networks (PCNs) 106
  protein motors 99
  protein nano-machines 99
  protein regulation 105
  protein switches 105
  synthetic and mechanistic engineering 103

q
quantitative elastography 217

r
radiofrequency ablation (RFA) 204–205
Raman-based cell characterization
  biomedical applications 157–160
  FTIR spectroscopy 149
  subcellular analysis 153–156
  whole-cell analysis 151–153
renal replacement devices 13, 15–16
rotating film oxygenator 58

s
scaffolds 4
Schizophrenia 188–189
secreted molecule feature 165–167
self-organization 41
signal processing methods
  connectivity analysis methods 182
  frequency domain methods 183–184
  preprocessing 181–182
  time domain methods 183
silicon-based membranes (SNM) 17
single-pass albumin dialysis 30–31
single photon emission computed tomography (SPECT) 125–126
stem cells 38
subcellular analysis 153–156
superparamagnetic iron oxide (SPIO) 121
supersonic shear wave imaging (SSI) 218
support vector machine (SVM) 149
synthetic probiotic therapy 5

t
Tellegen's theorem 98–99
thymidine kinase 1 (TK1) 124
time domain methods 182
tissue destruction
  cold temperature for 202–203
  hyperthermal tissue destruction 200–202
tissue engineering 3, 15
  and artificial organs 108–110
  for liver construction 40–41
TMS-EEG methodology
  artifacts and prevention 180–181
  clinical perspectives
    Alzheimer's disease 187–188
    disorders of consciousness 189
    Schizophrenia 188–189
  data acquisition 178–180
  oscillatory activity 186–187
  signal processing methods
    connectivity analysis methods 182

frequency domain methods 183–184
preprocessing 181–182
time domain methods 183
TMS-Evoked Potentials (TEPs) 185–186
TMS-evoked potentials (TEPs) 185–186
transcranial magnetic stimulation (TMS) 175–176. *see also* TMS-EEG methodology
transplantation costs 12
transport phenomena 62
tumors, thermal trearment of
hyperthermal treatments
cryoablation 207
HIFU ablation 203–204
laser ablation (LA) 206–208
microwave ablation (MWA) 205–206
radiofrequency ablation (RFA) 204–205
mathematical models 209–211
temperature monitoring
invasive (contact) thermometric techniques 212–215
non-invasive (contactless) thermometric techniques 215–218
tissue destruction
cold temperature for 202–203
hyperthermal tissue destruction 200–202

u

ultrafiltration technology 2
ultrasonography 119
ultrasound imaging 217

w

wearable artificial kidney (WAK) 15
whole-cell analysis 151–153
whole organ engineering 42–43
World Economic Forum 22

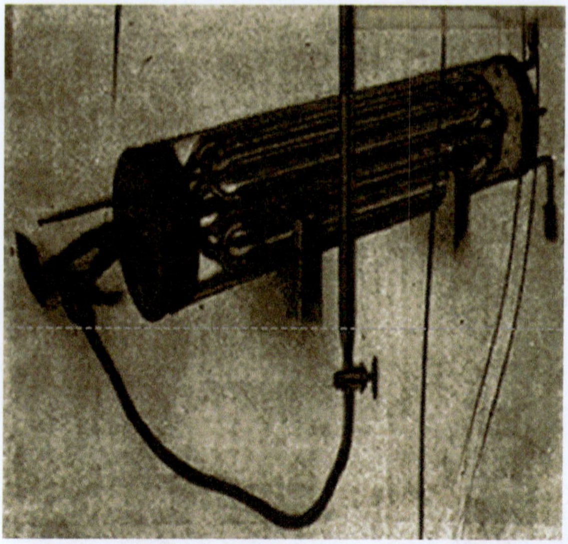

Figure 2.1 First experimental dialysis (1913). *Source:* Journal of Pharmacology and Experimental Therapeutics January 1914, 5 [3] 275–316, http://jpet.aspetjournals.org/content/5/3/275.

Biomedical Engineering Challenges: A Chemical Engineering Insight, First Edition.
Edited by Vincenzo Piemonte, Angelo Basile, Taichi Ito, and Luigi Marrelli.

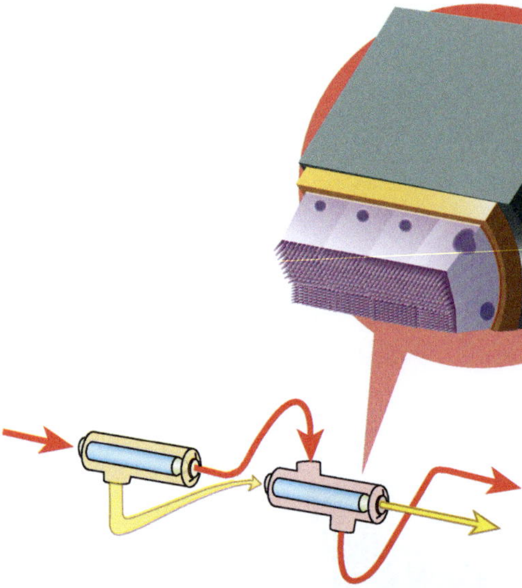

Figure 2.2 Schematic of an extracorporeal circuit of a bioartificial kidney used to treat patients with acute renal failure. The first cartridge is a hemofiltration cartridge in series with a renal tubule cell device (RAD). The ultrafiltrate is delivered to the luminal compartment of the RAD which contains the cells and the post-filtered blood is pumped into the extracapillary space of the RAD. The processed luminal ultrafiltrate from the RAD is discarded to waste and the processed blood is returned to the patient. *Source:* Principles of Tissue Engineering, Robert Lanza, Robert Langer, Joseph Vacanti (2013) 1142.

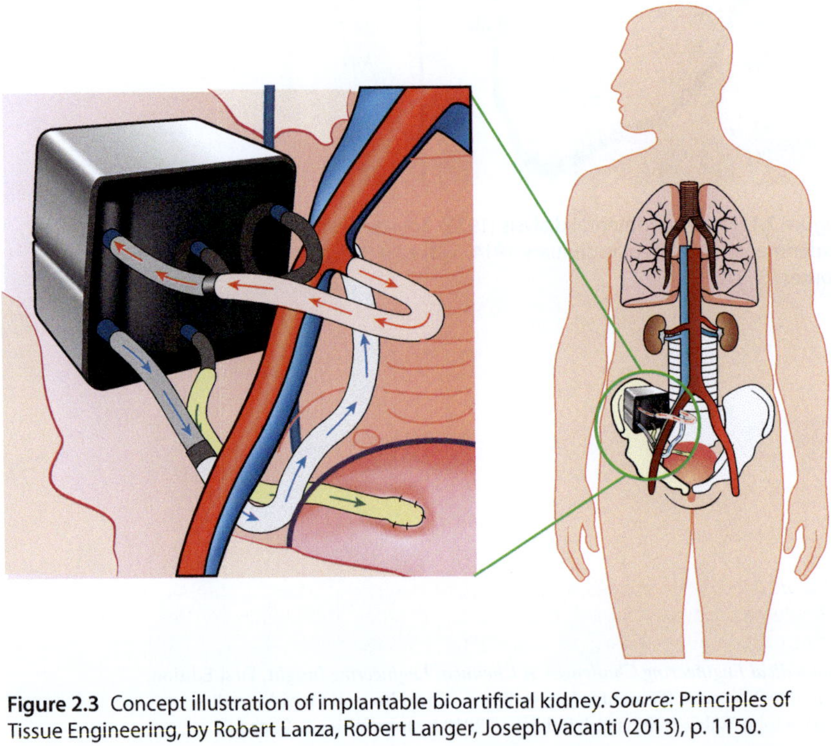

Figure 2.3 Concept illustration of implantable bioartificial kidney. *Source:* Principles of Tissue Engineering, by Robert Lanza, Robert Langer, Joseph Vacanti (2013), p. 1150.

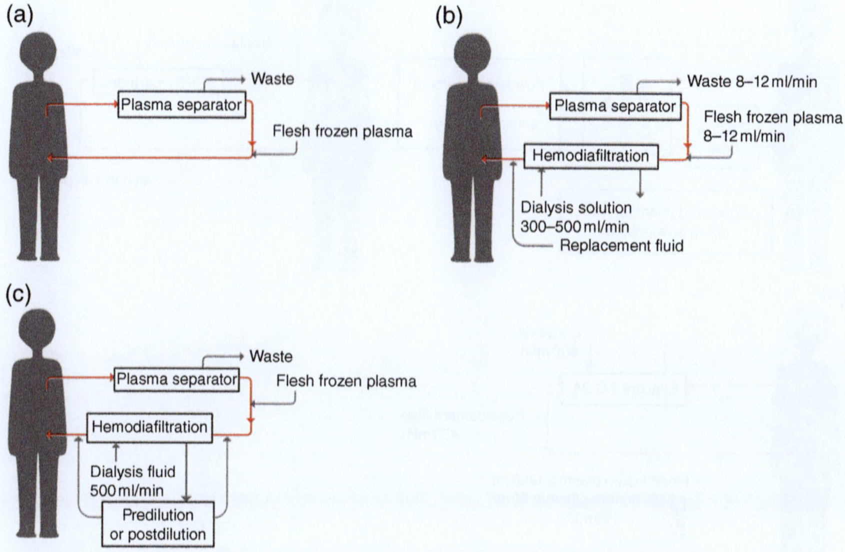

Figure 3.1 System images of non-biological artificial livers using plasma exchange and hemodiafiltration. (a) Plasma exchange, (b) plasma exchange and high-flow hemodiafiltration, and (c) plasma exchange with online hemodiafiltration.

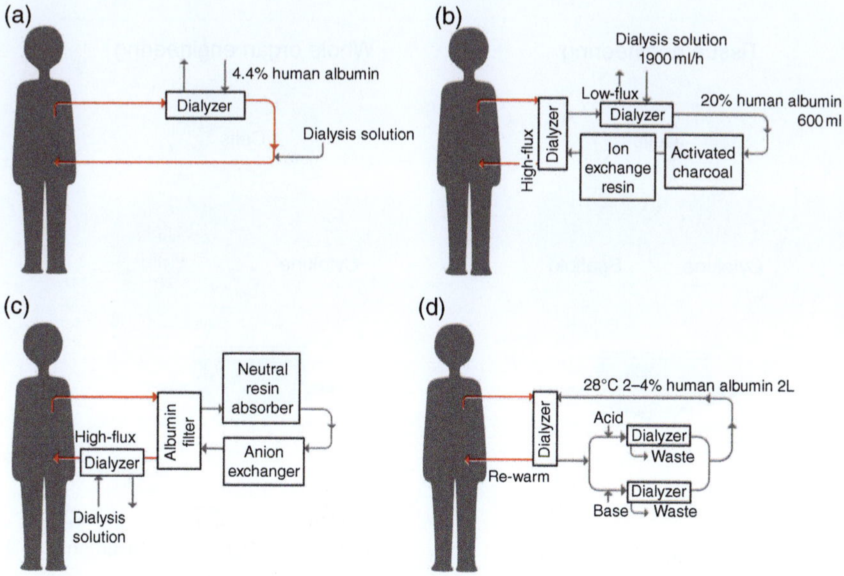

Figure 3.2 System images of non-biological artificial livers using albumin dialysis. (a) Single-pass albumin dialysis, (b) molecular adsorbent recirculating system, (c) fractionated plasma separation and adsorption, and (d) Hepa Wash.

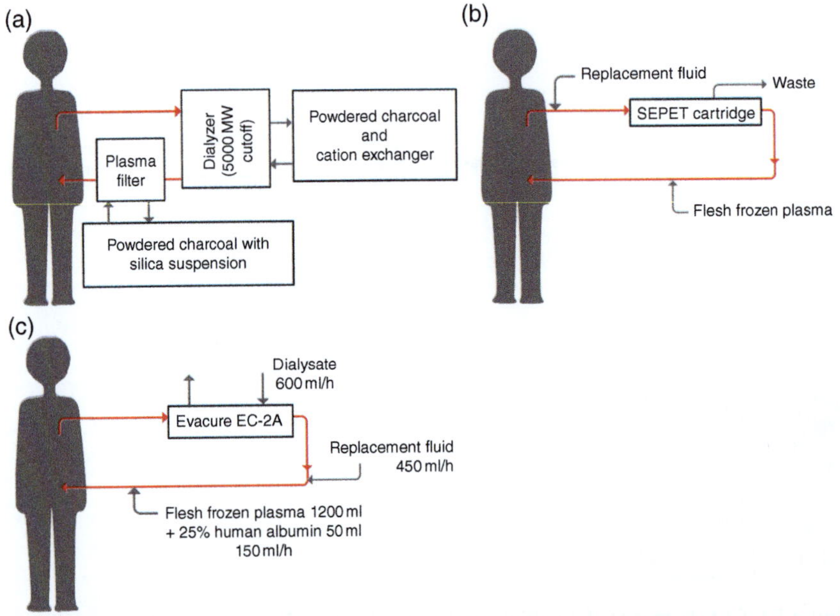

Figure 3.3 System images of non-biological artificial livers using selective plasma filtration. (a) Biologic-detoxifilter/plasma filter, (b) selective plasma-exchange therapy, and (c) plasma filtration with dialysis.

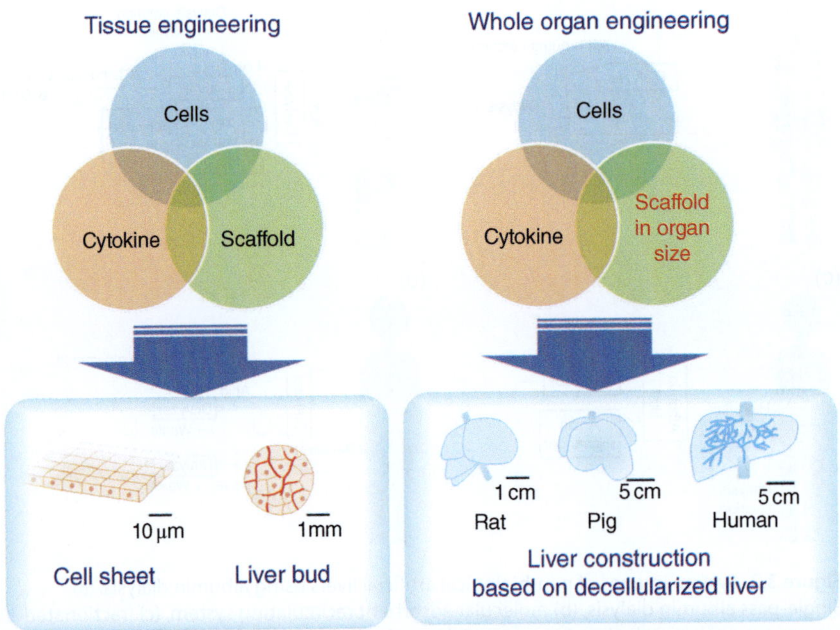

Figure 3.4 Concepts of tissue engineering and whole-organ engineering.

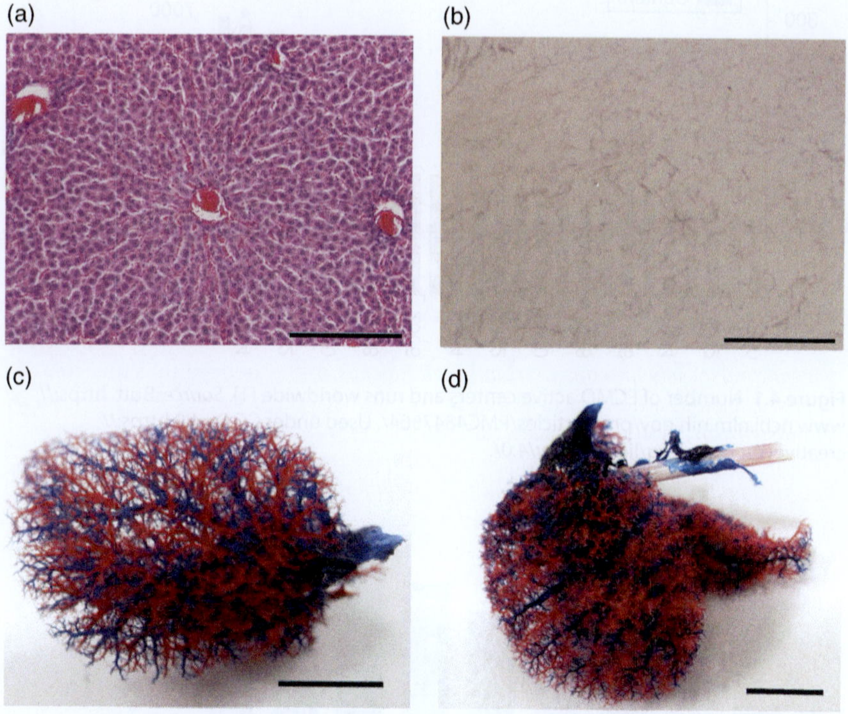

Figure 3.5 Histological analysis using hematoxylin and eosin staining (a, b) and molded resin of vascular structure (c, d). native liver (a, c) and decellularized liver (b, d). Scale bars indicate 100 μm (a, b) and 5 mm (c, d), respectively (reproduced from Shirakigawa et al., 2013 with permission from Elsevier) (Shirakigawa et al., 2013).

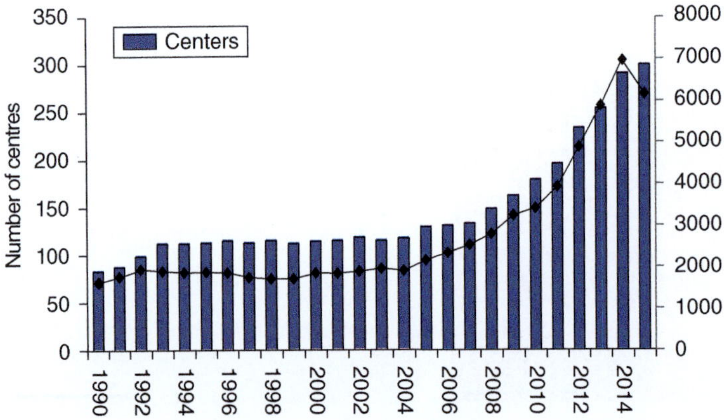

Figure 4.1 Number of ECMO active centers and runs worldwide [1]. *Source:* Butt, https://www.ncbi.nlm.nih.gov/pmc/articles/PMC4847564/. Used under CC-By 4.0 https://creativecommons.org/licenses/by/4.0/.

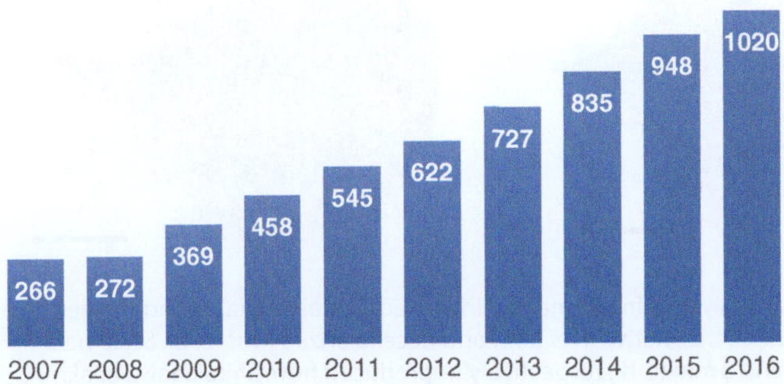

Figure 4.2 Number of papers regarding ECMO during years (*Source:* PubMed, language English).

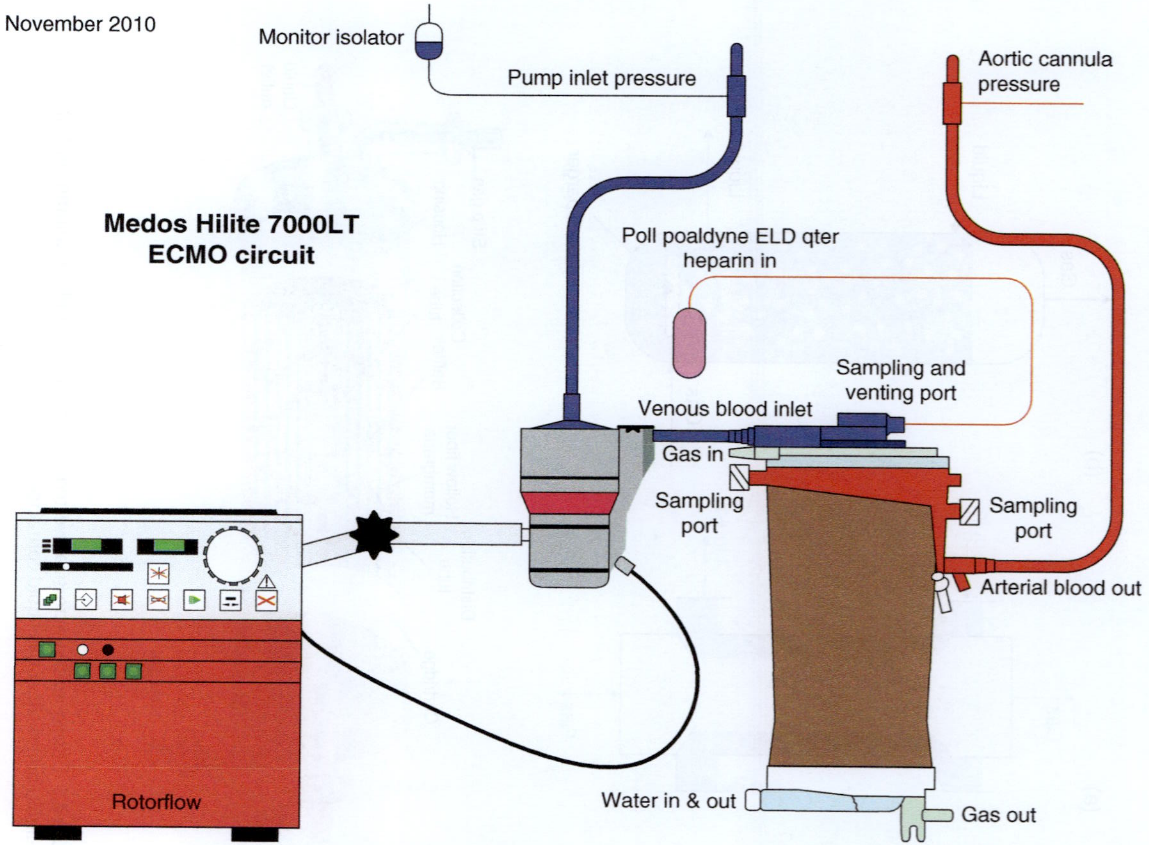

Figure 4.4 A modern ECMO circuit (reprinted with from Lequier, L., Horton, S., McMullan, D., and Bartlett, R.H. (2013) Extracorporeal membrane oxygenation circuitry. *Pediatr. Crit. care* ..., **14**, 1–10).

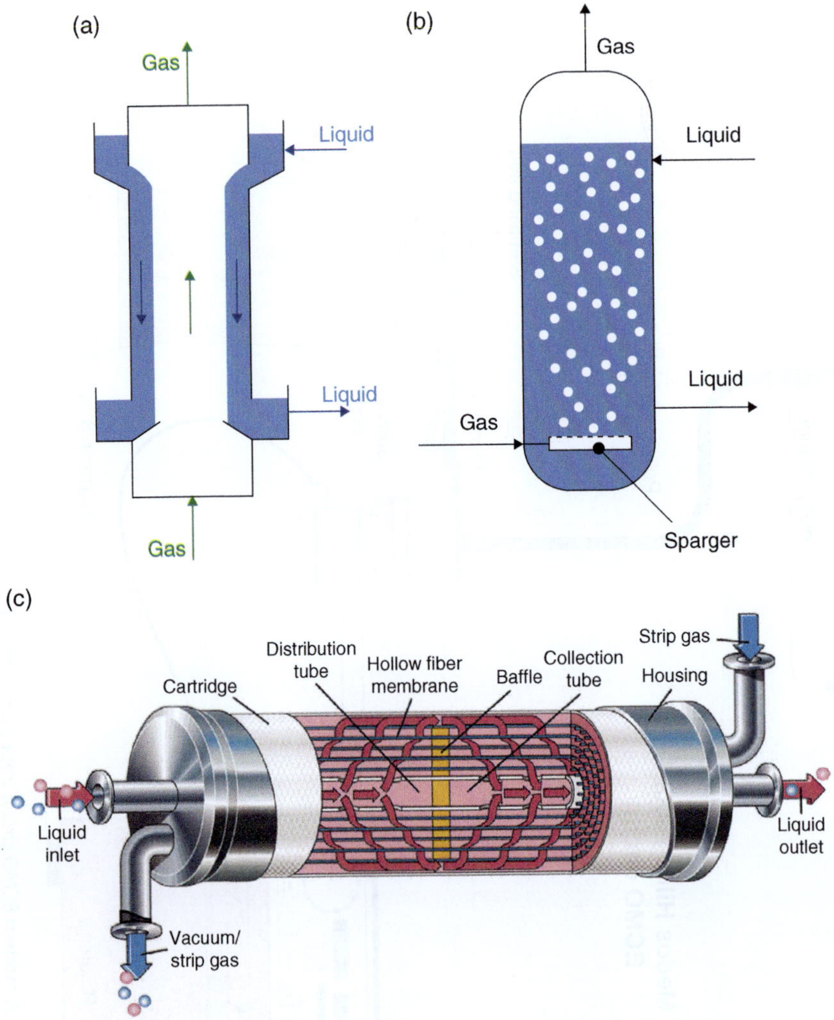

Figure 4.6 Gas-liquid contactors: (a) wetted wall column; (b) bubble column; and (c) gas-liquid hollow-fiber gas-liquid contactor.

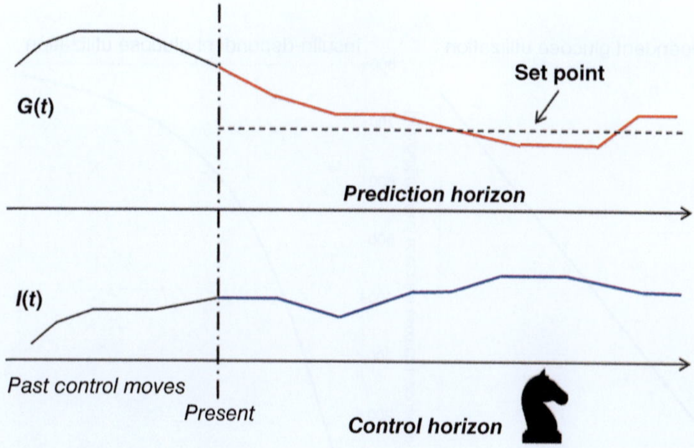

Figure 5.2 Basic concept of MPC during a chess game consisting of predicting and regulating glucose level.

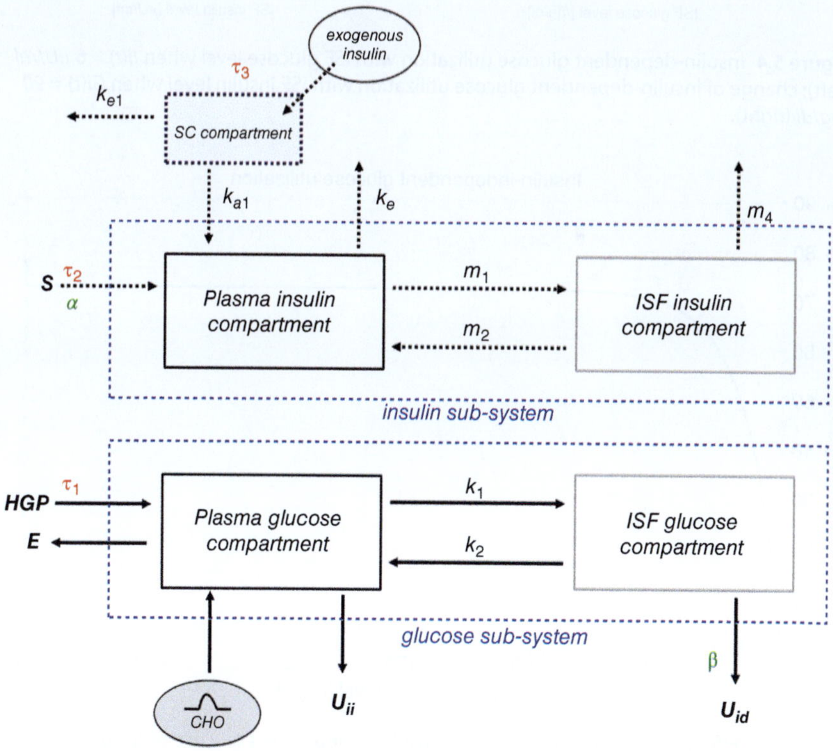

Figure 5.3 Block diagram of the model presented and tested by Piemonte et al. [10].

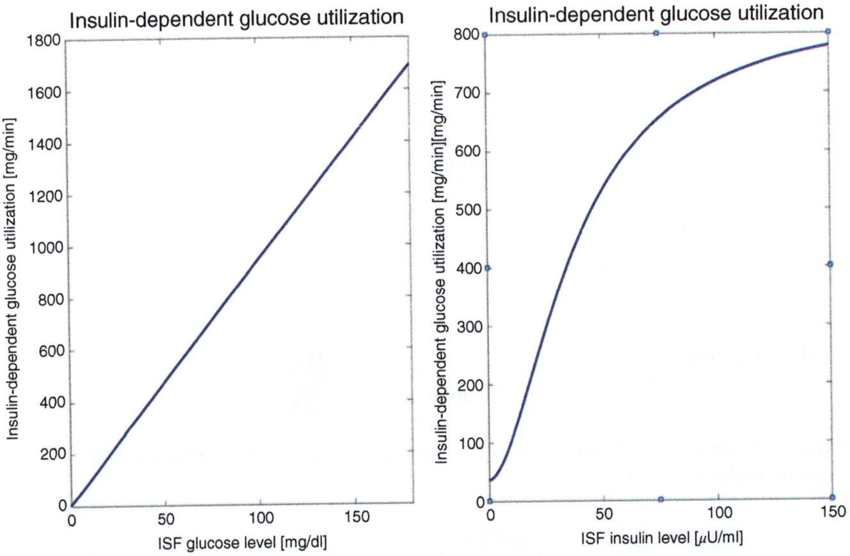

Figure 5.4 Insulin-dependent glucose utilization with ISF glucose level when $Ii(t) = 6\ \mu U/ml$ (left); change of insulin-dependent glucose utilization with ISF insulin level when $Gi(t) = 90\ mg/dl$ (right).

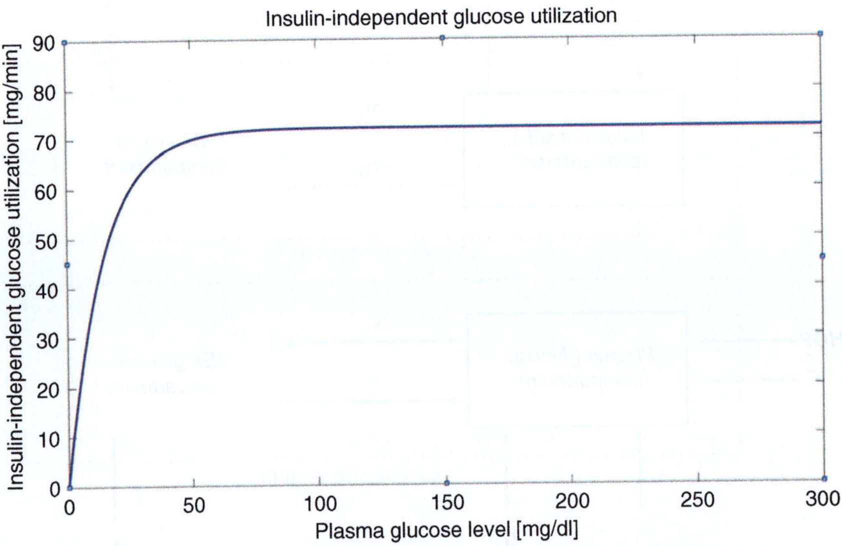

Figure 5.5 Effect of plasma glucose level on insulin-independent glucose utilization.

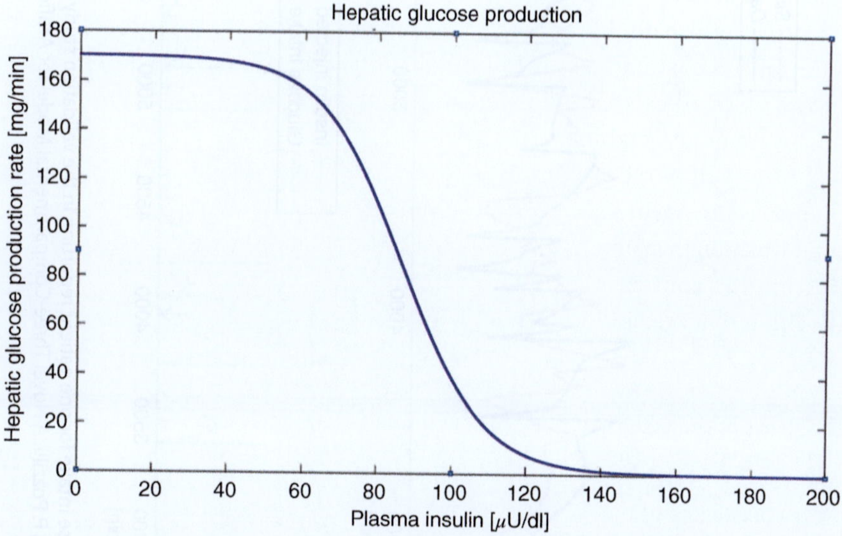

Figure 5.6 Change of hepatic glucose production with plasma insulin level.

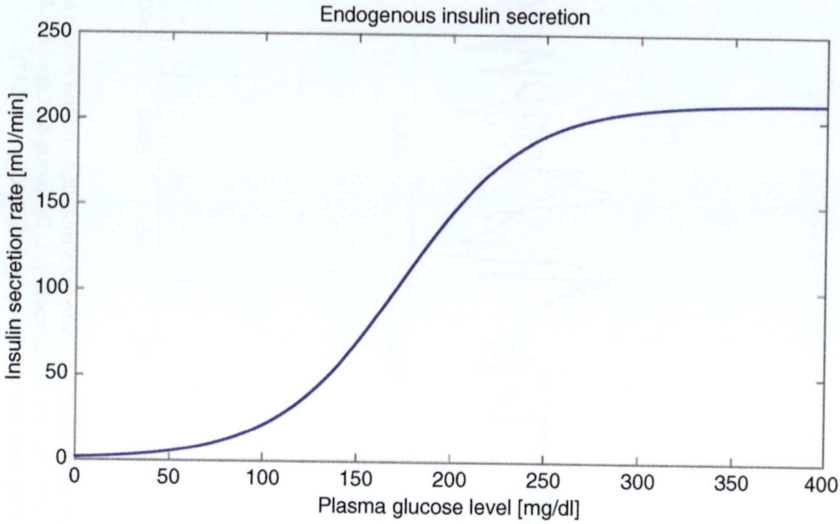

Figure 5.7 Insulin secretion rate with plasma glucose level.

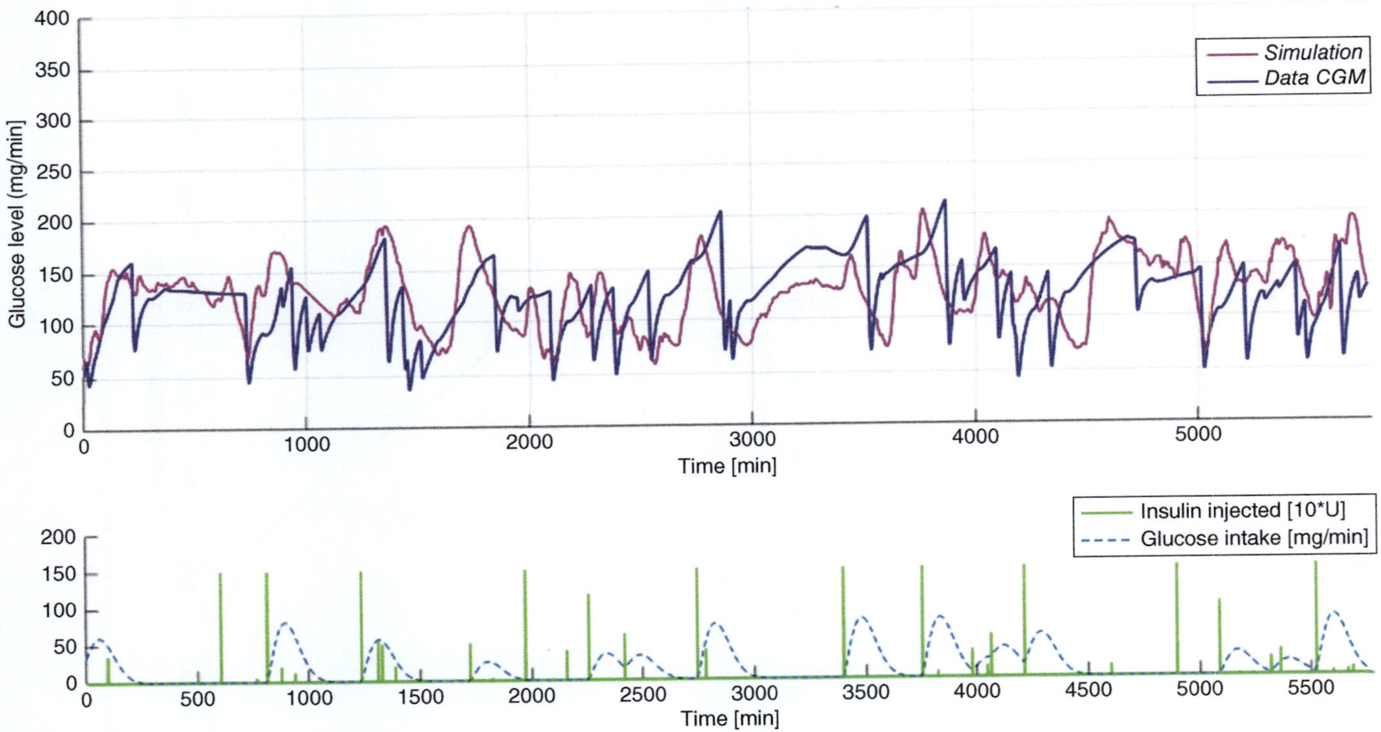

Figure 5.8 Glucose level profiles with the simulation and the insulin and glucose intake for one patient reporting in the validation study by Piemonte et al. [10] (*Source:* V. Piemonte V., L. De Santis, M. Capocelli, A. Maurizi, P. Pozzilli. A Novel Three-Compartmental Model for Artificial Pancreas: Development and Validation. J Artificial Organs, 2017).

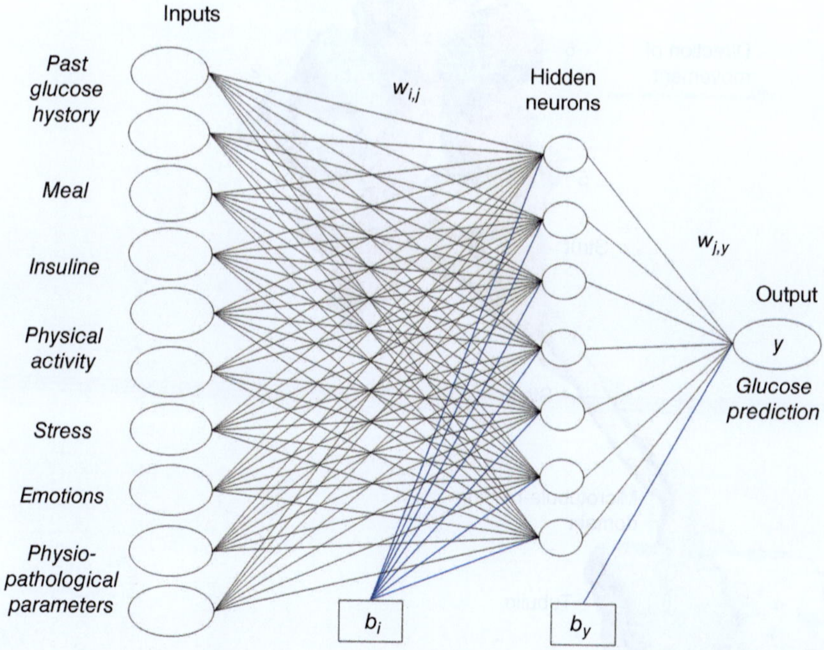

Figure 5.9 Conceptual scheme of an ideal glucose predictor able to exploit most of the inputs influencing glucose concentration.

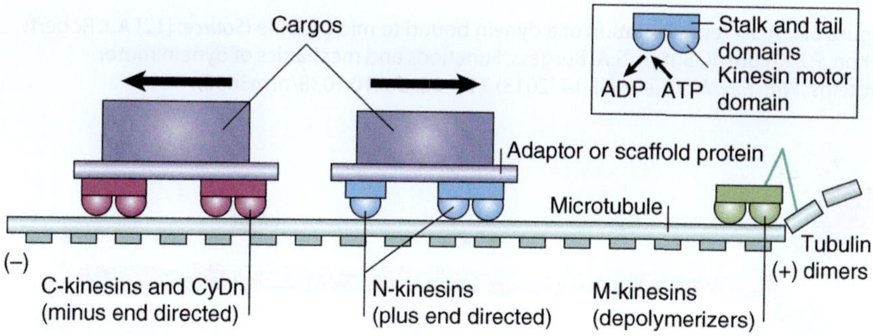

Figure 6.2 A sketch of kinesin functioning (*Source:* [7] N. Hirokawa, Y. Noda, Y. Tanaka, S. Niwa, Kinesin superfamily motor proteins and intracellular transport., *Nat. Rev. Mol. Cell Biol.* 10 (2009) 682–696. doi:10.1038/nrm2774).

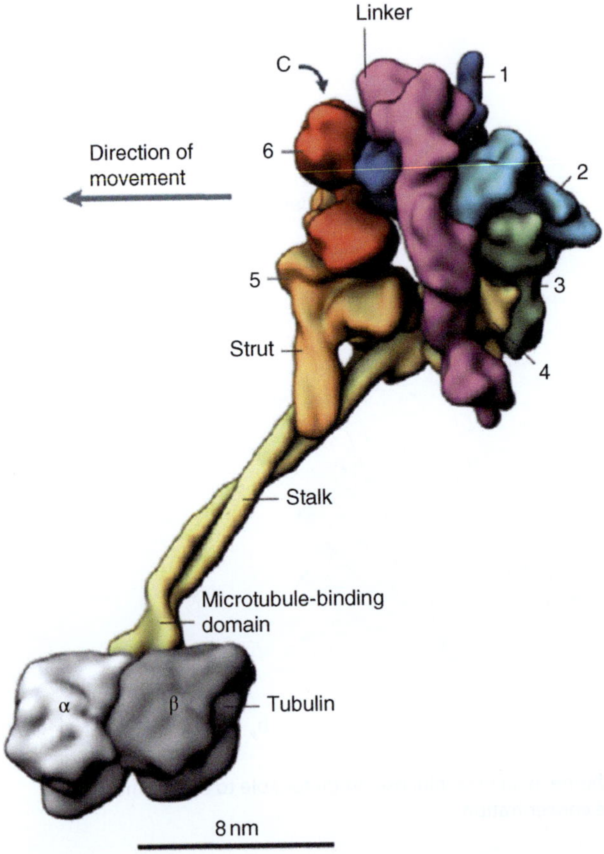

Figure 6.3 A 3D representation of a dynein bound to microtubule (*Source:* [12] A.J. Roberts, T. Kon, P.J. Knight, K. Sutoh, S.A. Burgess, Functions and mechanics of dynein motor proteins., *Nat. Rev. Mol. Cell Biol.* 14 (2013) 713–26. doi:10.1038/nrm3667).

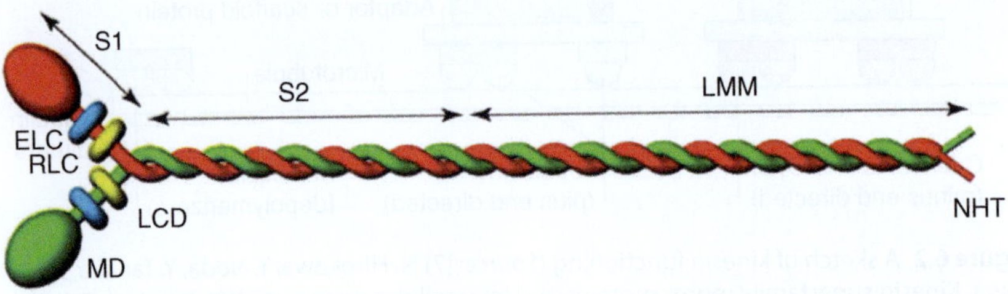

Figure 6.4 Myosin structure: the long tail (S2 + LMM) is a coiled coil, head S1 comprises the motor domain (MD) and the light chain domain (LCD) in turning including the essential light chain (ELC) and the regulatory LC (RLC) (*Source:* [16] M. Vicente-Manzanares, X. Ma, R.S. Adelstein, A.R. Horwitz, Non-muscle myosin II takes centre stage in cell adhesion and migration., *Nat. Rev. Mol. Cell Biol.* 10 (2009) 778–90. doi:10.1038/nrm2786).

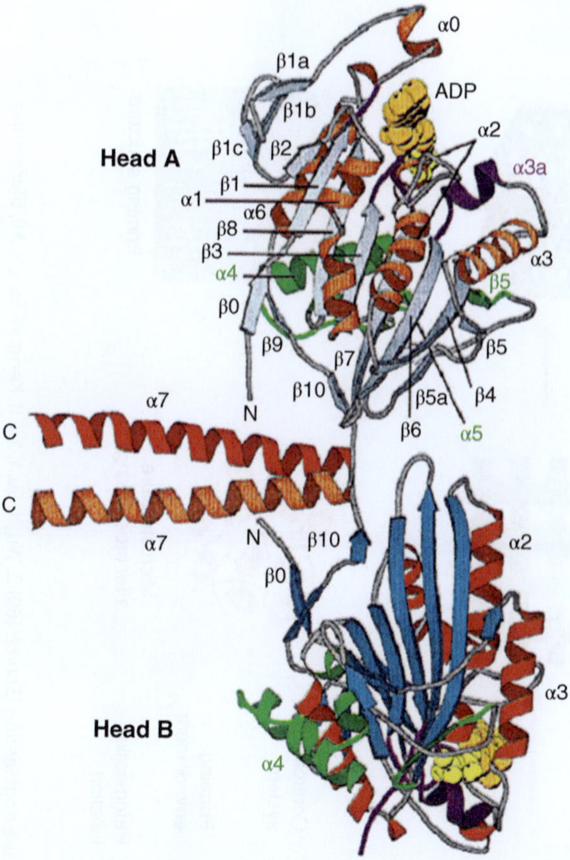

Figure 6.5 Ribbon structure of kinesin (*Source:* [17] F. Kozielski, S. Sack, A. Marx, M. Thormählen, E. Schönbrunn, V. Biou, A. Thompson, E.M. Mandelkow, E. Mandelkow, The crystal structure of dimeric kinesin and implications for microtubule-dependent motility, *Cell.* 91 (1997) 985–994. doi:10.1016/S0092-8674(00)80489-4).

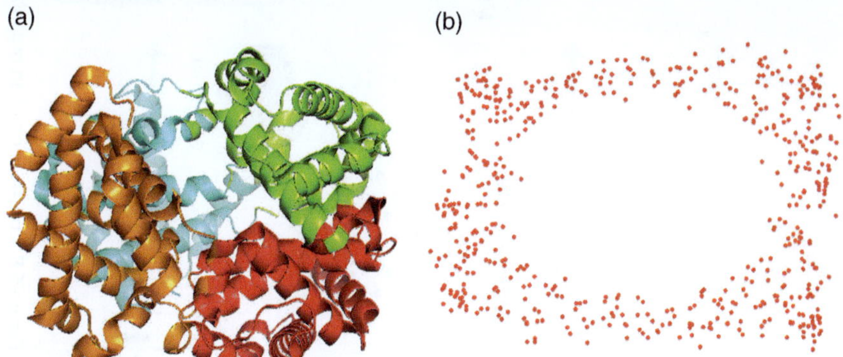

Figure 6.7 Protein contact networks. Panel (a) the ribbon representation of the T-state of human hemoglobin; Panel (b) the corresponding Protein Contact Network.

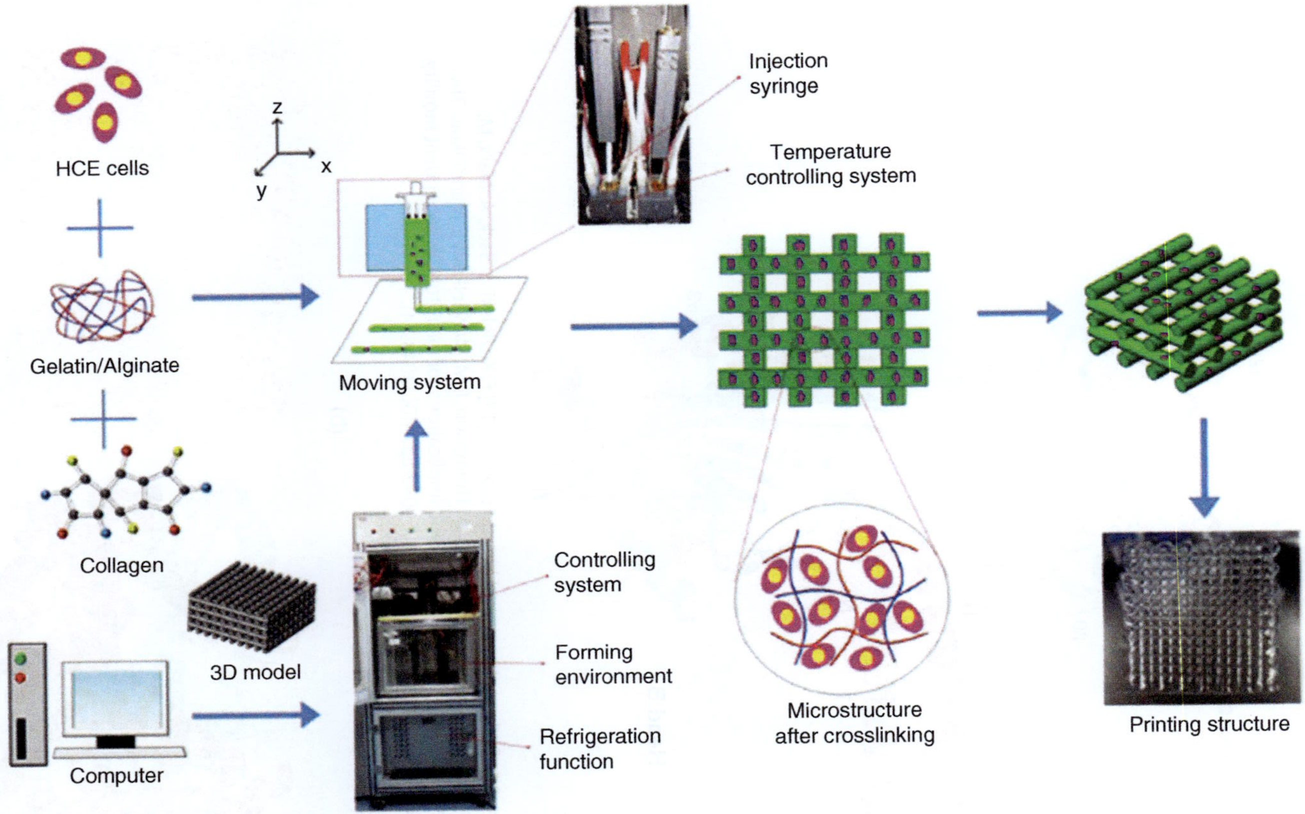

Figure 6.8 Construction of an organized scaffold for tissue engineering (*Source:* [60] Z. Wu, X. Su, Y. Xu, B. Kong, W. Sun, S. Mi, Bioprinting three-dimensional cell-laden tissue constructs with controllable degradation, *Sci. Rep.* 6 (2016) 24474. doi:10.1038/srep24474).

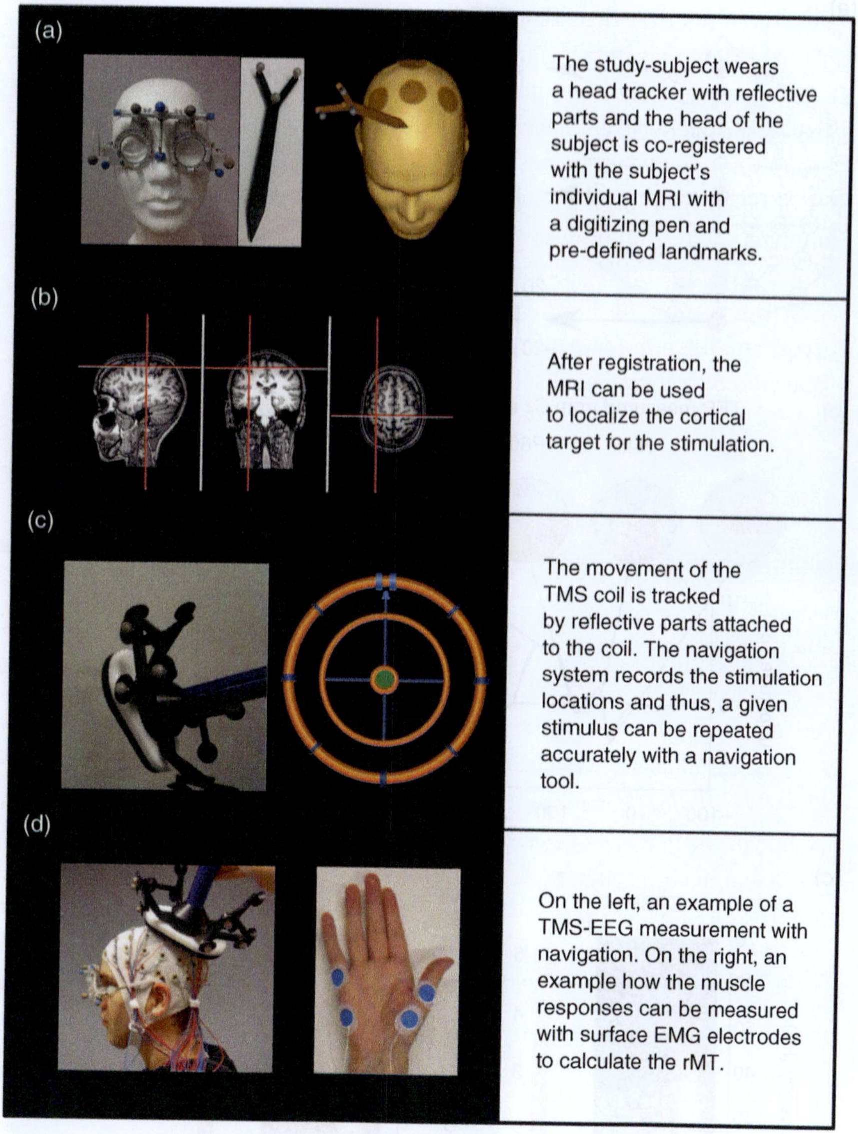

Figure 9.1 The methodology of navigated transcranial magnetic stimulation (an example from one commercial vendor) combined with electroencephalography (TMS-EEG). (a) In order to perform accurate navigation, the subject wears a head tracker with spherical reflective parts. These reflective parts are monitored by an infrared camera. With the help of the head tracker, the subject is co-registered with the subject-specific magnetic resonance image (MRI). **(b)** After co-registration, the MRI can be used to localize the stimulation target. **(c)** The movement of the stimulation coil is also followed by reflective parts. The navigation system records the location, tilt and induced electric field direction of a given TMS pulse and therefore, pulses can be repeated. **(d)** An example of a TMS-EEG measurement setup. If the applied stimulation intensity is normalized according to the resting motor threshold (rMT), TMS pulses are first focused on the motor cortex and evoked motor responses are measured from the target muscle using surface electromyography (EMG) electrodes to calculate the rMT.

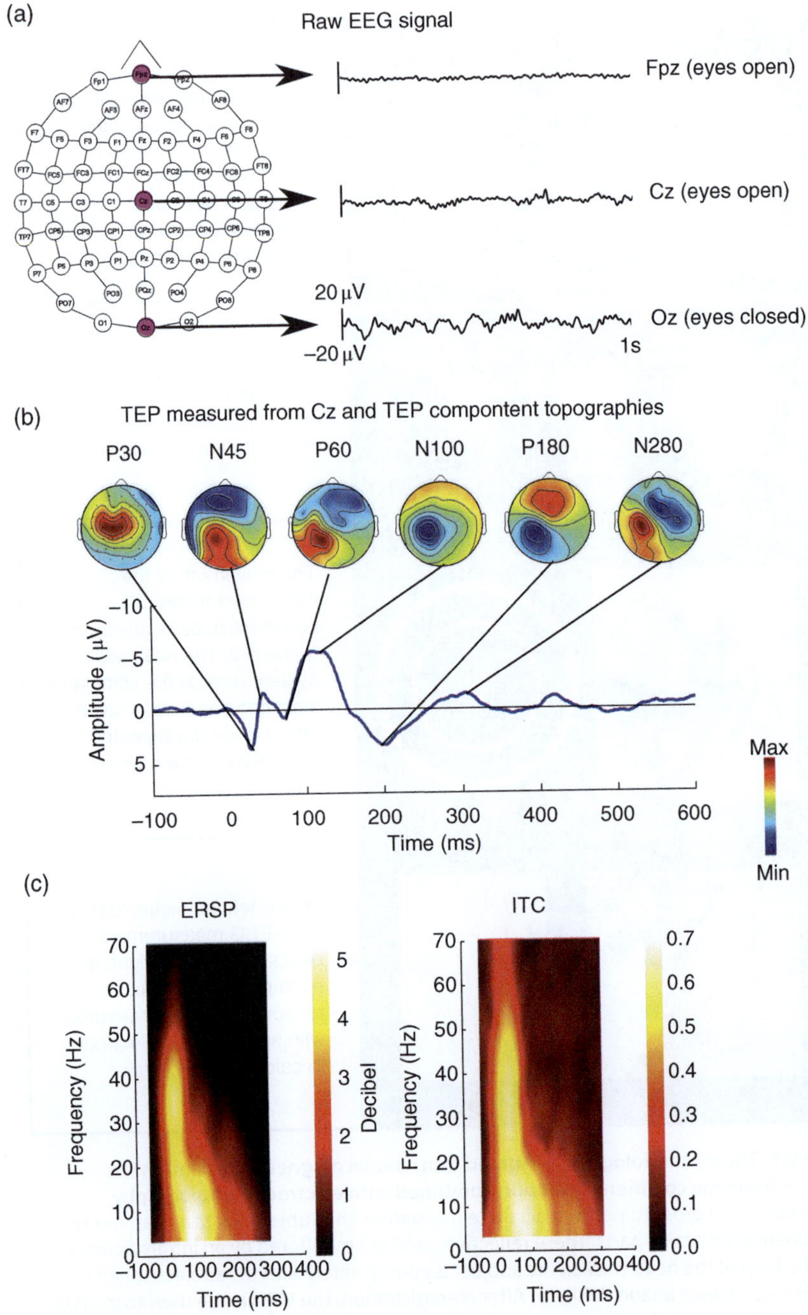

Figure 9.2 EEG signal from a representative subject. (a) An electroencephalography (EEG) electrode montage. Raw EEG signal measured from frontal (Fpz), central (Cz), and posterior (Oz) electrodes is shown. **(b)** A typical TMS evoked potential (TEP) induced by motor cortex stimulation. The TEP is measured from the Cz electrode. The topographical maps present the spread of neural activity at each TEP component (P30 ms, N45 ms, P60 ms, N100 ms, P180 ms, and N280 ms). **(c)** Event related spectral perturbation (ERSP) and inter-trial coherence (ITC) of motor cortex TEPs. In B and C, the TMS pulse was given at a time 0 ms.

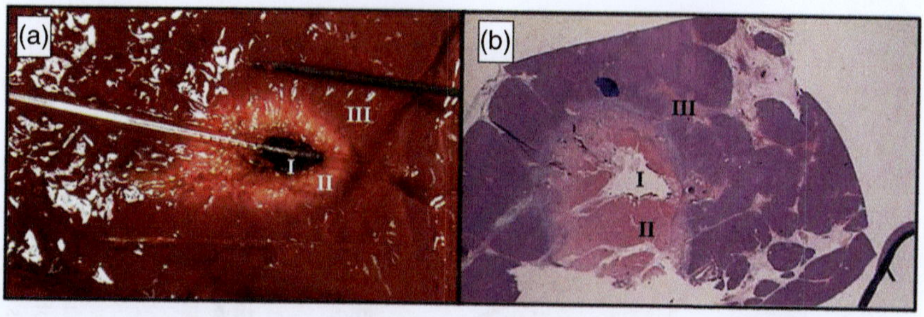

Figure 10.1 Effects of thermal ablation on tissue: (a) Macroscopic effect of diode laser-induced thermal ablation of healthy ex vivo animal liver; (b) Specimen from a histopathological examination (zoom × 10) 24 h after Nd:YAG laser ablation on in vivo animal pancreas. Although the laser settings differed between the two treatments, it is possible to distinguish similar effects: (I) the central zone of irreversible damage (in both cases, cells vaporized as a result of the raised temperature), (II) a zone of coagulated necrosis, and (III) the surrounding uninjured parenchyma. Also, a thin (1–2 mm) and watershed zone of early inflammatory response surrounding coagulated tissue is detectable [11]. (*Source:* Di Matteo, F. et al. EUS-guided Nd: YAG laser ablation of normal pancreatic tissue: a pilot study in a pig model. *Gastrointest Endosc* 2010, 72, 358–363.)

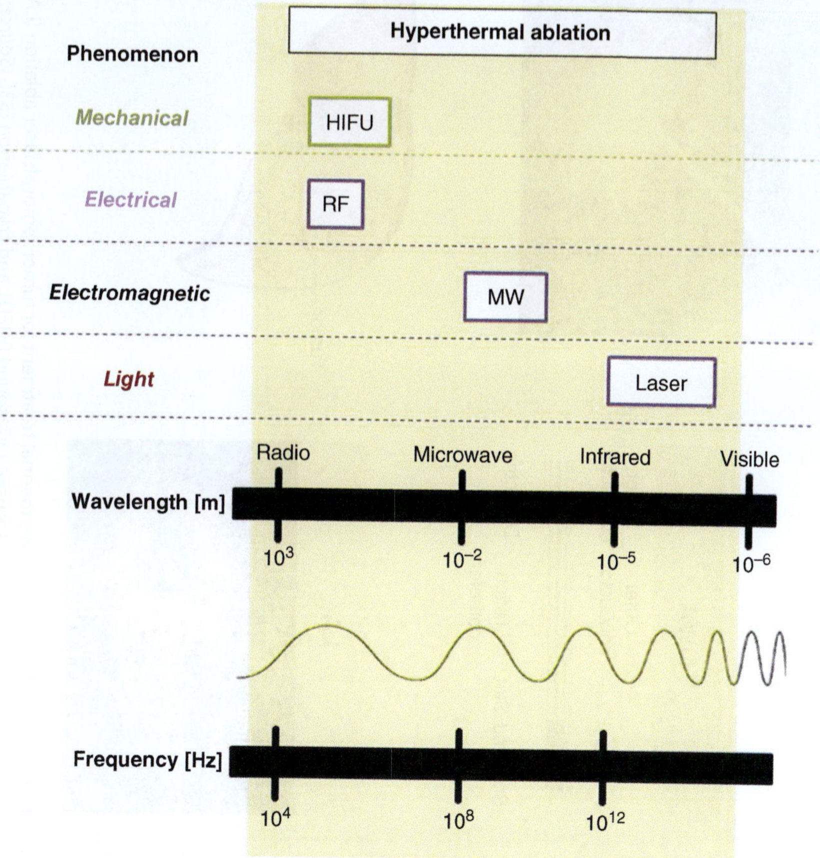

Figure 10.2 Different kinds of hyperthermal treatment whose energy source is characterized by different frequencies (or wavelengths).

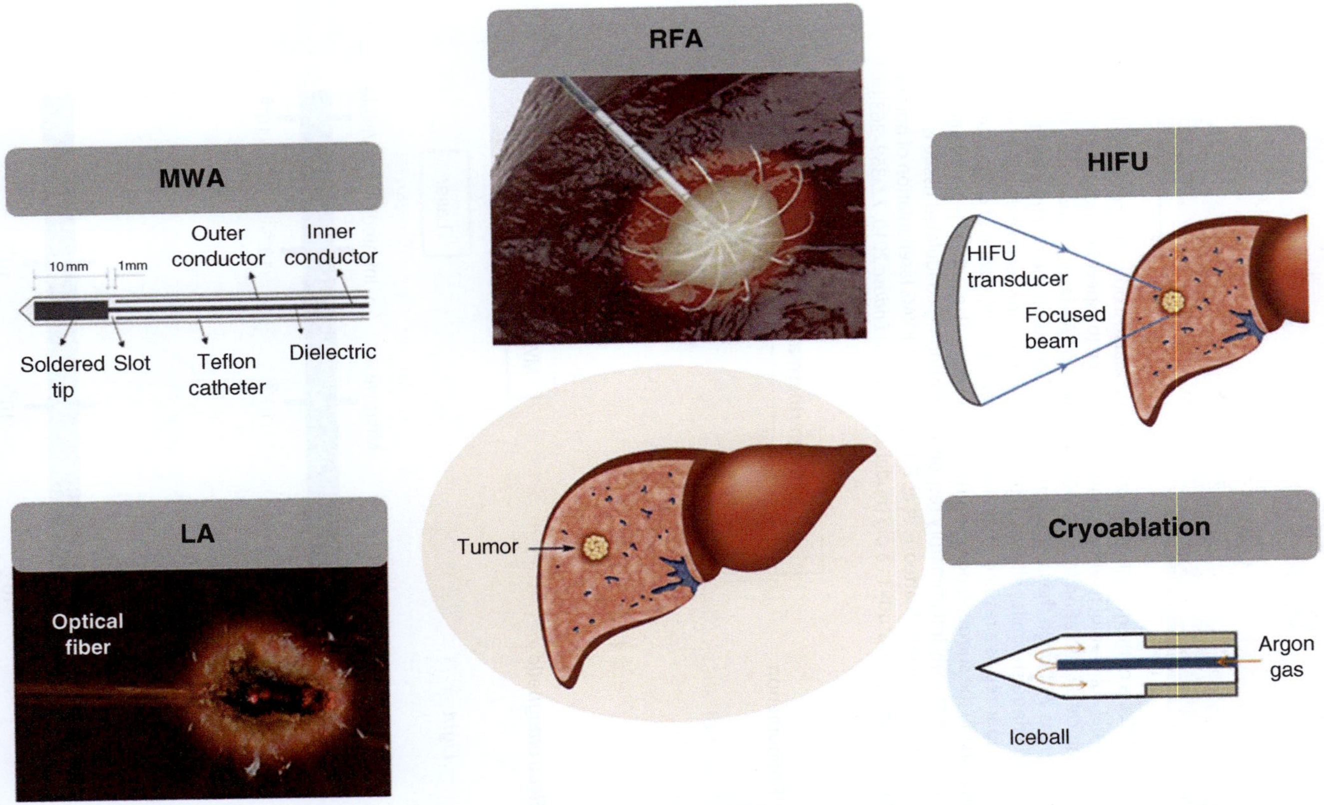

Figure 10.3 Minimally invasive thermal treatment for tumor removal: laser ablation (LA), microwave ablation (MWA), radiofrequency ablation (RFA), high-intensity focused ultrasound (HIFU), and cryoablation [53]. (*Source*: Schena, E. et al. Fiber Optic Sensors for Temperature Monitoring during Thermal Treatments: An Overview. Sensors 2016, 16, 1144.)

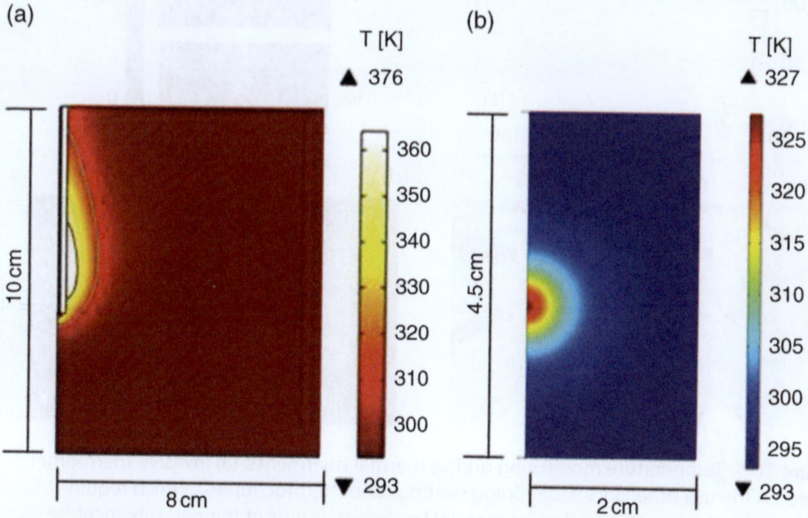

Figure 10.4 Finite Element Modeling (FEM) simulations of temperature distribution induced by (a) MWA and (b) LA within liver tissue.

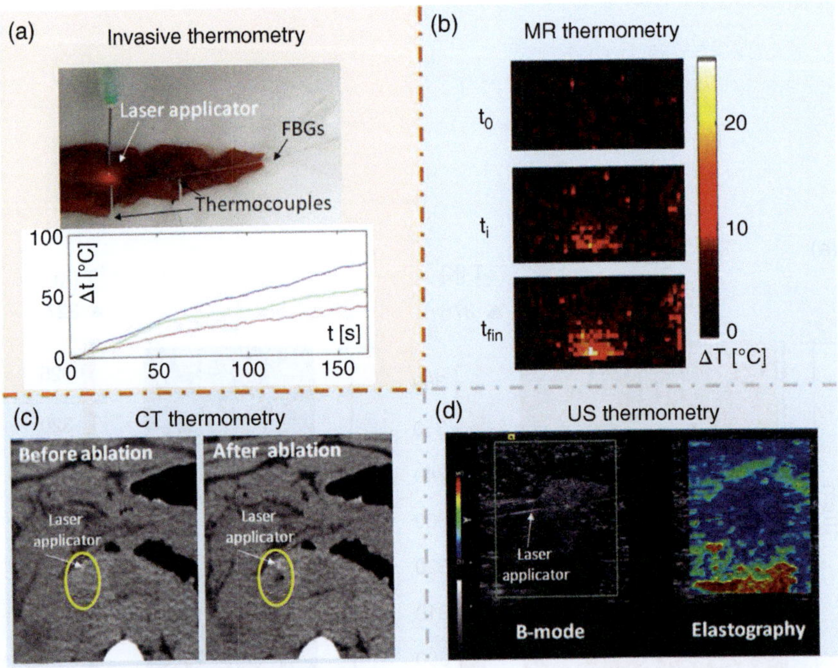

Figure 10.5 Temperature monitoring during thermal treatments. (a) Invasive thermometry, based on the use of sensors (Fiber Bragg Gratings and thermocouples), which require contact with the target intended for thermal treatment; trends of temperature increase measured by FBG sensors placed inside the ex vivo organ undergoing LAs. Non-invasive thermometry based on: (b) MR imaging: MR scans are acquired at different times during the ablation procedure, and the temperature increase in each pixel of the image is obtained by comparing the successive scans and the change of signal intensity in each pixel; (c) CT-scans, where the change of tissue temperature is related to the change of CT-number of the image (when the temperature increases, the CT-number decreases, hence the pixel values turn from gray to dark gray), as clearly visible from the images obtained before and at the end of the ablation; (d) US imaging, where the increase of tissue temperature can be monitored by means of the classical B-mode, resulting into an increase of echogenicity, or, more efficiently, by means of elastography, where the variation of tissue stiffness can be used as an indirect parameter to estimate tissue temperature.

Printed and bound by CPI Group (UK) Ltd, Croydon, CR0 4YY

05/10/2026

14975859-0001